ESSENTIALS OF HUMAN BIOCHEMISTRY

Essentials of Human Biochemistry

C R Paterson MA, DM, BSc, FRCPath
Senior Lecturer in Biochemical Medicine, University of Dundee
Honorary Consultant, Tayside Area Health Board

CHURCHILL LIVINGSTONE
EDINBURGH LONDON MELBOURNE AND NEW YORK 1987

CHURCHILL LIVINGSTONE
Medical Division of Longman Group UK Limited

Distributed in the United States of America by
Churchill Livingstone Inc., 1560 Broadway, New York,
N.Y. 10036, and by associated companies, branches
and representatives throughout the world.

First published 1983
    Reprinted 1983 (Pitman Publishing Ltd)
    Reprinted 1987 (Churchill Livingstone)

ISBN   0-443-03895-3

British Library Cataloguing in Publication Data
Essentials of human biochemistry.

    1. Biological chemistry
    I. Paterson, C. R.
    612′.015 QP514.2

Library of Congress Cataloging in Publication Data
Main entry under title:

Essentials of human biochemistry.
    Bibliography: p.
    Includes index.
    1. Biological chemistry. I. Paterson, Colin Ralston.

QP514.2.E85 1983 612 83-3991

Produced by Longman Singapore Publishers Pte Ltd
Printed in Singapore

Contents

Preface

THIS book was written primarily to provide medical students with a clear account of those aspects of human biochemistry which are important directly or indirectly for the intelligent practice of medicine. Several excellent books are currently available for undergraduates studying biochemistry but none is entirely appropriate for those whose ultimate goal is clinical medicine. For example material relevant to plants or bacteria, but not to man, is included and, more important, several subjects of great physiological or clinical importance are omitted. This book reflects my conviction that biochemistry is important for the understanding of medicine and that it should be seen to be important.

The stimulus for the preparation of this book came from my involvement with the ninth edition of the *Textbook of Physiology and Biochemistry* ('BDS') published in 1976. I am indebted to Prof. G. H. Bell and Dr. D. Emslie-Smith, and also to Mr. R. Duncan of Churchill Livingstone, for permission to include here a small part of the material originally prepared for that edition. I was preceded, as editor of the biochemistry sections of BDS, by the late Prof. J. N. Davidson of Glasgow University. Although none of his text remains here, I am glad to acknowledge the inspiration he provided in his earlier editions.

I owe a great debt to many people for their help in the preparation of this work. The contributors listed on the next page helped in different ways. Some wrote a first draft which I then edited to ensure brevity and uniformity of style; others criticised and improved first drafts which I had written. All went to a great deal of trouble to ensure the accuracy and relevance of the material included in their chapters. I am indebted to correspondents in many parts of the world for their kindness in providing me with a unique collection of clinical photographs.

I thank Professor G. H. Bell for reviewing the entire text and making many valuable suggestions for improving the style and content. I am also indebted for help in various ways to Dr. M. A. Bakry, Dr. P. E. G. Mitchell, Dr. R. F. Oliver, Dr. G. R. Tudhope, Dr. R. H. Sankhala, Mrs. P. Mole and Miss E. Paterson.

This edition called for a large number of new drawings for which I thank Miss M. Benstead, Miss M. M. Sneddon and Mr. J. Burns. Mrs. M. Lawson and Mrs. M. Alexander did the lettering for the figures including many which must have been very trying. For skilled secretarial help I thank Mrs. M. Geekie, Mrs. J. Murant and Mrs. L. Reid. I owe a great debt to my wife and family for their tolerance of such a large distraction.

While I have been privileged to have the assistance of so many helpful people, I must take responsibility for any errors which remain. I hope that readers will write to let me know of any they detect, or of any improvements which they would like to see.

Dundee, September 1982   C R Paterson

Additional acknowledgements

I am indebted to many correspondents for their letters and for pointing out errors or ambiguities which it has proved possible to correct on the occasion of the book's reprinting. I am particularly indebted to Professor G. H. Bell, to Professor A. G. Campbell, to Dr. J. Hawkins, to Mrs. P. Mole and to Professor F. Vella for their comments. I have also used the opportunity provided by the preparation of a reprint to bring the reading lists up to date.

Dundee, November 1984   C R Paterson

Contributors

Dr. G. C. Barr, Department of Biochemistry, University of Dundee *Chapter 7.*

Miss M. C. K. Browning, Department of Biochemical Medicine, University of Dundee *Chapters 4 & 22.*

Dr. D. A. Chignell, Department of Chemistry, Wheaton College, Illinois *Chapter 2.*

Prof. P. Cohen, Department of Biochemistry, University of Dundee *Chapter 12.*

Dr. G. A. J. Goodlad, Department of Biochemistry and Microbiology, University of St Andrews *Chapters 5, 13, 14 & 15.*

Dr. R. Griffiths, Department of Biochemistry and Microbiology, University of St Andrews *Chapter 8.*

Dr. D. G. Hardie, Department of Biochemistry, University of Dundee *Chapters 12 & 22.*

Dr. M. I. S. Hunter, Department of Biochemistry and Microbiology, University of St Andrews *Chapters 4, 6, 11 & 18.*

Dr. T. E. Isles, Department of Biochemical Medicine, University of Dundee *Chapter 12.*

Prof. H. Lehmann, Department of Biochemistry, University of Cambridge *Chapter 16.*

Dr. D. G. Nicholls, Department of Psychiatry, University of Dundee *Chapters 7, 9 & 10.*

Dr. D. Thirkell, Department of Biochemistry and Microbiology, University of St Andrews *Chapters 4 & 11.*

Dr. R. Y. Thomson, Department of Biochemistry, University of Glasgow *Chapters 1, 7 & 9.*

Dr. D. B. Walsh, Department of Biochemical Medicine, University of Dundee *Chapters 9 & 19.*

1 Water & solutions

About 70 per cent of the body is water; its importance is indicated by the fact that the loss of as little as 10 to 20 per cent is fatal. The role of water, as a structural molecule and as a participant in biochemical reactions, is often overlooked. This chapter outlines the unique properties of water and the way in which it plays its essential function in the body.

Properties of water

While the O–H bonds of water are covalent, the shared electrons are attracted more strongly towards the oxygen nucleus with its greater number of protons. This leads to a slight negative charge in this region, and a slight positive charge in the region of the hydrogen nuclei. These charges give the molecule an electrical polarity and water is thus a *polar* molecule. When water molecules are close to one another, these attractions (hydrogen bonds) lead to their alignment as shown in figure 1.1.

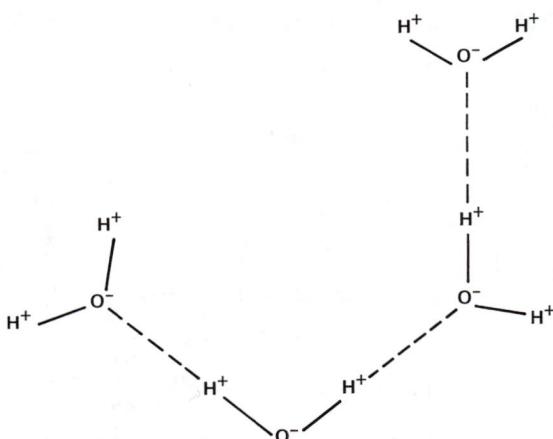

Fig. 1.1 The role of hydrogen bonds (shown by the dotted lines) in the structure of water

Although hydrogen bonds are weaker than covalent bonds they are considerably stronger than the van der Waals forces which exist between uncharged molecules. An illustration of the significance of these bonds is provided by the high boiling point of water, a measure of the amount of energy required to break the bonds. In comparison H_2S, which does not possess hydrogen bonds, boils at $-61\ °C$.

Solutions

When a solid dissolves in a liquid, three processes take place: (1) the bonds between the molecules or ions of the solid are broken, (2) the bonds between the molecules of the liquid are broken and (3) a form of bonding occurs between the molecules of the solvent and the molecules or ions of the solute.

Thus water, a highly polar liquid, is a good solvent for ions and polar molecules. For example, water can dissolve sodium chloride, a solid whose inter-ionic bonds are electrostatic bonds between full positive and negative charges (fig. 1.2). When sodium chloride dissolves in water the ions become arranged so that sodium ions are adjacent to the negatively charged oxygen atoms while the chloride ions are surrounded by positively charged hydrogen atoms (fig. 1.3).

Compounds which are neither ionised nor polar do not readily dissolve in water. Non-polar compounds such as benzene and ether, in which most of the bonds are non-specific van der Waals forces, are completely miscible with one another but not with water. Some solvents such as ethanol have properties intermediate between those of water and benzene.

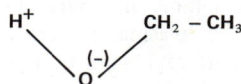

1

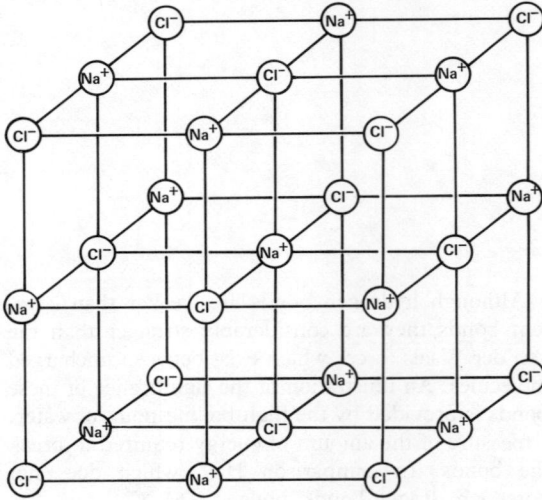

Fig. 1.2 The regular arrangement of atoms of sodium and chloride in a crystal of sodium chloride

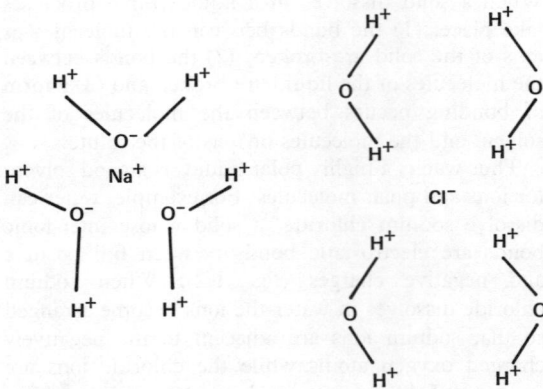

Fig. 1.3 Arrangement of water molecules around sodium and chloride ions in solution

Solubilisation of non-polar substances. While many of the molecules in tissues are polar certain non-polar compounds, notably lipids, must be carried in solution around the body to take part in biochemical reactions. Such non-polar molecules may be made soluble by being attached to polar molecules which are themselves soluble in water. One example is provided by the plasma proteins, in particular albumin, which are capable of transporting insoluble substances like bilirubin (p. 199) and many drugs.

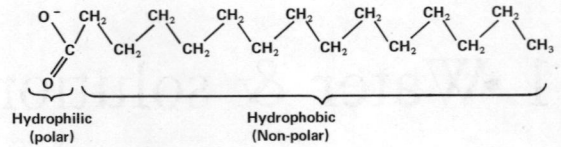

Fig. 1.4 Fatty acid molecule to show the polar carboxyl group and the long non-polar hydrocarbon chain

The albumin molecule has many charged side-chains and is water-soluble. It also has non-polar areas which provide the binding sites for the non-polar molecules.

Another way in which non-polar substances can be made water-soluble is the emulsification of fats to form *micelles*. Fatty acid molecules have long non-polar hydrocarbon chains and polar carboxyl groups at one end (fig. 1.4). The hydrocarbon chain attracts other hydrocarbon chains or other non-polar molecules by van der Waals forces; it does not attract polar molecules such as water and is described as *hydrophobic*. The carboxyl group attracts water and other polar molecules and is described as *hydrophilic*.

Many similar molecules with polar and non-polar components are recognised. The bile acids described in Chapter 4 are important examples (fig. 1.5).

Bile acids are secreted in the bile as micelles with their hydrophobic surfaces toward the interior and the hydrophilic surfaces on the outside (fig. 1.6a). On entering the duodenum fatty acids are taken up to give *mixed micelles*; these also contain monoglycerides and, in the hydrophobic interior, non-polar molecules such as triglycerides, cholesterol and fat-soluble

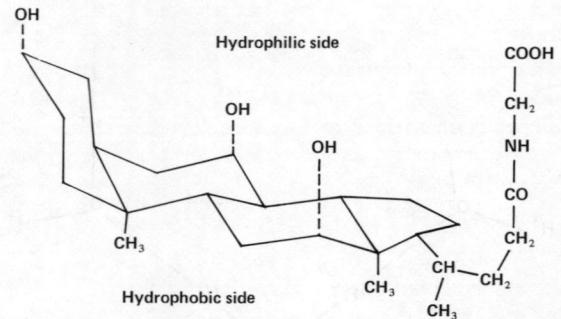

Fig. 1.5 Molecule of glycocholic acid to show that all the hydroxyl groups are on one side of the rigid ring structure, while the other side is predominantly hydrophobic

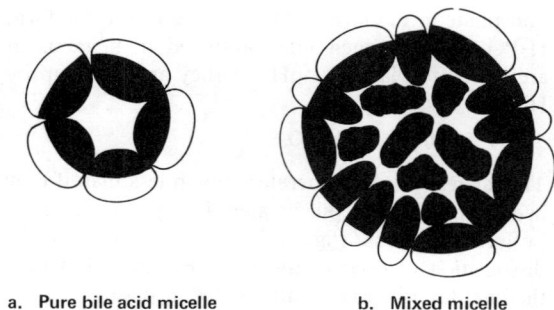

a. Pure bile acid micelle   b. Mixed micelle

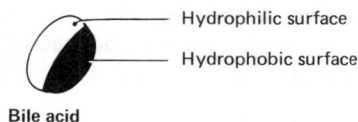

— Hydrophilic surface

— Hydrophobic surface

Bile acid

— Hydrophilic carboxyl group

— Hydrophobic hydrocarbon tail

Fatty acid

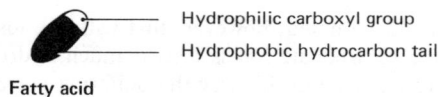

Triglycerides and fat-soluble vitamins

Fig. 1.6 Cross-sections of (a) a pure bile acid micelle and (b) a mixed micelle containing bile acids, fatty acids, triglycerides and fat-soluble vitamins

vitamins (fig. 1.6b). Thus all these substances can be carried in the aqueous environment of the intestinal lumen in particles small enough to be dealt with by the mucosal cells.

Ions in aqueous solution. The extent to which the ions of a solute dissociate in aqueous solution depends on their attraction to each other relative to their attraction to the water molecules. The dissociation of a substance in water can be represented as shown here for acetic acid:

$$CH_3COOH \rightleftharpoons CH_3COO^- + H^+$$

or, more accurately:

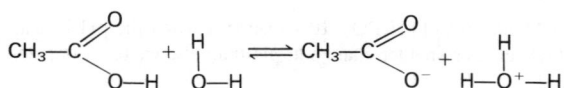

Water itself ionises:

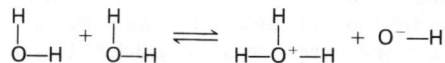

The dissociation constant is very small; at 20 °C

$$[H^+] \times [OH^-] = 10^{-14}$$

where $[H^+]$ is the concentration of hydrogen ions and $[OH^-]$ is that of hydroxyl ions, both being measured as mol/litre. In pure water the concentrations of hydrogen and hydroxyl ions are theoretically equal at 10^{-7} mol/l (100 nmol/l).

Hydrogen ion concentration

The range of hydrogen ion concentrations which is possible in aqueous solutions is so wide, between 10^{-14} mol/l for concentrated alkalis and 1 mol/l for concentrated acids, that the pH scale is used. pH is the negative logarithm, to base 10, of the hydrogen ion concentration in mol/l. Thus pH 0 = 1 mol H^+/l, pH 7 = 10^{-7} mol/l (100 nmol/l) and pH 14 = 10^{-14} mol/l.

The concentration of hydrogen ion in most body fluids is very low compared with that of other ions such as sodium, potassium or chloride. Gastric juice, with a hydrogen ion concentration as high as 100 mmol/l (pH 1.0), is exceptional; the cells of the gastric mucosa are able to secrete hydrogen ions at a concentration more than one million times higher than that of plasma.

The pH of arterial blood is normally in the range 7.36 to 7.44; in severe illness values as low as 6.9 or as high as 7.8 may be found. The intracellular pH is about 7.2 in red blood cells and 6.9 in muscle.

The pH of the urine varies widely, between 4.8 and 8.0. Urine is more commonly acid than alkaline because it is the principal route for the excretion of hydrogen ions produced in metabolism.

Buffering of hydrogen ions. Normal functioning of the tissues requires that the hydrogen ion concentration of the extracellular fluid should remain within narrow limits. Metabolism in man leads to the production of some 50 mmol of hydrogen ions daily. These have to be removed from the body and the main methods for their excretion are described later (Chap. 19).

3

There is inevitably a delay between the production of hydrogen ions in the tissues and their disposal. While in the blood, hydrogen ions must be buffered to prevent the development of patches of acidity which would affect enzyme activity and the solubility of other ions.

The general equation for the buffering of hydrogen ions is

$$H^+ + Buffer^- \rightleftharpoons H.Buffer$$

The extent to which a buffer can accept hydrogen ions depends not only on the concentration of the buffer but also on its pK_a, the pH at which 50 per cent of the buffer is in the H.Buffer form. In order to take up H^+ ions and to discharge them at the excretory organs, the buffers of the body must have a pK_a near the plasma's normal pH of 7.4. A buffer is most effective when its pK_a is nearing the prevailing pH.

For example the equilibrium of phosphate in plasma is

$$H_2PO_4^- \rightleftharpoons HPO_4^{2-} + H^+$$

Since the pK_a of this reaction is 6.7 the concentration of HPO_4^{2-} and $H_2PO_4^-$ are equal at pH 6.7 while at pH 7.4 there is approximately five times as much

phosphate in the form HPO_4^{2-} as there is in the form $H_2PO_4^-$. If hydrogen ions produced in metabolism are added to plasma at pH 7.4 they are taken up by HPO_4^{2-}:

$$H^+ + HPO_4^{2-} \rightleftharpoons H_2PO_4^-$$

The change in pH is therefore much less than if they had been added to pure water. Phosphate has three buffering regions (fig. 1.7) corresponding to the dissociation of each of the three H^+ ions, but only the one described is of importance in plasma.

The other major buffers in plasma are bicarbonate and protein. Bicarbonate is of particular importance because it also provides a means for disposing of surplus hydrogen ions as carbonic acid; the CO_2 is lost in the expired air:

$$H^+ + HCO_3^- \rightleftharpoons H_2CO_3$$

$$H_2CO_3 \rightleftharpoons H_2O + CO_2$$

It should be noted, however, that with the loss of CO_2, the bicarbonate ion is lost; if much hydrogen ion were removed in this way the buffering power of the plasma would be reduced.

Protein buffers are of interest in that each is confined to one body compartment and therefore cannot affect the buffering of other areas of the body.

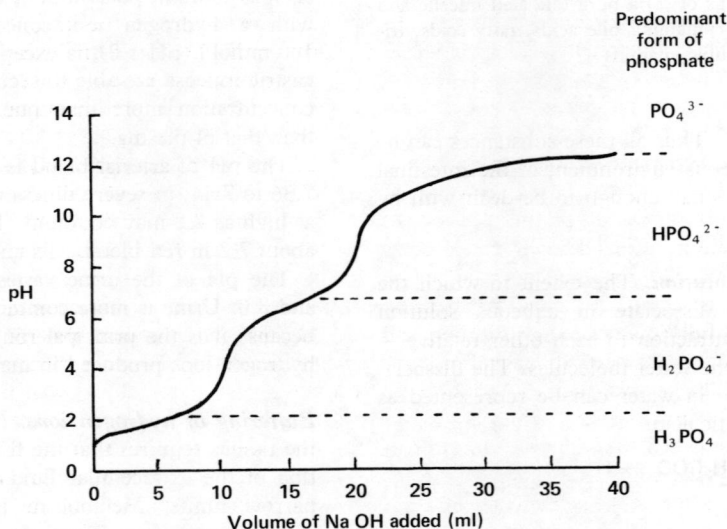

Fig. 1.7 Effect of successive small additions of 2.0 mol/l NaOH to 20 ml 1.0 mol/l H_3PO_4. It is seen that when the pH is near the pK_a values (indicated by the dotted lines) each addition of NaOH causes a smaller change in pH than elsewhere. (Courtesy of J. Burns)

They are able, because their constituent amino acids have different pK_a values, to buffer over a wider range of pH than bicarbonate or phosphate alone.

The most important buffer of the blood is not in the plasma but in the red cells. Haemoglobin has a large number of ionising amino acid side-chains. The haemoglobin molecule has the interesting property of altering its affinity for oxygen as the pH changes. Thus in an area of high hydrogen ion concentration, such as occurs in the tissues, oxygen is released from the molecule, and in a region of low hydrogen ion concentration, such as the capillary bed of the lung, oxygen is taken up (p. 204).

Osmotic pressure

While the concentration of individual ions in a body fluid is important, the overall concentration of solutes also needs to be noted because of its effect on the osmotic pressure. One particle of any solute, an ion or a molecule, whatever its size or weight, has the same effect on the osmotic pressure of a solution. Similarly it has the same effect on the depression of the freezing point of the solvent.

While the biochemical importance of overall solute concentration lies in its effect on osmotic pressure, the simplest method for estimating it is the measurement of the depression of freezing point. Instruments for this are available commercially and give reproducible readings of *osmolality* in moles per kilogram of water. In normal people plasma has an osmolality between 281 and 297 mmol/kg. No direct methods are available for the measurement of *osmolarity* which is expressed in mmol/l.

The theoretical osmolality of plasma, calculated from the concentrations of its known constituents (Table 1.1) is about 325 mmol/kg but the actual plasma osmolality is about 291 mmol/kg because electrolytes are not fully dissociated in solution and because some substances are partly bound to protein.

In normal people the osmolality of the urine varies between 50 and 1400 mmol/kg; low values are found early in renal disease when the kidney is not able to produce a concentrated urine. If an osmometer is not available, a crude measure of the solute concentration in a urine sample is given by the specific gravity; in normal people this is between 1.001 and 1.050.

The importance of osmolality is that, since water can penetrate membranes whereas other molecules

Table 1.1 Composition of mixed venous plasma

	Mean values for fasting adults	
	mg/l	mmol/l
Cations		
Sodium	3 265	142.0
Potassium	164	4.2
Calcium	96	2.4
Magnesium	21	0.9
Anions		
Chloride	3 568	100.5
Bicarbonate	1 586	26.0
Phosphate (mainly HPO_4^{2-})	96	1.0
Sulphate	48	0.5
Organic anions	210	6.0
Proteins	66 300	2.0
Uncharged molecules		
Urea	270	4.5
Glucose	900	5.0
Others	3 000	10.0
Total solute	79 524	305

A total solute concentration of 305 mmol/l is equivalent to 325 mmol/kg water.

and ions cannot, the transfer of water from one compartment to another depends on the osmotic pressure within each.

One example of the importance of osmolality is provided by the capillaries. The hydrostatic pressure exerted on the plasma by the heart is sufficient to drive water out of the capillaries and into the interstitial fluid but, largely due to the presence of proteins in the plasma, water does not normally accumulate in the interstitial space. Most of the solutes of the plasma are free to move across the capillary endothelium but the proteins are confined to the plasma. There is therefore an osmotic gradient across the endothelium known as the *colloid osmotic pressure*. Although this is small (about 2.0 mmol/kg equivalent to about 30 mmHg or 4.0 kPa) it is sufficient to promote the return of water from the extracellular space into the venous side of the capillaries (fig. 1.8). When the plasma albumin level is low in disease (p. 20), the colloid osmotic pressure of the plasma is low and water may accumulate in the extracellular space (*oedema*).

Interstitial fluid

Hydrostatic
pressure 4.3 kPa

Hydrostatic pressure
1.6 kPa

Lumen of capillary

Colloid osmotic
pressure 3.1 kPa

Colloid osmotic
pressure 3.1 kPa

Interstitial fluid

Arterial end:
net outflow of fluid

Venous end:
net inflow of fluid

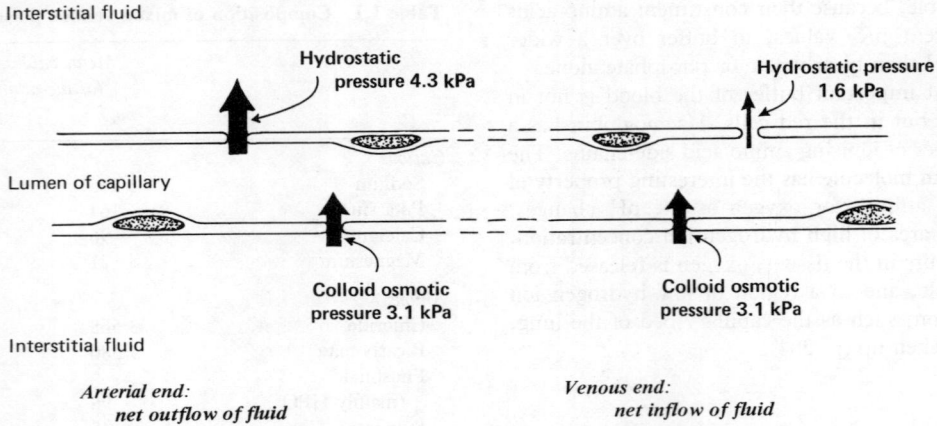

Fig. 1.8 Typical figures for the pressures affecting the flow of fluid through the wall of a capillary. This is a simplified explanation; the capillary endothelium is not completely impermeable to proteins and the hydrostatic pressure within the capillary varies with many factors including the height of the vessel in relation to the heart and the state of the arterioles

A further example of the significance of osmotic pressure is provided by the renal medulla which is an area of high osmolality due to the concentration of sodium and urea by the loop of Henle. Urine passing down the collecting ducts becomes more concentrated as water is drawn back into the renal medulla by osmotic pressure.

Further reading

Chang, R. (1977) *Physical Chemistry with Applications to Biological Systems.* New York: Macmillan

Maxwell, M. H. and Kleeman, C. R. (1980) *Clinical Disorders of Fluid and Electrolyte Metabolism*, 3rd edn. New York: McGraw-Hill

CRP RYT

2 The proteins

A PROTEIN is a polymer of amino acids folded into a unique three-dimensional structure. Just as many styles of house can be built from the same basic building materials, the vast range of structures found among proteins is achieved by combining a small number of basic units (twenty amino acids) in different sequences in a protein chain. In the human body, proteins are found as *enzymes* which catalyse metabolic reactions (Chap. 8), as *structural proteins* which provide mechanical support in the skin, connective tissues, tendons and bones (Chap. 17) and as *transport proteins* which bind substances in the blood and move substances across membranes (Chap. 18). Proteins are also responsible for muscle contraction (myosin and actin), defence against foreign substances (immunoglobulins) and the detection of light in the retina (visual pigments). Proteins are found as hormones, such as insulin and growth hormone, as receptors on a cell surface and as integral parts of chromosomes and ribosomes.

The fundamental units from which all proteins are constructed are amino acids; many proteins contain nothing else but some proteins have additional components as integral parts of their molecules. Examples of these are the metallo-proteins such as haemoglobin (Chap. 16) and the cytochromes (Chap. 10), the glycoproteins containing carbohydrates (Chap. 3), and the lipoproteins responsible for lipid transport in the plasma (Chap. 11). Proteins are synthesised from a pool of twenty different amino acids whose structures are given in an appendix (p. 25). Some proteins contain additional amino acids (Table 2.1) which are formed by enzymatic modification of one of the original twenty after the initial protein chain has been synthesised.

The wide variety of functions of proteins is matched by their wide variety of physical and chemical characteristics. For example their molecular weights vary from 5000 to several million, their shapes vary from globular to linear and their surfaces are capable of binding a variety of substances. Proteins have a finite life span and are degraded eventually to their constituent amino acids, to be replaced by newly synthesised proteins. The dynamic equilibrium between the body proteins and the amino acid pool is described in Chapter 15.

Table 2.1 Modified amino acids found in proteins

Original amino acid	Modified amino acid	Occurrence
Cysteine	Cystine	Covalent cross-links in many proteins
Proline	4-Hydroxyproline (Hyp)	Collagen (Chap. 17)
Lysine	5-Hydroxylysine (Hyl)	Collagen
Lysine	Desmosine, isodesmosine	Elastin (Chap. 17)
Lysine	Mono-, di- and trimethyl lysines	Histones, myosin
Histidine	3-Methylhistidine	Certain types of myosin (Chap. 15)
Tyrosine, histidine	Iodinated derivatives such as mono- and di-iodotyrosine, monoiodohistidine, tri-iodothyronine	Thyroglobulin (Chap. 22)
Serine	Phosphoserine	Many regulatory enzymes (Chap. 12)
Glutamate	γ-carboxy-glutamate	Prothrombin (p. 172)
Glutamate	Pyroglutamate	Certain hypothalamic hormones
Arginine	Ornithine	Certain binding proteins

Amino acids

Amino acids have the general formula

$$\begin{array}{c} COOH \\ | \\ R—C—H \\ | \\ NH_2 \end{array}$$

The central carbon atom, known as the α-carbon atom, has attached to it an amino group ($-NH_2$), a carboxyl group ($-COOH$), a hydrogen atom and a side-chain usually designated by the letter R. Except in the case of glycine (in which R = H), the α-carbon atom is joined to four different groups and is thus *asymmetric*. Amino acids other than glycine therefore exist in two possible isomeric forms, a D-form and an L-form which are mirror images.

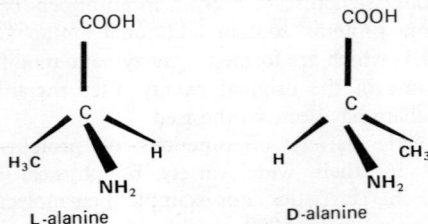

L-alanine   D-alanine

Amino acids of both types are found in microorganisms but in higher animals and plants only L-amino acids are found; only L-amino acids are used in the synthesis of proteins.

In addition to those found in proteins many other amino acids are found in nature. Some of these are of great physiological importance (Table 2.2).

Properties of the amino acids

Affinity for water. Amino acids may be classified into groups according to the affinity of their side-chains for water (Table 2.3). Side-chains with a high affinity for water are described as *hydrophilic;* some are charged and attract water dipoles while the uncharged hydrophilic side-chains readily form hydrogen bonds with water molecules. In contrast, *hydrophobic* side-chains do not readily interact with water and in aqueous solution such side-chains tend to attract each other, to the exclusion of water.

Table 2.3 Classification of the amino acids according to the affinity of their side-chains for water

Side-chain uncharged at pH 6.0

Hydrophilic

Glycine	Asparagine
Serine	Glutamine
Threonine	Cysteine
Methionine	

Hydrophobic

Alanine	Proline
Valine	Phenylalanine
Leucine	Tyrosine
Isoleucine	Tryptophan

Side-chain charged at pH 6.0 (all hydrophilic)

Basic

Lysine	Histidine
Arginine	

Acidic

Aspartate	Glutamate

Table 2.2 Some amino acids of physiological importance not found in proteins

Amino acid	Structure	Occurrence, function etc.
β-alanine	$H_3N^+—CH_2—CH_2—COO^-$	A component of coenzyme A (p. 54)
γ-Aminobutyric acid (GABA)	$H_3N^+—CH_2—CH_2—CH_2—COO^-$	Inhibitory transmitter in the central nervous system
δ-Amino-laevulinic acid (δALA)	$H_3N^+—CH_2—CO—CH_2—CH_2—COO^-$	An intermediate in the synthesis of haem (Chap. 16)
Creatine	$\begin{array}{c} NH\ CH_3 \\ \|\ \ \| \\ H_3N^+—C—N—CH_2—COO^- \end{array}$	An important constituent of muscle (Chap. 15)
Taurine	$H_3N^+—CH_2—CH_2—SO_3^-$	A constituent of muscle and a component of taurocholate, a bile salt (Chap. 15)

Ionisation. Both the α-amino and the α-carboxyl groups of amino acids ionise in aqueous solutions. The extent to which either group is ionised depends on the pH of the solution.

At pH values of about 6.0, the un-ionised molecule is in equilibrium with a molecule in which both α-groups are ionised.

$$H_2N-\overset{\overset{\displaystyle R}{|}}{CH}-COOH \rightleftharpoons H_3N^+-\overset{\overset{\displaystyle R}{|}}{CH}-COO^-$$

The equilibrium is strongly in favour of the ionised molecule which at this pH has little or no *net* charge and is known as a *dipolar ion* or *zwitterion*. At low values of pH the carboxyl group ceases to be ionised while at a high pH the amino group loses a proton:

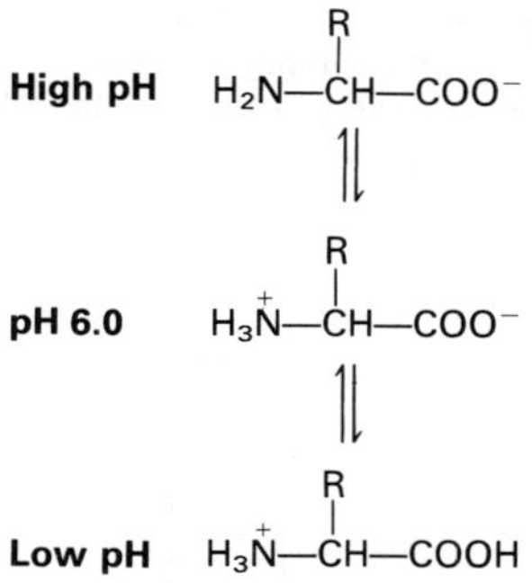

High pH $\quad H_2N-\overset{\overset{\displaystyle R}{|}}{CH}-COO^-$

$\qquad\qquad\qquad \Updownarrow$

pH 6.0 $\quad H_3\overset{+}{N}-\overset{\overset{\displaystyle R}{|}}{CH}-COO^-$

$\qquad\qquad\qquad \Updownarrow$

Low pH $\quad H_3\overset{+}{N}-\overset{\overset{\displaystyle R}{|}}{CH}-COOH$

Every amino acid therefore has at least two pK_a values, one for the amino group and one for the α-carboxyl group. In the case of glycine the values are 9.60 and 2.34, respectively (fig. 2.1). However, pK_a values vary between amino acids and this variation must reflect differences between the side-chains. Table 2.4 shows the pK_a values for the

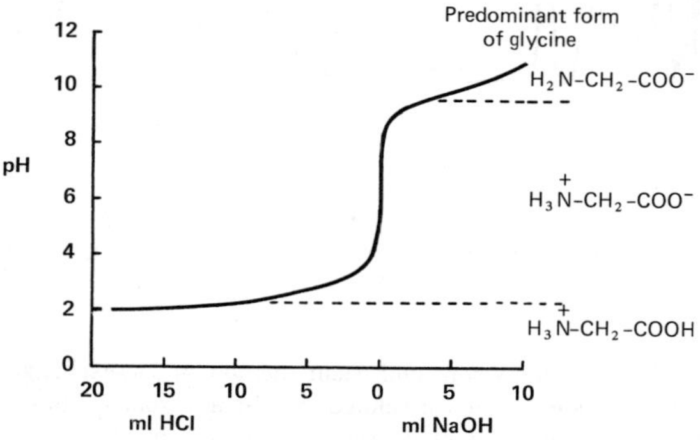

Table 2.4 pK_a values for some amino acids

	α-carboxyl	α-amino
Glycine	2.34	9.60
Alanine	2.34	9.69
Tryptophan	2.38	9.39
Serine	2.21	9.15

α-amino and α-carboxyl groups on three amino acids. Amino acids with an ionising side-chain have an additional pK_a value; those for lysine and glutamate, for example, are shown in Table 2.5.

Table 2.5 pK_a values for some amino acids with ionising side-chains

	α-carboxyl	α-amino	Side-chain
Lysine	2.18	8.95	10.53
Glutamate	2.13	9.76	4.31
Aspartate	1.88	9.60	3.65

At any particular pH value there are differences in the net charge on different amino acid molecules. Such differences form the basis of two methods for separating amino acids, namely ion-exchange chromatography and electrophoresis.

Side-chain reactivity. One amino acid, cysteine, has a side-chain containing a sulphydryl (–SH) group and the sulphydryl groups of two cysteine molecules can be oxidised together to give a disulphide 'bridge' (–S–S–) between the two molecules. The resulting amino acid is known as cystine.

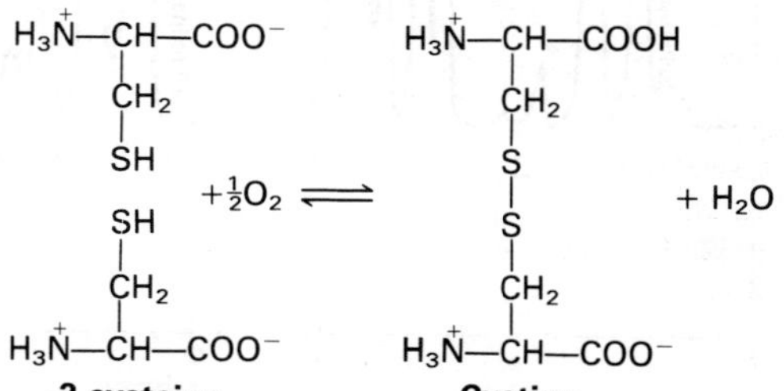

2 cysteine molecules $\quad +\tfrac{1}{2}O_2 \rightleftharpoons$ **Cystine** $\quad + H_2O$

Disulphide bridges have an important role in the structure of proteins. Insulin, for example, has three disulphide bridges (fig. 12.11, p. 139).

Fig. 2.1 Titration curve for glycine (Courtesy of J. Burns)

Other important reactions of side-chains include the modifications of lysine and hydroxylysine which form the cross-links between collagen molecules (p. 212) and the desmosine formed from lysine which links the peptide chains of elastin (p. 214).

Amino acids in plasma and urine

The amino acids of the plasma form a pool to which are added those derived from the diet and from the breakdown of tissue proteins, and from which are taken those needed for protein synthesis (Chap. 14). The total amino acid content of the plasma is 2.4 to 5.0 mmol/l (3.4 to 7.0 mg N/100 ml). Techniques for the separation and measurement of the individual amino acids include partition chromatography, ion-exchange chromatography (fig. 2.2), electrophoresis and various combinations of these techniques (fig. 2.3). Abnormally high values for particular amino

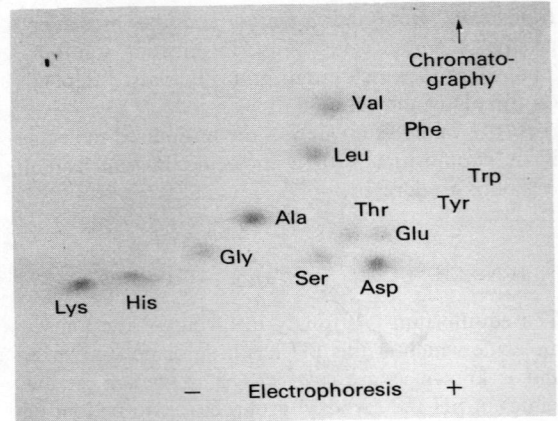

Fig. 2.3 Separation of amino acids in normal plasma on a thin layer of cellulose by chromatography followed by electrophoresis at right angles. (Courtesy of Patricia Macfarlane and Sheila Sharp)

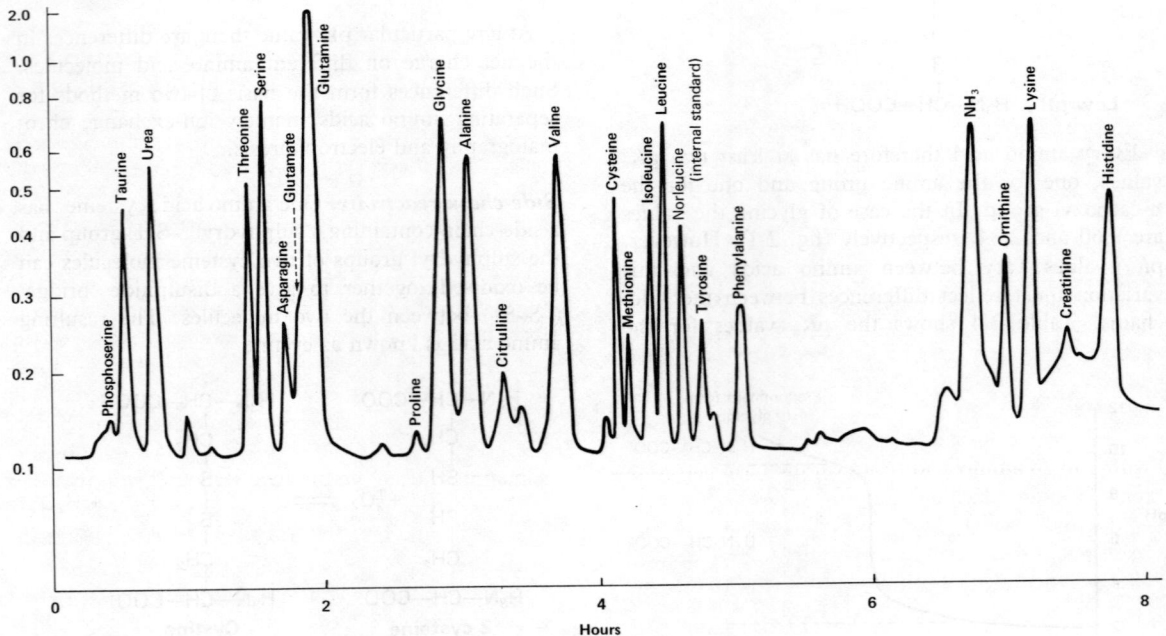

Fig. 2.2 Amino acid analysis of a plasma sample from a normal child by the Moore and Stein method of ion-exchange chromatography (Spackman *et al.* (1958) *Analytical Chemistry* **30**, 1190). The amino acids are eluted in sequence from a column of ion-exchange resin by a buffer of progressively changing composition. The amino acids are demonstrated by the colour formation with ninhydrin. (Courtesy of Iris Mackie)

acids are found in some inborn errors of metabolism (Chap. 15).

Normal subjects excrete some 85 mmol free amino acids daily in the urine. High values for particular amino acids in urine are found in patients with high levels in plasma (*overflow aminoaciduria*) and in patients with defects of renal tubular reabsorption of amino acids (*renal aminoaciduria*((Chap. 18).

Peptides and polypeptides

A peptide is the condensation product of two or more amino acids. In the *peptide bond* which links two amino acids the α-amino group of one amino acid is condensed with the α-carboxyl group of the next with the loss of a molecule of water:

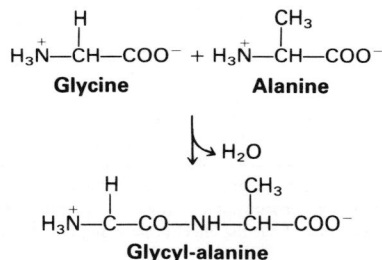

$$\underset{\textbf{Glycine}}{H_3\overset{+}{N}-\underset{\underset{H}{|}}{C}H-COO^-} + \underset{\textbf{Alanine}}{H_3\overset{+}{N}-\underset{\underset{H}{|}}{\overset{\overset{CH_3}{|}}{C}}H-COO^-}$$

$$\searrow H_2O$$

$$\underset{\textbf{Glycyl-alanine}}{H_3\overset{+}{N}-\underset{\underset{H}{|}}{C}-CO-NH-\underset{\underset{}{}}{\overset{\overset{CH_3}{|}}{C}}H-COO^-}$$

A peptide, such as glycyl-alanine, formed from two amino acids is known as a *dipeptide* while compounds derived from three or four amino acids are known as *tripeptides* and *tetrapeptides* respectively. Since it would be laborious to write out the structural formulae of the larger peptides in full shortened forms are used. These replace the side-chain and its associated –NH–CH–CO– group with a three-letter code to represent the amino acid from which it was derived (p. 25). The dipeptide glycyl-alanine thus becomes *Gly-Ala*.

The term *peptide* is generally used for a product with 2 to 20 amino acid residues, the term *polypeptide* for a compound containing 20 to 50 residues while compounds with more than 50 residues in their native state are usually called *proteins*. Many peptides and polypeptides of physiological importance have been isolated and some are included in Table 2.6.

Determination of amino acid composition

A knowledge of the amino acid composition of a peptide or polypeptide is an important first step in identifying its structure. For this determination it must first be hydrolysed to its constituent amino acids, for example by incubation with concentrated hydrochloric acid. The amino acid mixture produced is usually separated by ion-exchange chromatography.

Determination of amino acid sequence

Determination of the sequence of amino acids is equally important. For example the amino acid composition of vasopressin (Table 2.6) is *Arg, Gly, Asn, Gln, Cys$_2$, Pro, Phe, Tyr*, but this gives no indication as to which of the many thousand possible combinations is the biologically active hormone.

An important instrument used for determining amino acid sequence is the *sequenator*. Amino acids are chemically removed one after another from the amino end of the peptide chain, and identified. The process is repeated until the full sequence is obtained, or until the side-reactions become so significant that the identity of the amino acid is not clear. Runs of up to 50 residues have been successfully completed. No one method for the determination of an amino acid sequence gives all the information needed. Typically, a polypeptide is first broken down enzymically or chemically into smaller fragments whose sequence can be determined separately. The results are then pieced together to give the full sequence.

Knowledge of amino acid sequences is important in a number of ways. For example oxytocin and vasopressin, two peptide hormones of the posterior pituitary gland, have similar structures, differing from each other only in two amino acid residues (fig. 2.4). They have quite different physiological functions; oxytocin causes contraction of the uterus in labour while vasopressin promotes water reabsorption in the renal tubules. Oxytocin in high doses is sometimes used to induce labour but with its structural similarity to vasopressin it has a vasopressin-like action which can cause serious side effects due to excessive water retention at these doses.

Determination of amino acid sequences may have a practical value in other ways. Many peptide hormones are secreted only in minute amounts and it may be impossible to obtain enough animal material for clinical use. The hypothalamic hormone LHRH from the pig is biologically active in man but many thousand pigs are needed to produce one milligram

11

Table 2.6 Amino acid composition of some peptide and polypeptide hormones

Amino acids	TRH	Met-enkephalin	Oxytocin	Vasopressin	LHRH	ACTH	Insulin
Lysine						4	1
Arginine				1	1	3	1
Histidine	1				1	1	2
Aspartic acid						1	
Glutamic acid	1				1	5	4
Glycine		2	1	1	2	3	4
Serine					1	3	3
Threonine							1
Asparagine			1	1		1	3
Glutamine			1	1			3
Cysteine			2	2			6
Methionine		1				1	
Alanine						3	3
Valine						3	5
Leucine			1		1	1	6
Isoleucine			1				1
Proline	1		1	1	1	4	1
Phenylalanine		1		1		3	3
Tyrosine		1	1	1	1	2	4
Tryptophan					1	1	
TOTAL	3	5	9	9	10	39	51

TRH (thyrotropin releasing hormone) and LHRH (luteinising hormone releasing hormone) are hypothalamic hormones which stimulate the anterior pituitary. Oxytocin and vasopressin are hormones of the posterior pituitary. ACTH (adrenocorticotrophic hormone) is produced by the anterior pituitary and stimulates the adrenal cortex. Insulin (Chap. 12) is produced by the pancreas and has the effect of lowering blood glucose. Met-enkephalin is a morphine-like peptide secreted during episodes of pain.

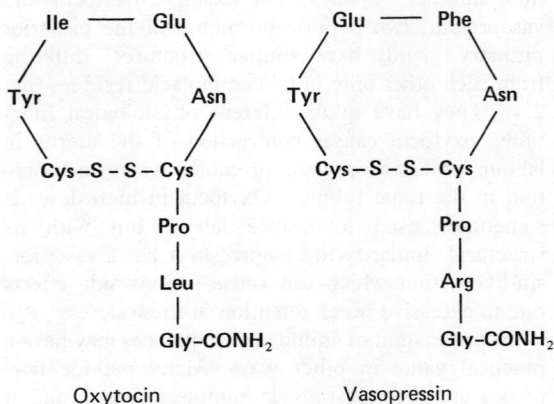

Fig. 2.4 The amino acid sequences of human oxytocin and vasopressin. The CONH$_2$ indicates a carboxy-terminal in which the OH group has been replaced by NH$_2$

of the pure hormone. The use of pigs as the source of the hormone would be very expensive but knowledge of the amino acid sequence has allowed the hormone to be synthesised chemically.

Another example of the value of sequence studies is provided by the hormone gastrin which stimulates gastric secretion. It was found that the fragment containing the five amino acids at the amino end was biologically active. This knowledge permitted the production of a synthetic analogue, *pentagastrin* (fig. 2.5) which is used in the investigation of the secretory capacity of the stomach.

Insulin was the first biologically active polypeptide to have its amino acid sequence determined. It consists of two polypeptide chains joined by two disulphide bonds and contains 51 amino acid residues. The insulin molecule is synthesised as an inactive precursor containing 84 residues in a single

$\overset{+}{\text{N}}\text{H}_3$-Phe-Asp-Met-Trp-Gly-Tyr-Ala-(Glu)$_5$-Leu-Trp-Pro-Gly-Glu
Human gastrin I

$\overset{+}{\text{N}}\text{H}_3$-Phe-Asp-Met-Trp-Gly-$\beta$-ala-BOC
Pentagastrin

Fig. 2.5 The five residues at the amino end of the gastrin molecule are common to all forms of gastrin including those from a wide variety of species. A synthetic analogue, pentagastrin, is effective in stimulating gastric secretion. BOC denotes a butyloxycarbonyl group

polypeptide chain (fig. 12.11 p. 139). This *proinsulin* molecule is converted to insulin by the removal of a 33-residue section in the middle of the chain. Many other biologically active peptides, polypeptides and proteins are activated in the same way, so that the active molecule is rapidly available when required without the delay involved in synthesis from amino acids.

Separation methods

Peptides, like amino acids, may be separated by chromatography or by electrophoresis. In one method for detecting abnormalities in protein structure a mixture of peptides is obtained either by enzyme hydrolysis of by chemical cleavage, for example by cyanogen bromide (CNBr) which attacks the peptide bond on the carboxyl side of methionine residues. The mixture can be separated by electrophoresis in one direction and by chromatography at right angles. After staining the peptides appear as an array of spots ('peptide map') characteristic of the protein digested. An example of the method, applied to an abnormal haemoglobin (Chap. 16), is given in figure 2.6.

Proteins

Proteins may be classified in several ways. One method is based on the form in which the proteins are found in the tissues. Thus *soluble proteins* are single proteins with a defined function in the cytoplasm or the plasma. These are mostly globular molecules.

Other proteins are associated as *multi-protein assemblies*, functional units with several components. These include the proteins of flagella, the actin–myosin complex of muscle and multi-enzyme complexes such as that responsible for the synthesis of fatty acids (Chap. 11).

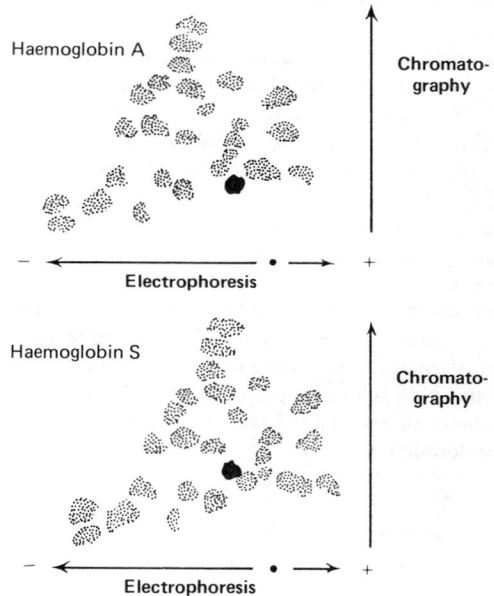

Fig. 2.6 'Peptide maps' produced by trypsin hydrolysis of haemoglobin A and haemoglobin S. Only one peptide (black) is displaced. This contains the abnormal residue (valine) which differs from the glutamate normally present at this position in being less polar and less negatively charged. (After Baglioni, C. (1961) *Biochimica et Biophysica Acta* **48**, 392)

Some proteins are associated with membranes (p. 63). These may be integral components of the membrane, such as the proteins responsible for the transport of solutes, or may be attached to the surface of a membrane such as the cytochromes of the respiratory chain (Chap. 9).

Proteins may also be classified by their chemical composition as *simple proteins* which contain only amino acids, and *conjugated proteins* which have additional components (Table 2.7).

13

Table 2.7 Composition of conjugated proteins

Glycoprotein	Protein + carbohydrate
Lipoprotein	Protein + lipid
Flavoprotein	Protein + flavin
Haemoprotein	Protein + haem
Metalloprotein	Protein + metal ion(s)

Structure of proteins

The full amino acid sequence of a number of proteins has now been determined. This information has given insight into the cause of some inherited diseases including the disorders of the globin component of haemoglobin, the *haemoglobinopathies* (Chap. 16).

The recording of the structure of a polypeptide chain simply as ---Arg-Glu-Gly-Pro-Lys--- ignores the position of the individual groups in space. Even if the chain were extended fully, it would not be linear. The α-carbon atom has a tetrahedral arrangement of bonds around it and while rotation around each bond in the chain might be expected this is limited by the tendency of one of the bonds to assume the character of a double bond:

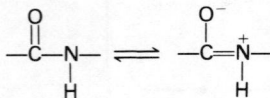

As a result the four atoms shown are confined to one plane, indicated in figure 2.7. There is therefore a limit to the extent to which the chain can fold.

Many soluble proteins have a compact globular structure which is maintained in two ways. First the

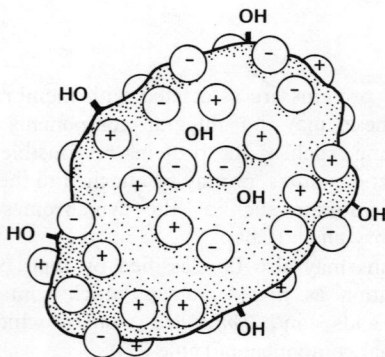

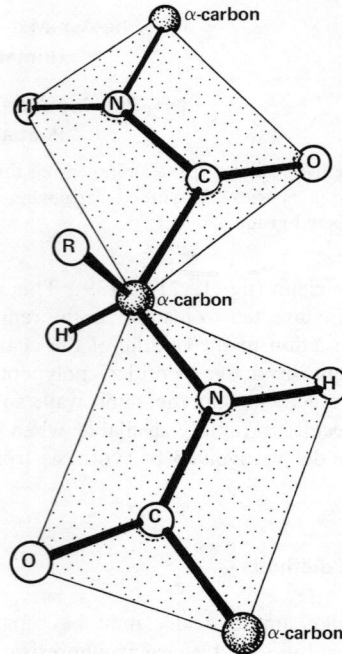

Fig. 2.7 In a polypeptide chain rotation is only possible at the α-carbon atom. The 'double bond' character of the peptide linkage confines to one plane the groups attached to the C and N atoms involved. (After Dickerson, R. E. and Geis, I. (1969). *The Structure and Action of Proteins*. New York: Harper and Row)

protein is folded so that the hydrophobic side-chains are kept away from the aqueous environment by being buried inside the molecule while the hydrophilic side-chains are on the surface. In this respect the protein molecule resembles a micelle in which the hydrophilic heads project into the aqueous medium while the hydrophobic tails are together in the centre (p. 3).

Secondly the three-dimensional structure of a protein is maintained by bonds between residues, notably covalent disulphide bridges (p. 9) and hydrogen bonds between amide and carboxyl groups:

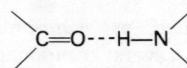

Hydrogen bonds play an important part in two special types of protein structure, the helix and the pleated sheet. In the *helix* the polypeptide chain

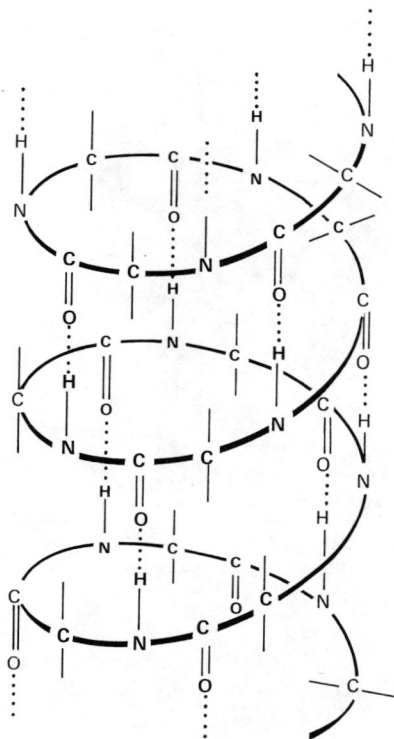

Fig. 2.8 An α-helix with a pitch of 0.54 nm and 3.6 residues per turn. Hydrogen bonds (dashed lines) link the amide groups of one turn with the carboxyl groups on the next. In the helices of most proteins the spiral is right handed as shown; an exception is collagen (Chap. 17) with a left handed α-helix

forms a spiral with a series of regularly spaced hydrogen bonds holding the structure together. The α-helix allows the greatest number of hydrogen bonds to be made (fig. 2.8). Proteins such as haemoglobin and myoglobin (fig. 16.11, p. 202) consist almost entirely of lengths of α-helix joined by 'hinge regions' in which proline seems to play an important part. Other proteins with substantial lengths of α-helix are the structural proteins collagen and α-keratin.

In the *pleated sheet* two sections of the same polypeptide chain, or two polypeptides, lie close to each other so that hydrogen bonds form between them (fig. 2.9). The structure is 'pleated' because of the limitations to the folding of polypeptide chains and because it allows the side-chains to be accommodated above and below the plane of the sheet. Bulky side-chains distort the structure and in some globular proteins the pleated structure is barely recognisable. In contrast some fibrous proteins such as β-keratin have large amounts of undistorted pleated sheet.

Drawings of proteins as in figures 2.8 and 2.9 might suggest a loose structure with much space between the chains. However the three-dimensional structure of many proteins has now been determined by X-ray crystallography; if a model of a protein is constructed so that each atom is depicted with its 'true volume' of electron shells, the structure is seen to be very compact (fig. 2.10).

Subunit structure. An individual polypeptide chain in a globular protein seldom contains more than 1000

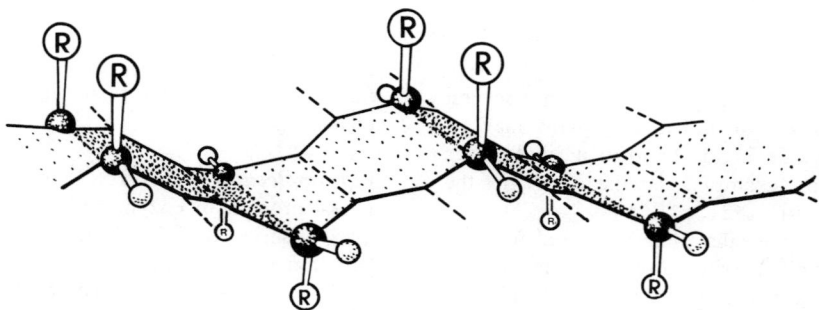

Fig. 2.9 The β-pleated sheet with side-chains (R) on each side. The dashed lines denote hydrogen bonds. Two forms of pleated sheet are recognised; in one the two chains run in the same direction (parallel) while in the other they run in opposite directions (antiparallel). (After Dickerson, R. E. and Geis, I. (1969) *The Structure and Action of Proteins.* New York: Harper and Row)

15

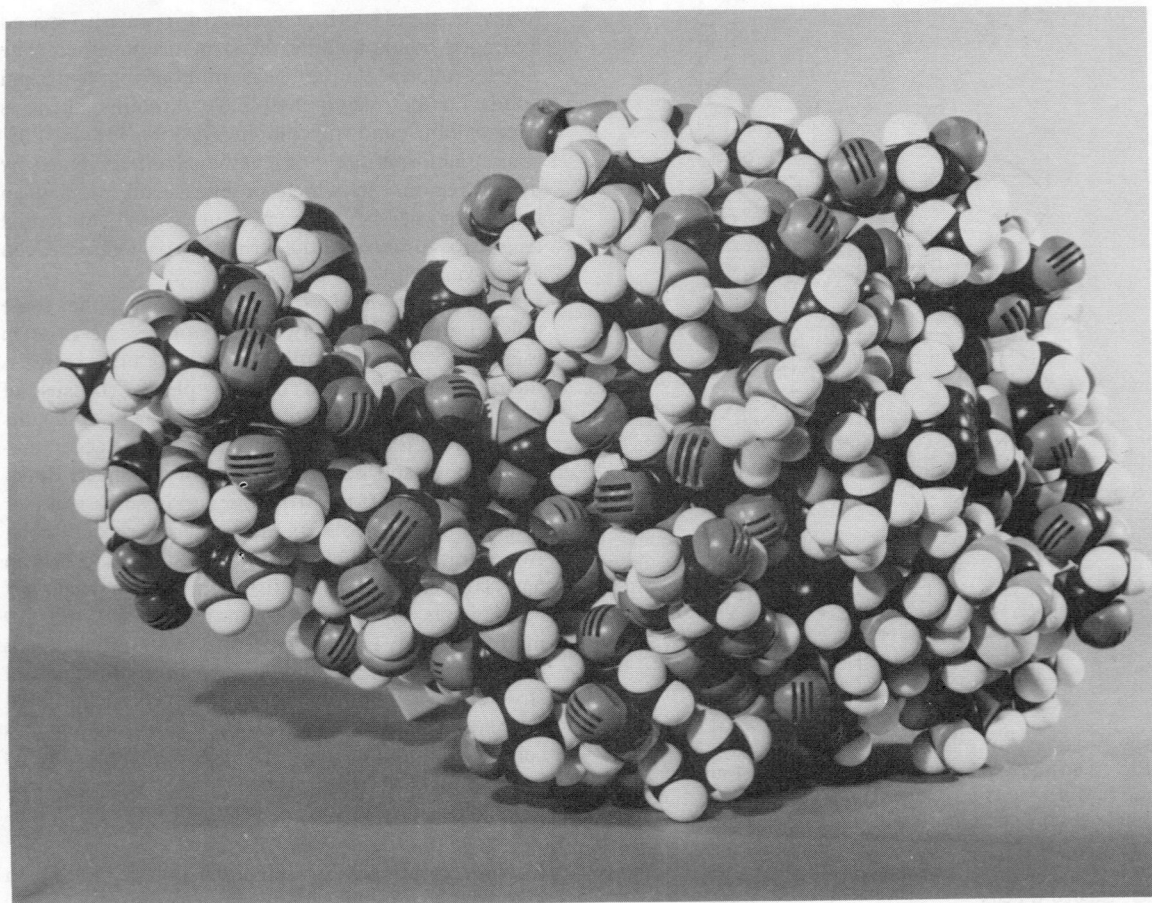

Fig. 2.10 'Space-filling' model of the molecule of the enzyme lysozyme (muramidase) to show its compact structure (Courtesy of J. A. Rupley)

residues but many proteins contain several polypeptide chains. The individual chains may be linked by disulphide bonds to form a single large molecule as, for example, is the case in the immunoglobulins (fig. 2.21, p. 23)

Several protein molecules may be linked together to form a larger unit. Ferritin, an iron-storage protein (p. 250), is an example. It consists of 24 subunits linked together to form a hollow rounded cube (fig. 2.11). Iron is stored in the central cavity; a single molecule can accommodate as many as 4300 iron atoms as ferric oxide. Further examples of proteins

with subunits are haemoglobin (Chap. 16) and collagen (p. 209).

Supramolecular structure. Complexes exist where several proteins or enzymes are associated to accomplish a series of related activities. The pyruvate dehydrogenase complex (p. 97) has a molecular weight of 4 600 000 and combines three enzymic activities. The respiratory electron transport chain is a membrane-associated complex with several protein components in which electrons are passed from one component to the next (Chap. 10).

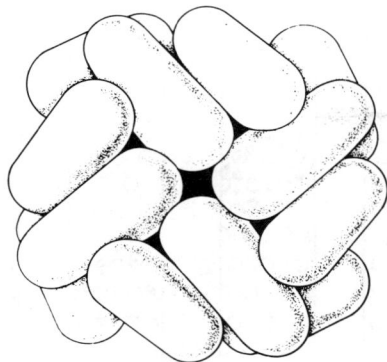

Fig. 2.11 Arrangement of subunits to make up a molecule of apoferritin. (From Harrison, P. M. *et al.* (1980) In *Iron in Biochemistry and Medicine II*, eds Jacobs, A. and Worwood, M., p. 131. London: Academic Press, by courtesy of authors and publisher)

Properties of proteins

All molecules of a particular protein are identical; this explains the formation of crystals, composed of identical, regularly repeating units, from a protein solution. Each protein has its own unique amino acid sequence, its own three-dimensional structure and its own characteristic polarity and charge.

The production of crystals is an essential step in the determination of the three-dimensional structure of a soluble protein. Diffraction of X-rays by the regularly repeating units in the crystal has provided the information needed for the elucidation of the complete structure of many protein molecules.

Binding. Proteins bind with other substances including small molecules and macromolecules (polysaccharides, polynucleotides or other proteins). The binding of identical protein molecules gives rise to multi-subunit proteins, multi-protein complexes, or rods, sheets or cylinders in the case of structural proteins. The binding process is highly specific. For example enzymes bind specific substrates and then promote a reaction at a specific position in their molecules (Chap. 8).

Denaturation. Denaturation consists of the unfolding of the polypeptide chain and the loss of the compact structure. It is usually irreversible and can result from the application of heat, extremes of pH or the action of detergents. The previously buried hydrophobic side-chains become exposed to the aqueous medium and the solubility of the protein falls. Denaturation therefore frequently causes precipitation of the protein. When food is cooked its protein is denatured so that more peptide bonds are exposed to the digestive enzymes of the gastrointestinal tract.

A protein in its native state is folded into the conformation of its greatest stability. An unfolded protein may not be able to re-fold because of the length and flexibility of its own chain. Ribonuclease is an exception. This 124 residue protein renatures very rapidly under appropriate conditions, and full activity can be restored.

When a protein is synthesised on the ribosome (p. 165), the conditions for folding are very different from those found in free solution. In some cases, the new chain folds as it is being made and before it separates from the ribosome. Sometimes this is ensured by having the growing chain feed directly into a membrane where the hydrophobic surroundings promote a specific form of folding. The normal folding of a protein may be ensured by its initial production as a precursor (such as proinsulin) which may ensure that the ultimate active molecule has the correct conformation.

Charge. The various charged side-chains contribute to the overall charge of the protein. At low pH a protein carries a net positive charge since most of the side-chains are either positively charged or neutral. The opposite is true at high pH. For each individual protein, therefore, there is a pH at which the net charge on the protein is zero; this is known as the *isoelectric point*. At this pH the protein does not move in an electric field and the repulsion between individual molecules is least (fig. 2.12).

Separation methods

Methods for the isolation of a pure protein from a mixture may make use of four properties: solubility, charge, size and specific binding properties.

Solubility. When the pH of a protein solution is brought to the isoelectric point, the protein may precipitate. Proteins are also precipitated from solution at high ionic strengths. Since precipitation occurs at different ionic strengths for different

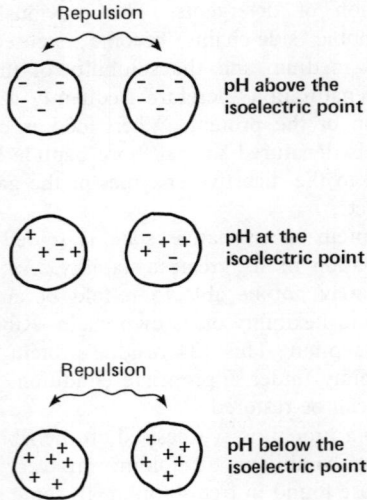

Repulsion

pH above the
isoelectric point

pH at the
isoelectric point

Repulsion

pH below the
isoelectric point

Fig. 2.12 The repulsion between molecules of a protein in solution is least when the pH is at the isoelectric point

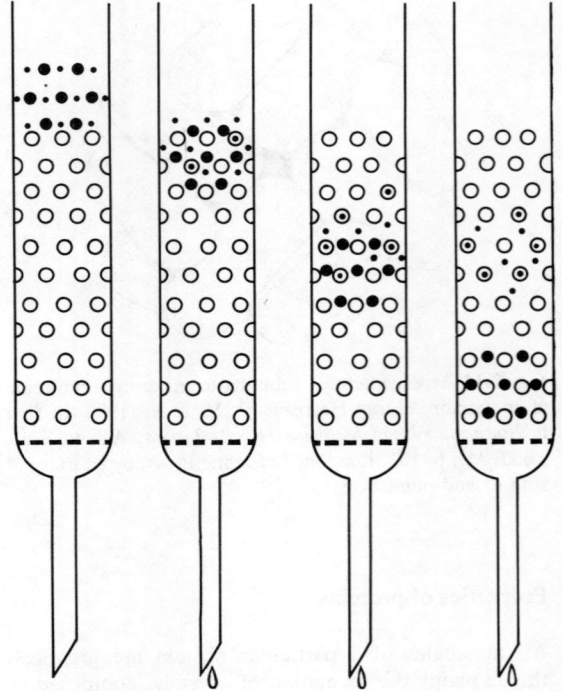

Fig. 2.13 Gel filtration with beads of cross-linked dextran (Sephadex) for the separation of proteins according to their molecular size. The beads are packed into a column and a mixture of proteins is layered on to the surface. As the proteins move down the column the larger molecules, unable to penetrate the beads, move faster than the small molecules which can. (After Lehninger, A. L. (1975) *Biochemistry*, New York: Worth)

proteins, partial separation of the individual components of a mixture of proteins becomes possible. Plasma proteins were first separated with different concentrations of ammonium sulphate into 'albumin' and 'globulin' fractions.

Charge. Ion-exchange chromatography is commonly used for protein separations; the procedure is similar to that used for amino acids but the conditions are designed to avoid the denaturation of the proteins. Electrophoresis is widely used for protein separations; its application to the plasma proteins is described later.

Size and shape. Proteins in solution subjected to a strong centrifugal force in an ultracentrifuge sediment at a rate which depends on their size and shape. Another useful separation method is *gel filtration* (fig. 2.13).

Specific binding properties. Enzymes are proteins with specific binding sites for particular molecules and this specificity may be used to select an enzyme from a protein mixture by *affinity chromatography* (fig. 2.14). Another method for separating proteins, involving both binding properties and electrophoresis, is *immuno-electrophoresis* in which proteins are first separated into broad groups by electrophoresis and then identified by their reaction with specific antisera (p. 19).

The plasma proteins

Plasma contains more than 100 proteins. Although the function of only a few is known it is likely that each has a specific role either in the circulation or in the interstitial fluid to which some can gain access. The plasma proteins include both simple proteins and conjugated proteins such as glycoproteins and lipoproteins.

Plasma proteins were at one time divided into only two fractions, albumin and globulin according to their solubility in salt solutions. It later became

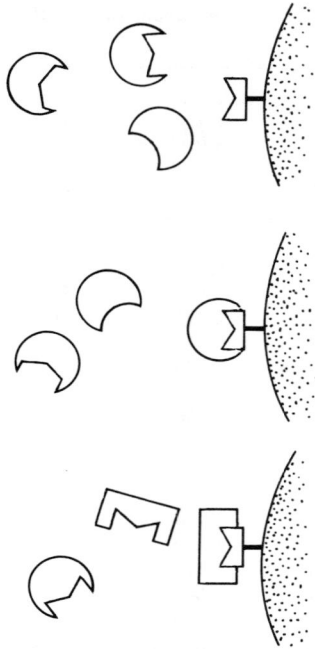

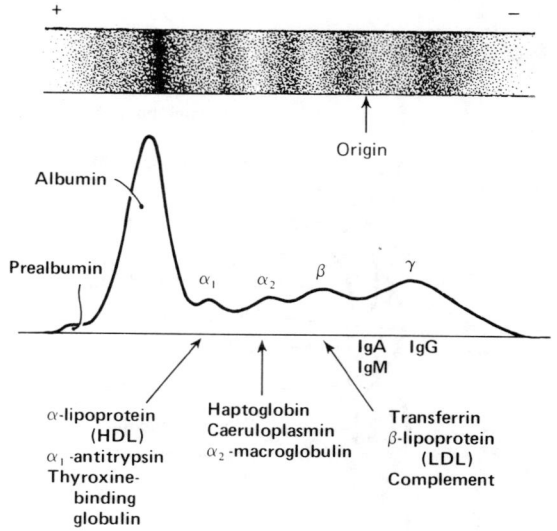

Fig. 2.15 Electrophoretic separation of the plasma proteins. With the exception of albumin, it can be seen that each band represents several distinct proteins

Fig. 2.14 Affinity chromatography. An analogue of an enzyme's normal substrate is bound by covalent bonds to a supporting material such an dextran beads. The beads are then packed into a column and the protein mixture is passed through it (a). Only the enzyme with a binding site of an appropriate shape binds to the beads (b) and may be displaced by changing the buffer conditions or by passing down the column a substance which competes for the binding site (c)

possible to separate the globulin fraction into five components by electrophoresis (fig. 2.15). It is now clear that each of these fractions includes several different proteins which can be demonstrated by the use of *immuno-electrophoresis* (fig. 2.16 and 2.17).

Table 2.8 lists the 13 most abundant proteins in plasma which together make up some 95 per cent of the total protein.

Fig. 2.16 Procedure for immuno-electrophoresis applied to an imaginary mixture of four proteins, which are incompletely separated by electrophoresis. After electrophoresis on an agar gel an antiserum containing antibodies to the four proteins is placed in the trough as shown. Precipitin lines develop where the concentrations of antigen and antibody are equivalent. (Courtesy of Margaret Gibb)

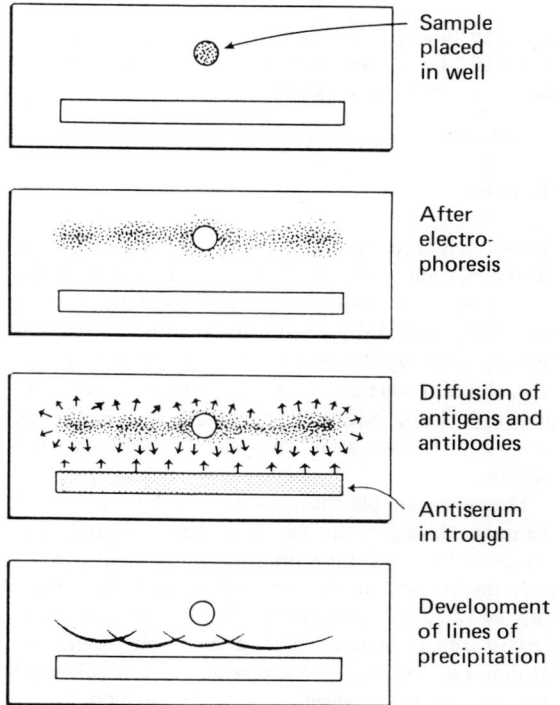

19

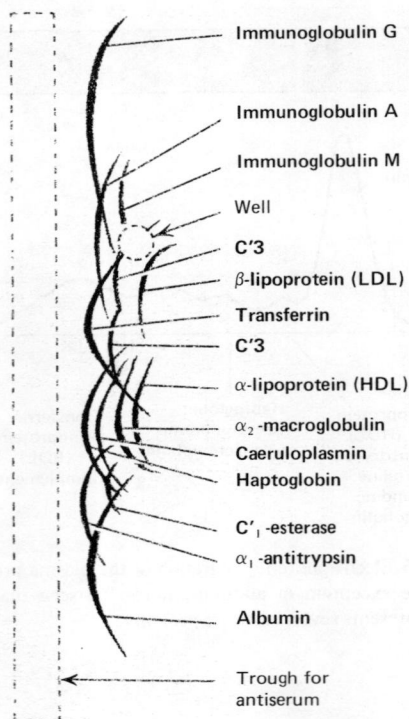

Immunoglobulin G

Immunoglobulin A

Immunoglobulin M

Well

C'3

β-lipoprotein (LDL)

Transferrin

C'3

α-lipoprotein (HDL)

α₂-macroglobulin

Caeruloplasmin

Haptoglobin

C'₁-esterase

α₁-antitrypsin

Albumin

Trough for antiserum

Fig. 2.17 Principal proteins in normal human plasma demonstrated by immuno-electrophoresis. (Courtesy of Margaret Gibb and P. E. G. Mitchell)

Table 2.8 The 13 most abundant protein fractions in normal plasma

	Molecular mass kdal	Electro- phoretic mobility	Plasma level in normal adult (g/l)
Albumin	66.3	'Albumin'	36–45
High density lipoprotein	175–360	α_1	2.5–3.8
Orosomucoid	44.1	α_1	0.7–1.0
α_1-Antitrypsin	52	α_1	2.1–5.0
α_2-Macroglobulin	820	α_2	2.2–3.8
Haptoglobin	100	α_2	0.3–2.0
Transferrin	90	β_1	1.2–2.0
Low density lipoprotein	2750	β_2	2.8–4.4
C3 fraction of complement	180	β_2	1.0–1.4
Fibrinogen	340	$\beta-\gamma$	1.5–4.0
Immunoglobulin A	175	γ	1.4–4.0
Immunoglobulin M	900	γ	0.5–2.0
Immunoglobulin G	150	γ	12–18

After Smith J. F. F. (1980). Plasma proteins. *British Journal of Clinical Equipment* **5**, 159–162.

Albumin

The principal function of albumin, the most plentiful of the plasma proteins, is the maintenance of the colloid osmotic pressure of the plasma. This plays an important role in the control of fluid exchanges between the capillaries and the interstitial fluid (p. 6). Albumin carries in the plasma the free fatty acids (p. 136), bilirubin (p. 199) and calcium (p. 238); it may also carry drugs such as penicillin and aspirin.

The levels of plasma albumin are low in several common disorders in which oedema occurs. The causes include defective albumin synthesis in chronic liver disease and in protein malnutrition (kwashiorkor, fig. 2.18), and excessive albumin loss notably in certain renal diseases and after severe burning. Albumin is absent from the plasma in a rare inherited disorder, *analbuminaemia* (fig. 2.19), which causes

surprisingly few symptoms probably because there is a substantial compensatory increase in the plasma levels of other proteins.

Lipoproteins

The lipoproteins are complexes of proteins with triglycerides, cholesterol and phospholipids. Their composition and function in the transport of lipids in the plasma is described in Chapter 11.

Orosomucoid

The function of this glycoprotein is not yet known; it may have a role in the defence against viral infections; its level in the plasma increases in many acute illnesses in which tissues are damaged.

α₁-antitrypsin and α₂-macroglobulin

These proteins are protease inhibitors which neutralise proteolytic enzymes released into the circulation

Fig. 2.18 Severe oedema due to a low plasma albumin in a child with kwashiorkor (protein malnutrition). (Courtesy of K. M. Waddell)

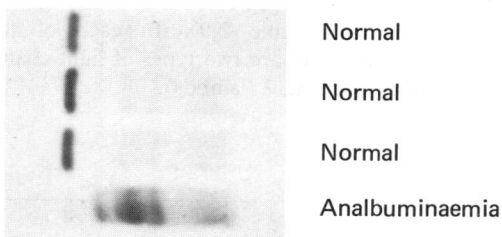

Normal

Normal

Normal

Analbuminaemia

Fig. 2.19 Cellulose acetate electrophoresis of serum from one patient with analbuminaemia compared with three normal controls. The striking increase in the concentrations of proteins other than albumin is seen. (From Di Guardo, C., et al. (1977) *Lab* **4,** 113 by courtesy of authors and editor)

during phagocytosis. In a rare inherited disorder, α_1-antitrypsin deficiency, progressive lung damage (emphysema) occurs.

Haptoglobin

This protein binds haemoglobin released into the plasma as a result of the breakdown of red cells in the circulation; the haptoglobin–haemoglobin complexes are rapidly removed from the circulation by monocytes. In patients with diseases causing increased red cell breakdown (*haemolysis*) the plasma levels of haptoglobin are greatly reduced.

Transferrin and carrier proteins

Two types of carrier protein are recognised: in one the protein enters the cell with its ligand (nutrient type) and in the other the ligand is taken up by itself (fig. 2.20). An example of the first type is transferrin which transports iron in the plasma (Chap. 21); an

21

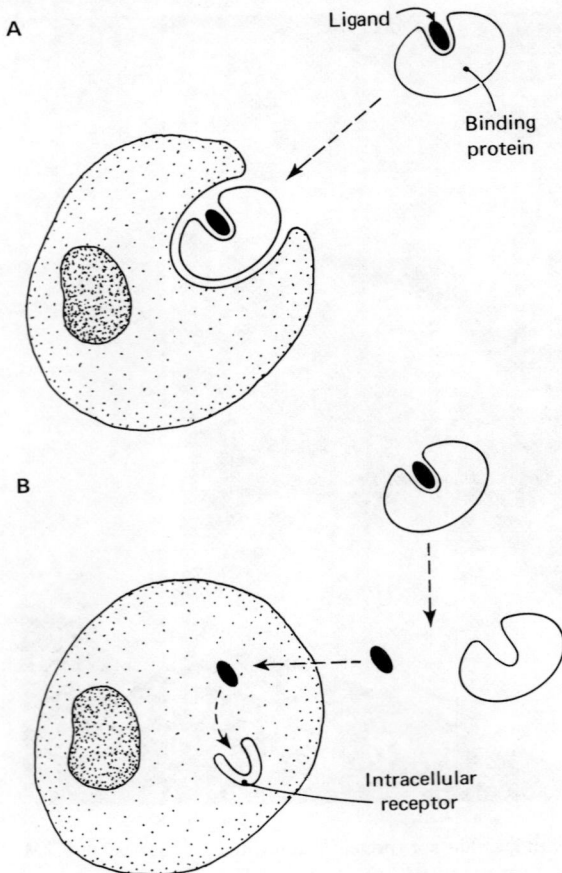

example of the second is transcortin which carries the hormone cortisol.

Complement

The term complement comprises a large number of proteins which act together in response to the reaction of an antibody with an antigen and mediate, for example, the destruction of bacteria. Complement also has a role in the attraction (chemotaxis) of leucocytes to areas of bacterial invasion.

Fibrinogen and the clotting factors

The clotting of blood involves a cascade of at least thirteen factors (twelve proteins and calcium). The orderly sequence of reactions leads to the conversion of fibrinogen to fibrin.

Immunoglobulins

The immunoglobulins (Ig) are antibodies; five distinct classes are recognised: IgG, IgA, IgM, IgD and IgE. The principal components of immunoglobulins are two types of polypeptide chains: heavy chains and light chains. The heavy chains are specific for each immunoglobulin class: γ (gamma) for IgG, μ (mu) for IgM, α (alpha) for IgA, δ (delta) for IgD, ε (epsilon) for IgE (Table 2.9). In each of the immunoglobulins there are two types of light chains known as κ (kappa) and λ (lambda).

Fig. 2.20 Two types of transport protein in the plasma. In the 'nutrient type' (A) the loaded transport protein is taken up complete by a cell. In the 'hormonal type' (B) the ligand is removed from the protein before entering a cell when it is bound to a receptor in the cytosol. (After Bouillon, R. and Van Baelen, H. (1981) *Calcified Tissue International* **33**, 451)

Table 2.9   The immunoglobulins

	Heavy chains	Light chains	Molecular mass (kdal)	Normal plasma level	Function
IgG	γ	κ, λ	150	8–16 g/l	Principal antibody in plasma
IgA	α	κ, λ	170	1.4–4.0 g/l	Immunity at mucosal surfaces e.g. gut and bronchial tree
IgM	μ	κ, λ	960	0.5–2.0 g/l	Early defence against bacteria and foreign cells (includes blood-group antigens)
IgD	δ	κ, λ	184	0–0.4 g/l	Not known
IgE	ε	κ, λ	188	0.1–1.3 mg/l	Allergic responses

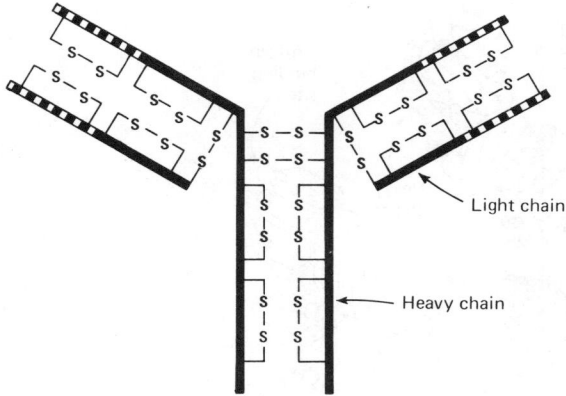

Fig. 2.21 Conventional representation of the structure of immunoglobulin G (IgG), with two heavy chains and two light chains linked by disulphide bridges. The dotted area of each chain has a variable amino acid composition and forms the antigen binding site. The structure of IgG is shown more precisely in figure 2.22

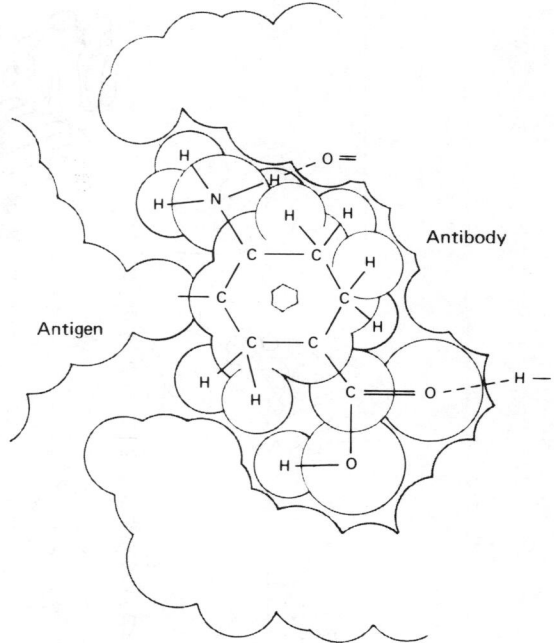

Fig. 2.23 Binding of antigen to antibody because their surfaces are complementary. (From Hansen, L. (1979) In *Plasma Proteins*, eds Blomback, B. and Hansen, H. Chichester: Wiley, by courtesy of author and publisher)

Immunoglobulin G. This is the principal antibody which binds to foreign antigens particularly in chronic or repeated infections. The IgG response to a second encounter with a particular antigen is substantially greater than on the first occasion. IgG is small enough to enter the interstitial fluid. The IgG molecule consists of two heavy chains (γ) and two light chains (κ or λ) bound together by disulphide bridges (figs. 2.21 and 2.22). The amino terminal

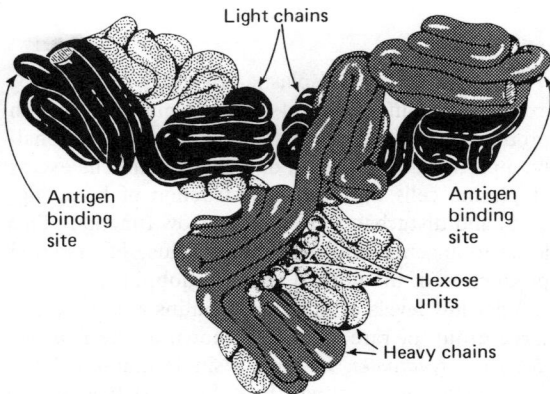

Fig. 2.22 Drawing of the structure of IgG to show the convoluted formation of the polypeptide chains. (Courtesy of R. S. H. Pumphrey)

part of each chain, which has a very variable amino acid composition is the site for binding to antigens (fig. 2.23).

Immunoglobulin A. IgA provides the principal antibody defence on mucosal surfaces such as those of the gut and the bronchial tree. It has initially a structure like that of IgG with two heavy chains (α) and two light chains (κ or λ). Two such molecules are secreted onto the surfaces as a dimer bound together by an additional protein known as a *secretory piece*.

Immunoglobulin M. IgM and to a smaller extent IgA provide the early response to infections particularly bacterial infections. The blood-group antibodies are also IgM. IgM is a pentamer with ten heavy chains and ten light chains (fig. 2.24). Since it is a very large molecule it is confined to the blood plasma and does not reach the interstitial fluid.

23

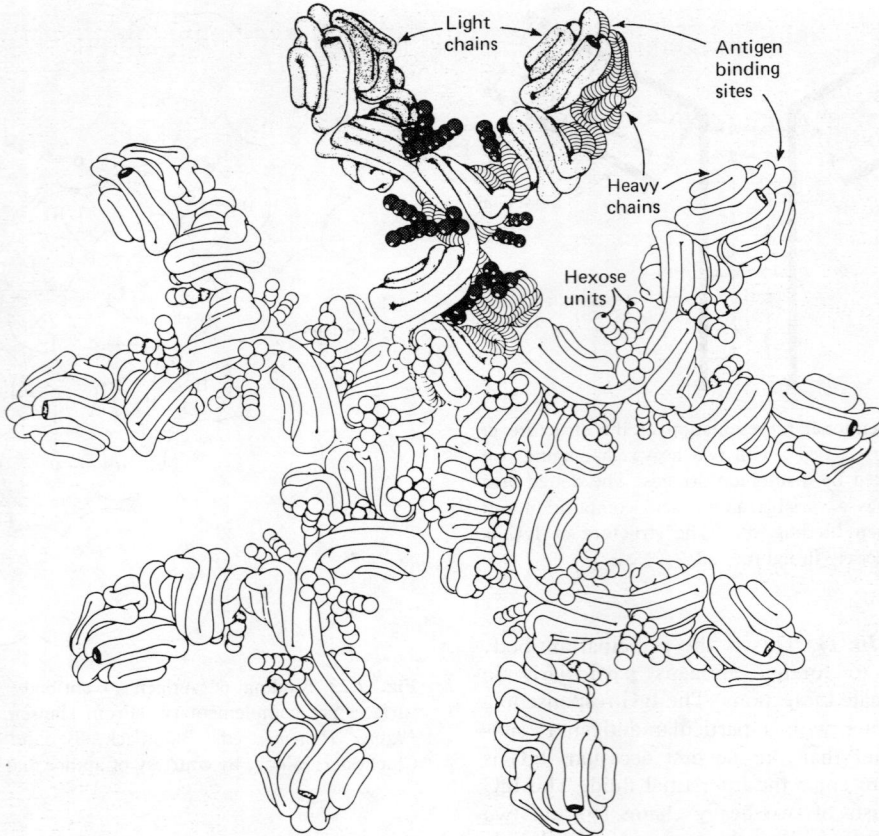

Fig. 2.24 Drawing of the structure of immunoglobulin M (IgM) (Courtesy of R. S. H. Pumphrey)

IgD and IgE. The function of IgD is not yet known. IgE mediates allergic and hypersensitivity reactions; high values are found in patients with asthma and hay fever.

Disorders of the immunoglobulins. High values for some or all of IgG, IgM or IgA are found in patients with chronic infections, chronic liver disease and other chronic disorders such as rheumatoid arthritis. A specific rise in IgA is found in infections of the lungs or intestine. The pattern of changes in immunoglobulin levels can give a guide to the nature of an illness.

Myeloma is a neoplastic disorder caused by excessive proliferation of the plasma cells which produce a particular immunoglobulin, or occasionally, a particular light chain or heavy chain. The excess of plasma cells may cause destruction of bone (fig. 2.25) and disturbance of bone marrow function. This leads to anaemia and infections because of impaired production of the normal immunoglobulins.

Very low levels of immunoglobulins are found in a large group of rare disorders known as the *immunodeficiency syndromes*. In some a single immunoglobulin is deficient, in others IgG, IgA and IgM are all lacking and in yet others defects are also found in cell-mediated immunity. In all these disorders the characteristic clinical problem is recurrent infections.

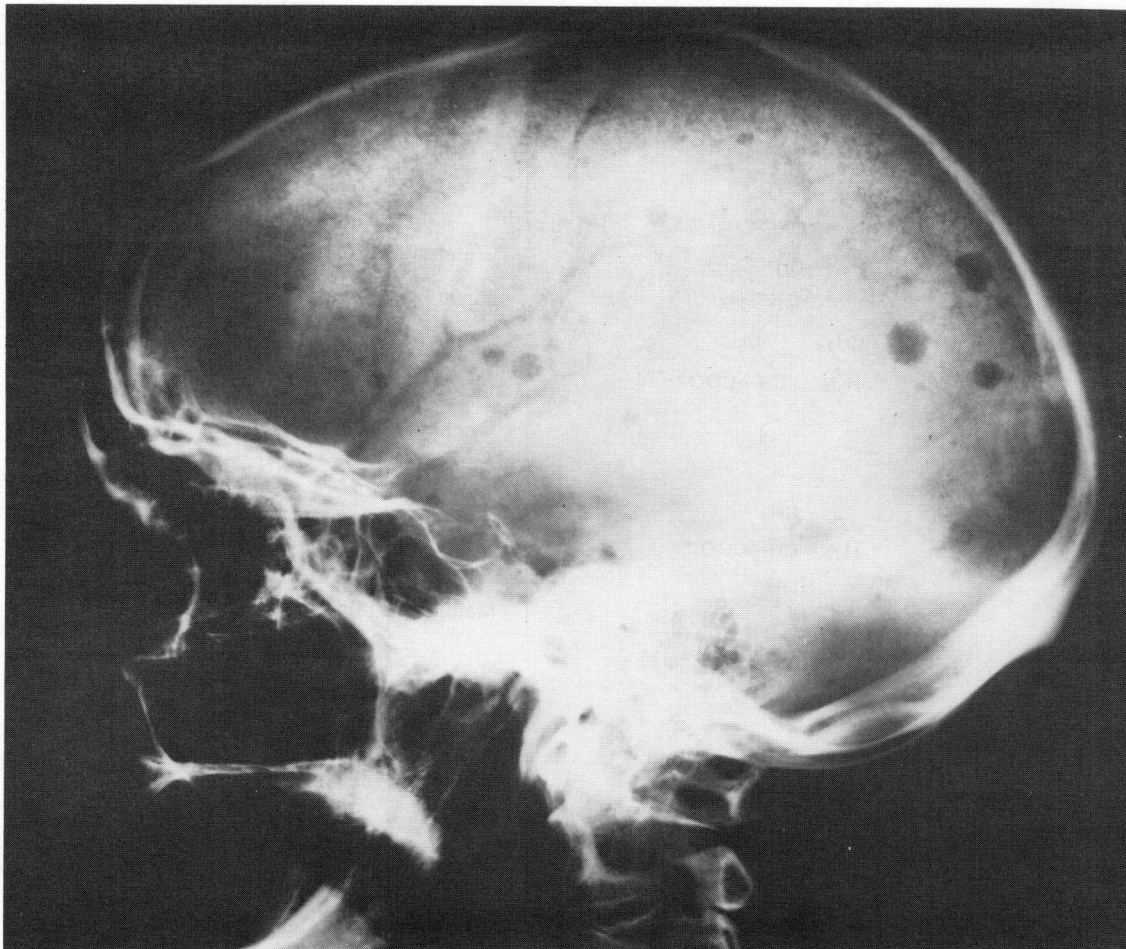

Fig. 2.25 X-ray of skull in a patient with myeloma

Further reading

Dickerson, R. E. and Geis, I. (1982) *Proteins: Structure, Function and Evolution*. New York: Benjamin

Hill, H. R. (1980) Laboratory aspects of immune deficiency in children. *Pediatric Clinics of North America* **27,** 805–830

Jaenicke, R. (1980) *Protein Folding*. New York: Elsevier

Keyser, J. W. (1979) *Human Plasma Proteins*. Chichester: Wiley

Neurath, H. and Hill, R. L. (1976) *The Proteins* 3rd edn. New York: Academic Press

Whicher, J. T. (1980) The interpretation of electrophoresis. *British Journal of Hospital Medicine* **24,** 348–360

APPENDIX—THE AMINO ACIDS

Hydrophobic amino acids

Alanine (Ala)

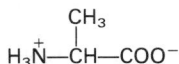

Valine (Val)

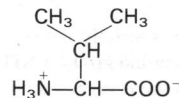

Leucine (Leu)

Isoleucine (Ile)

Proline (Pro)*

Phenylalanine (Phe)

Tyrosine (Tyr)

Tryptophan (Trp)

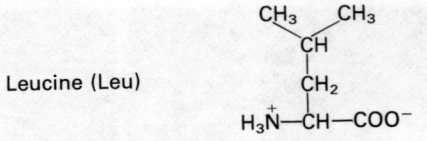

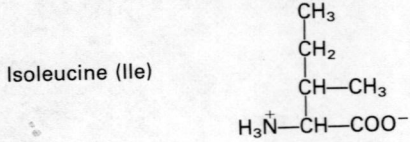

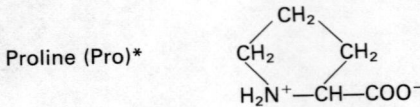

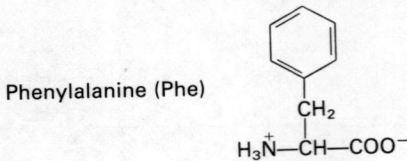

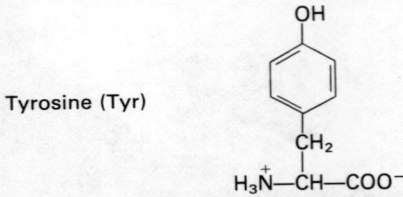

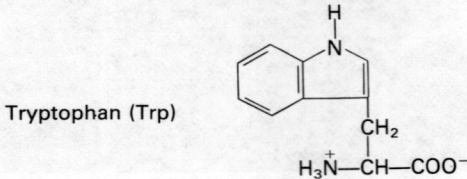

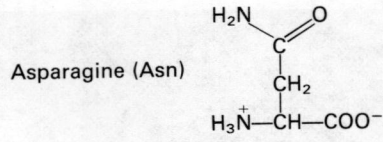

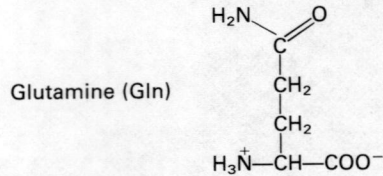

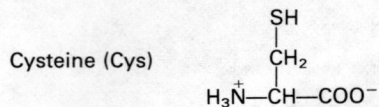

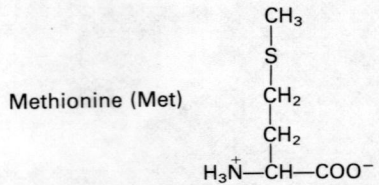

Hydrophilic amino acids—uncharged

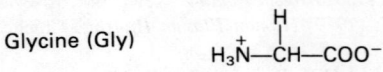

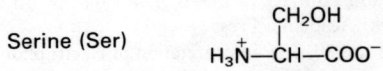

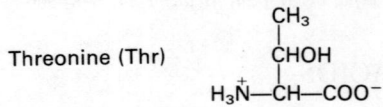

Glycine (Gly)

Serine (Ser)

Threonine (Thr)

Asparagine (Asn)

Glutamine (Gln)

Cysteine (Cys)

Methionine (Met)

Acidic amino acids

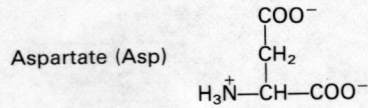

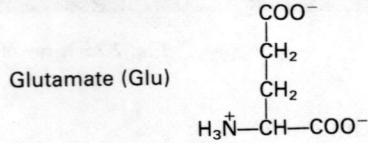

Aspartate (Asp)

Glutamate (Glu)

Basic amino acids

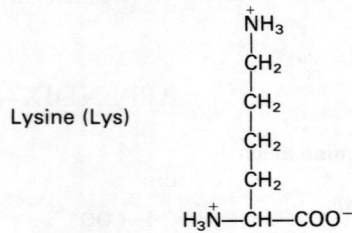

Lysine (Lys)

* Proline is strictly not an amino acid but an *imino acid* as it has an imino group ($-NH_2^+-$) instead of an $-NH_3^+$ group attached to the α-carbon atom.

Arginine (Arg)

$$H_3\overset{+}{N}\quad NH$$
$$C$$
$$NH$$
$$CH_2$$
$$CH_2$$
$$CH_2$$
$$H_3\overset{+}{N}—CH—COO^-$$

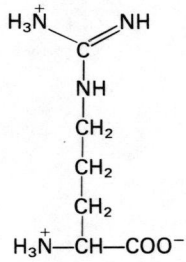

Histidine (His)

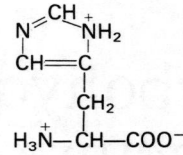

$$N=CH$$
$$\overset{+}{N}H_2$$
$$CH=C$$
$$CH_2$$
$$H_3\overset{+}{N}—CH—COO^-$$

DAC CRP

27

3 Carbohydrates: structure & function

CARBOHYDRATES make up about 10 per cent of the organic matter in a cell and are important as sources of energy and as energy stores. Carbohydrates are also components of large complex molecules such as DNA, RNA and proteoglycans.

The carbohydrates may be defined as substances with the general formula $C_m(H_2O)_n$. This definition is not completely accurate since some carbohydrates contain oxygen in a smaller proportion than this would imply and some carbohydrates contain nitrogen-, phosphorus- or sulphur- containing groups.

The simplest view of the smaller carbohydrates is to regard them as hydroxy-aldehydes or hydroxy-ketones. Thus the carbohydrates with three carbon atoms are glyceraldehyde and dihydroxyacetone.

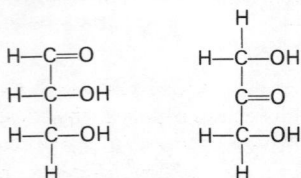

Glyceraldehyde   Dihydroxyacetone

Glyceraldehyde and dihydroxyacetone are known as *trioses*. Sugars with four carbon atoms are known as *tetroses*, with five as *pentoses*, with six as *hexoses*. Simple sugars like these are called *monosaccharides*.

When two monosaccharide units condense together with the elimination of one molecule of water the product is known as a *disaccharide*. If additional monosaccharide units are added an *oligosaccharide* is formed; molecules containing large numbers of monosaccharide units are known as *polysaccharides*.

Monosaccharides

Hexoses

The simplest hexose of the aldehyde type is:

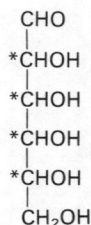

It has four asymmetric carbon atoms (marked with asterisks) so that sixteen variants are possible. Among them are

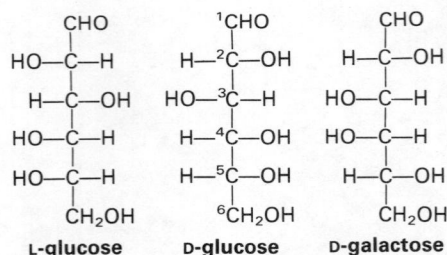

L-glucose   D-glucose   D-galactose

It can be seen that D-glucose and L-glucose are mirror images of each other; they have similar chemical and physical properties, but only D-glucose is of biological importance. D-glucose and D-galactose have distinctly different chemical properties.

Many of the chemical properties of glucose and other hexoses are difficult to explain on the basis of such structures. For example the aldehyde group is not as reactive as might be expected. In fact glucose exists to a large extent in a ring form in which the aldehyde group (carbon 1) is linked to carbon 5.

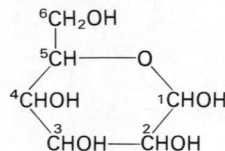

It can be seen that in such a structure carbon atom 1 is asymmetric and D-glucose exists in two forms with very different optical properties, α-D-glucose and β-D-glucose. These two forms are interconvertible in solution since both are in equilibrium with the open-chain aldehyde form present in small amounts. It is conventional, when depicting the structures of these compounds, to show the ring as if it were sticking out of the paper so that the hydrogen and hydroxyl groups are above and below the plane of the ring. The carbon atoms in the ring are often not shown. Thus the two forms of glucose just mentioned are illustrated

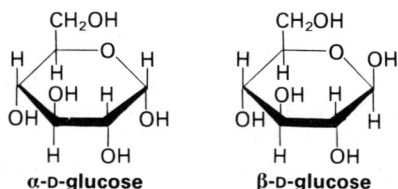

α-D-**glucose**   β-D-**glucose**

Even this illustration does not properly describe the structure of hexoses. The ring is not flat but 'puckered'.

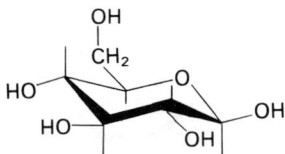

Glucose. D-glucose is of great importance as a source of energy for the tissues. It is an essential fuel for the brain, the erythrocytes and the skin. Glucose in plasma also makes a major contribution to the energy needs of exercising muscle (Chap. 12). The blood glucose level is maintained within the range 2.5 to 6.0 mmol/l (45 to 100 mg/100 ml) by a number of hormones (Chap. 12). Glucose molecules combine together to give a polysaccharide, *glycogen*, which provides a reserve of energy in liver and in muscle. Glucose also makes up the plant polysaccharide, starch, a major component of foodstuffs.

Since glucose can take on an open-chain aldehyde form it is a reducing agent capable, for example, of reducing cupric compounds to the cuprous state. This property is sometimes used in methods for detecting or estimating glucose or similar sugars in blood or urine.

Glucose is an alcohol and can form esters with acids. Glucose 1-phosphate and glucose 6-phosphate are of great importance in carbohydrate metabolism (Chap. 9).

Other hexoses. *Fructose* occurs naturally in fruit and in honey and is a component of the disaccharide sucrose (p. 31). Fructose is also the principal sugar in a mixture known as *isoglucose*, a sweetener produced commercially by the fermentation of starch from maize. In the body fructose occurs free in the seminal fluid where it is a source of energy for spermatozoa. Phosphate esters of fructose are of great importance in carbohydrate metabolism (p. 94). In the free state fructose has a six-membered (*pyranose*) ring as in glucose but in combination adopts a five-membered (*furanose*) ring:

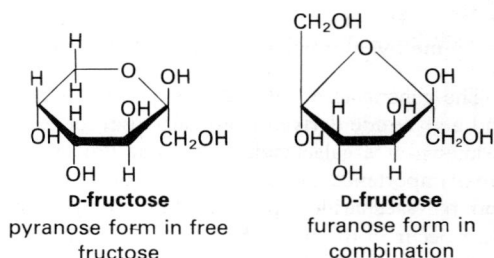

D-**fructose**
pyranose form in free
fructose

D-**fructose**
furanose form in
combination

Galactose is a constituent of the lactose in milk and a component of some complex polysaccharides, lipids and proteins. *Mannose* occurs in some complex polysaccharides and in certain proteins notably collagen in connective tissues.

Pentoses

The pentoses are monosaccharides with five carbon atoms. The best known are ribose and deoxyribose which are components of RNA and DNA respectively (Chap. 5). Like fructose, ribose exists in combination in a furanose form.

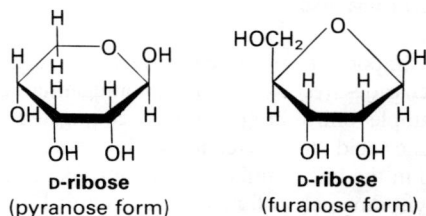

D-**ribose**
(pyranose form)

D-**ribose**
(furanose form)

In addition to their role in nucleic acids, a number of pentoses are important intermediates in the *pentose phosphate pathway* of glucose metabolism (p. 99). Some pentoses such as xylose are components of proteoglycans (p. 34).

Derivatives of monosaccharides

The glycosides. The hydroxyl groups of glucose may be replaced by other radicals with the formation of a glucoside. Two examples are

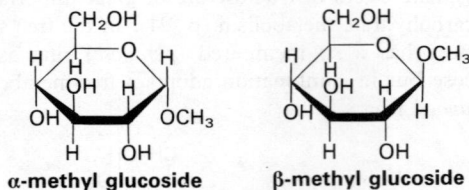

α-methyl glucoside   **β-methyl glucoside**

The general term for such compounds is *glycoside* and a glycoside derived from galactose, for example, is known as a galactoside. Glycoside linkages are of great importance in the formation of disaccharides and polysaccharides (p. 31). Many complex glycosides occur in the body. For example the glycolipids (p. 40) which form a major part of the myelin of nerve fibres contain galactosides and, to a smaller extent other glycosides. Many important drugs are glycosides including the antibiotic streptomycin, and digoxin which has an action on the heart.

Glucuronides. The carbon atom at position 6 in the glucose molecule can be oxidised to produce a carboxyl group. The resulting acid is known as glucuronic acid

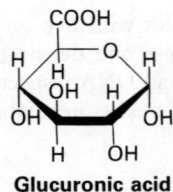

Glucuronic acid

and is important as a constituent of complex polysaccharides (p. 34) and as a conjugating agent. For example bilirubin (p. 200) is conjugated with glucuronic acid to make it soluble and readily excreted in the bile. Similarly many hormones, drugs and environmental pollutants are excreted in the urine or bile after being rendered soluble by conjugation with glucuronic acid. For example phenol, which is sometimes formed in the intestine as a result of the bacterial decomposition of aromatic amino acids, is excreted as a glucuronide.

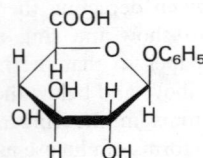

Glucuronic acid derivative of phenol

The hexosamines. When an amino group is introduced into a hexose the product is known as a hexosamine or an amino sugar. The hexosamine formed from glucose is glucosamine which is a major component of many complex polysaccharides, usually in the form of its acetyl derivative, *N*-acetyl glucosamine. The corresponding compound derived from galactose is *N*-acetyl galactosamine which is an important constituent of the glycoproteins (p. 35) and glycolipids (p. 40).

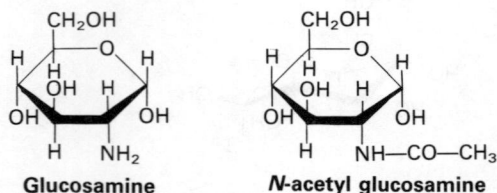

Glucosamine   ***N*-acetyl glucosamine**

Another important amino sugar is neuraminic acid, acetyl derivatives of which are called sialic acids. These are important constituents of glycoproteins in bone, cartilage and connective tissue, and of glycolipids in the nervous system.

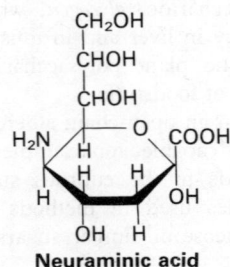

Neuraminic acid

Disaccharides

The three disaccharides of physiological importance are sucrose, maltose and lactose. Each comprise two hexose units joined by a glycoside linkage.

Sucrose consists of glucose and fructose

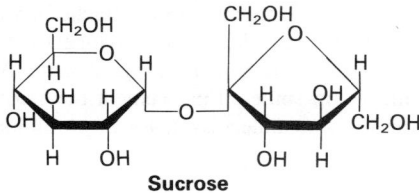

Sucrose

Sucrose ('sugar') is found in certain plants notably sugar cane and root vegetables, such as sugar beet. It forms an important part of the dietary carbohydrate in many countries.

Maltose consists of two glucose units. It is produced in the intestine from the breakdown of dietary polysaccharides such as starch.

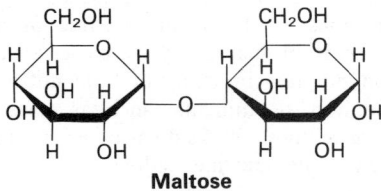

Maltose

Lactose is a constituent of milk and is synthesised in the breasts. It consists of one glucose and one galactose.

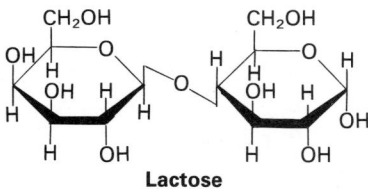

Lactose

Polysaccharides

Polysaccharides are formed by the condensation of large numbers of monosaccharide units with the elimination of water. Polysaccharides are usually insoluble in water. The principal polysaccharide in animal tissues is glycogen; in plants the predominant polysaccharides are starch and cellulose. All three are made up of glucose units.

Glycogen

A glycogen molecule has a mass of about 20 million and is large enough to be seen with the electron microscope (fig. 3.1). It consists of glucose units linked together to form a tree-like structure (fig. 3.2). The molecule starts with a single glucose unit and is enlarged by the addition of further units joined, as in maltose, by 1–4 glycoside linkages. Here and there the chain branches with a 1–6 linkage, details of which are shown in figure 3.3. The lengths of chain between branches are between 15 and 20 glucose units for the interior of the molecule and between 8 and 12 for the outer branches.

Glycogen stores in the liver and muscle represent only a small proportion of the body's energy reserves. However liver glycogen plays a vital role in the hour-to-hour control of the blood glucose; the enzymes responsible for its synthesis and breakdown respond within minutes to physiological changes (Chap. 12). Muscle glycogen constitutes a reserve of energy which is used in the early stages of exercise.

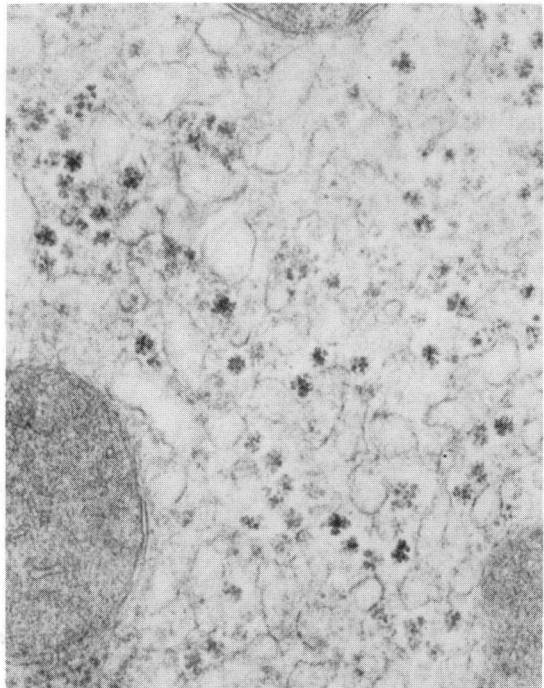

Fig. 3.1 Electron micrograph of human liver to show glycogen granules and smooth endoplasmic reticulum. (Courtesy of G. Milne)

31

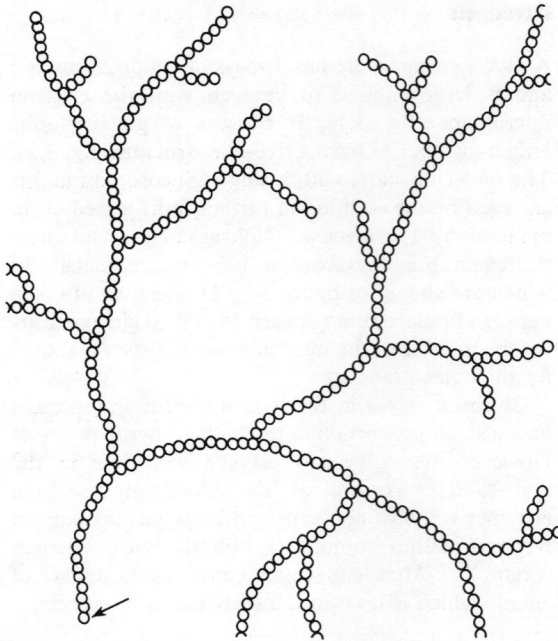

Fig. 3.2 Branched structure of a glycogen molecule. Each circle represents a glucose unit and the arrow indicates the unit with which synthesis commenced. The molecule includes branches which go outside the plane of the diagram, and is roughly spherical; a typical glycogen molecule may contain over 120 000 glucose units

Starch

Starch derived from plants represents about 60 per cent of the carbohydrate in the diet in western countries and a higher proportion in most developing countries. Starch has two components: amylose and amylopectin. The molecule of amylose is a straight chain of 25 to 2000 glucose units attached to each other by 1–4 glycoside linkages; amylopectin has a branched structure like that of glycogen. Amylopectin makes up some 80 per cent of the starch in the diet and has a molecular mass of 1 million or more.

Cell wall polysaccharides of plants

Cellulose is the major constituent of the cell walls of plants and, laid down in densely packed fibres, provides great strength. Cellulose differs from glycogen and starch in that its glucose units are attached to each other by β1–4 glycoside linkages (fig. 3.4). For this reason cellulose is not attacked by the digestive enzymes in the small intestine in man. Some cellulose is broken down by bacteria in the human colon but this probably of little nutritional value to the host. Ruminants, such as cows and sheep, can use cellulose for food since bacteria in the rumen produce the enzyme cellulase which does attack β1–4 linkages.

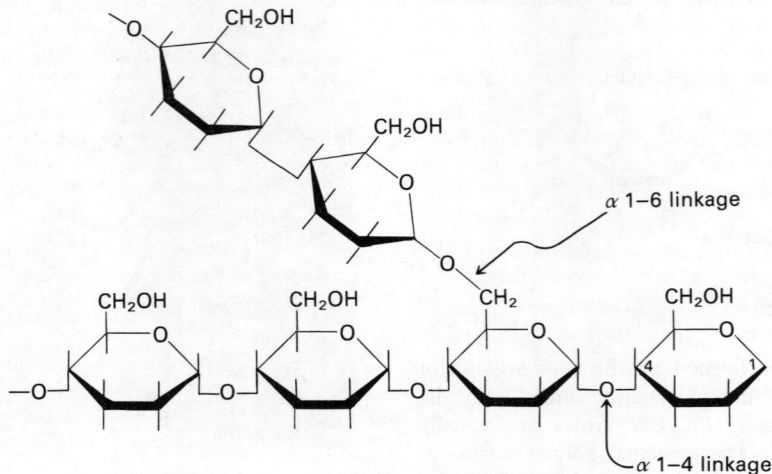

Fig. 3.3 A section of the structure of glycogen to show several 1–4 linkages and a branch point with a 1–6 linkage. The hydrogen atoms and hydroxyl groups have been omitted for the sake of clarity

32

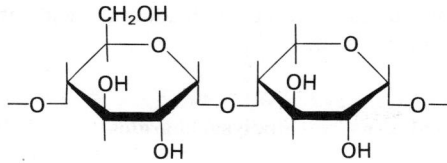

α 1–4 bonds of
glycogen and starch

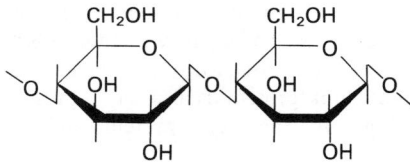

β 1–4 bonds
of cellulose

Fig. 3.4 Comparison of the α1–4 bonds of glycogen and starch with the β1–4 bonds of cellulose

Other cell wall polysaccharides of plants include hemicelluloses and pectins. *Hemicelluloses* have branched structures and consist mainly of xylose, together with other pentoses and hexoses, largely joined by β1–4 linkages as in cellulose. *Pectins* are a group of polysaccharides consisting mainly of galacturonic acid but also other hexoses and pentoses. They are present in large amounts in citrus fruit and apples, and are responsible for the 'setting' of jam. Pectins are broken down by bacteria in the gut.

All these substances are important in human nutrition. People, mainly in western countries, who have a refined diet with little cellulose or hemicellulose have a very solid stool and an increased incidence of constipation, appendicitis and diverticular disease of the colon. The effect of fibre from different sources depends on its composition. Fibre from wheat, for example, is resistant to bacterial action in the colon and causes an increase in stool bulk by its own bulk and that of the water it retains. Fibre from cabbage or apples is largely broken down and causes an increase in stool bulk by increasing the amount of bacteria.

Diets with a low pectin content are associated with high plasma cholesterol levels and a high incidence of vascular disease. The mechanism of the effect of pectin on plasma cholesterol is not yet known.

Proteoglycans and glycoproteins

Proteoglycans are enormous macromolecules consisting of proteins and complex carbohydrates known as *glycosaminoglycans*. The structure of a typical proteoglycan is shown in figure 3.5; such a molecule may have a molecular mass of 200 million. Proteoglycans are found in connective tissues, skin and bone, and are the major component of cartilage.

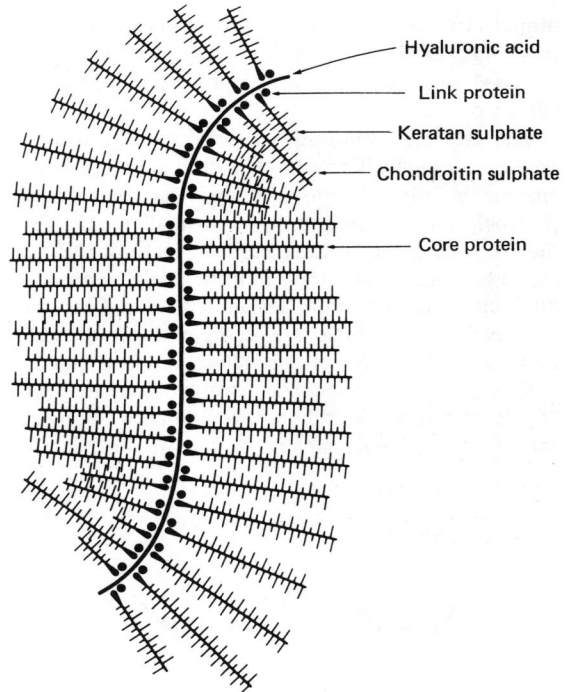

Fig. 3.5 The structure of a proteoglycan molecule. (From Rosenberg, L. E. (1975) In *Dynamics of Connective Tissue Macromolecules*, eds Burleigh, P. M. C. and Poole, A. R., pp. 105–128. Amsterdam: North Holland, by courtesy of author and publisher)

Glycosaminoglycans

A glycosaminoglycan is an unbranched polysaccharide consisting of alternate units of two different sugar residues. *Hyaluronic acid*, for example, contains alternating residues of *N*-acetyl glucosamine and glucuronic acid (fig. 3.6).

Hyaluronic acid forms the core of proteoglycans in cartilage, skin, the vitreous of the eye, and the

33

D-glucuronic acid   N-acetyl glucosamine

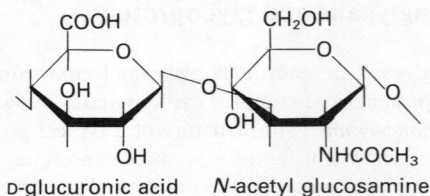

Fig. 3.6 The repeating unit of hyaluronic acid

umbilical cord. A single hyaluronic acid chain may be as long as 5 mm. Free hyaluronic acid occurs in synovial fluid and is largely responsible for its viscosity.

The enzyme hyaluronidase breaks down hyaluronic acid into smaller units. If a fluid containing this enzyme is injected into a tissue, the intercellular proteoglycans are broken down to smaller units and the injected fluid spreads easily. Hyaluronidase is found at a high concentration in the testis, in seminal fluid and in the venom of certain snakes and insects.

Chondroitin sulphates are composed of alternate units of sulphated *N*-acetylgalactosamine and glucuronic acid. *Keratan sulphate* consists of alternate units of galactose and *N*-acetyl glucosamine both sulphated (fig. 3.7). *Dermatan sulphate*, found in skin, is like

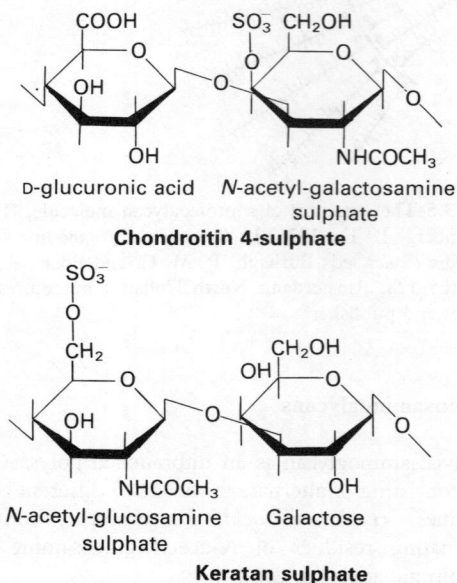

D-glucuronic acid   *N*-acetyl-galactosamine
                          sulphate

Chondroitin 4-sulphate

N-acetyl-glucosamine   Galactose
sulphate

Keratan sulphate

Fig. 3.7 The repeating units of chondroitin 4-sulphate and keratan sulphate

chondroitin sulphate but with iduronic acid instead of glucuronic acid.

Proteoglycans (mucopolysaccharides)

The structure of a typical proteoglycan was given earlier (fig. 3.5). The chondroitin sulphate and keratan sulphate are linked through specialised residues including galactose and xylose to serine or threonine residues in the protein core of each subunit. The protein has a partly globular portion at the end at which it is attached to the hyaluronic acid. An additional 'link protein' is also found in this area, attached both to the core protein and to the hyaluronic acid.

Since both keratan sulphate and chondroitin sulphate have numerous closely spaced negatively charged sulphate and carboxyl groups, a proteoglycan assumes a stiffly extended form in solution. Unlike a globular protein, a proteoglycan therefore tends to occupy a large volume in solution. Once a proteoglycan has expanded into the volume available to it considerable force is needed to deform it and squeeze some solvent out. On release of the force it tends to resume its previous volume. This property underlies the elasticity of cartilage particularly articular cartilage.

Proteoglycans are also involved in the lubrication of joints. Joint surfaces move over each other with less friction than ice sliding on ice. Articular surfaces have complexes of hyaluronic acid and proteins extending into the synovial space (fig. 3.8). The water trapped between the complexes is compressed and moved around as joints move but unless pressure

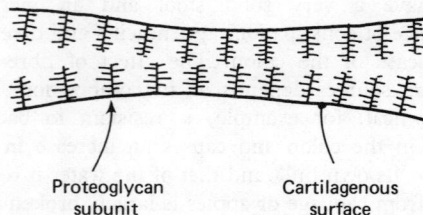

Proteoglycan   Cartilagenous
subunit   surface

Fig. 3.8 Boundary lubrication of articular cartilage. The load is carried on a layer of proteoglycan side-chains which trap synovial fluid. (After Armstrong, C. G. and Mow, V. C. (1980) In *Scientific Foundations of Orthopaedics and Traumatology*, ed. Owen, R. *et al.*, p. 223. London: Heinemann)

is heavy and prolonged the articular surfaces themselves do not come into contact.

Disorders of proteoglycan metabolism

Proteoglycans have half-lives between two and ten days; their breakdown involves an appreciable number of different enzymes. Deficiency of these enzymes causes disorders known as the *mucopolysaccharidoses*. All are rare; the best known are the Hunter and Hurler syndromes with defective dermatan sulphate metabolism (fig. 3.9), and Morquio's syndrome in which keratan sulphate metabolism is affected (fig. 3.10).

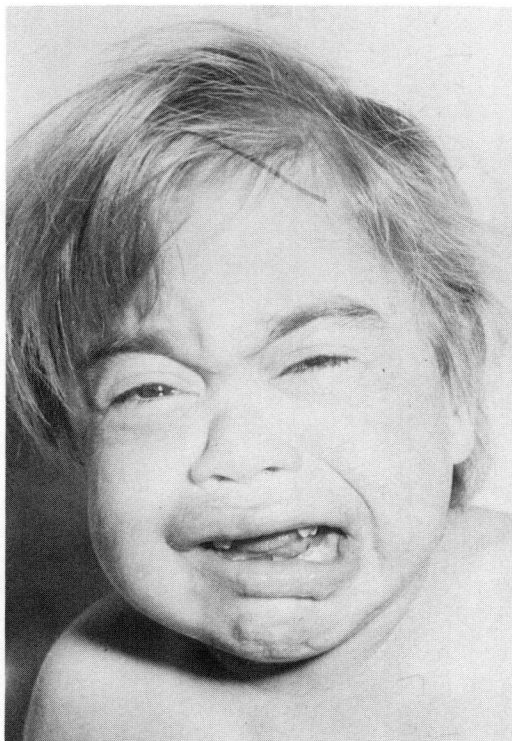

Fig. 3.9 A child with Hurler's syndrome (mucopolysaccharidosis I). The disorder is characterised by severe mental retardation, short stature, changes in the face and bones, clouding of the cornea and enlargement of the liver and spleen. The tissues contain vast amounts of dermatan sulphate which is also excreted in the urine. (From Beighton, P. (1978) *Inherited Disorders of the Skeleton*. Edinburgh: Churchill Livingstone, by courtesy of author and publisher)

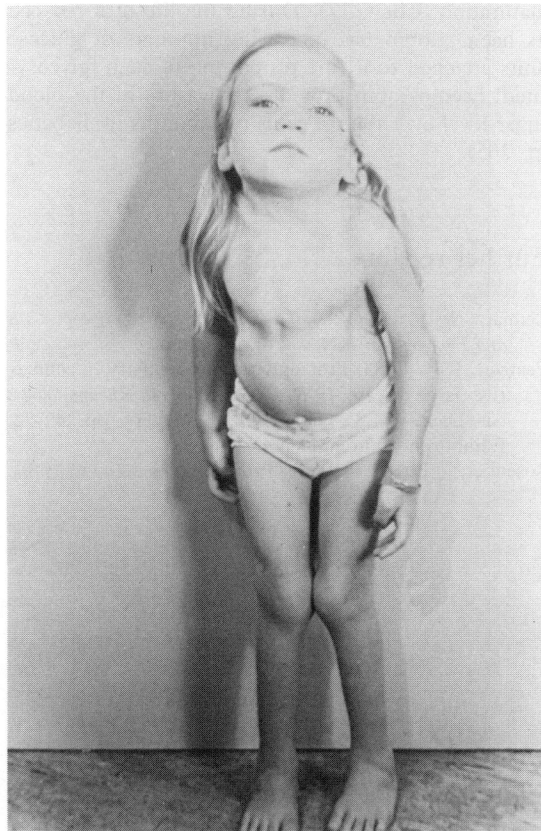

Fig. 3.10 A child with Morquio's syndrome (mucopolysaccharidosis IV). This disorder caused by the deposition of large amounts of keratan sulphate is consistent with normal intelligence but severe changes in the bones. Keratan sulphate is excreted in the urine. (From Beighton, P. (1978) *Inherited Disorders of the Skeleton*. Edinburgh: Churchill Livingstone, by courtesy of author and publisher)

Glycoproteins

The glycoproteins are proteins to which are attached relatively small amounts of carbohydrate, generally as monosaccharides or short chains of two or three monosaccharide units. Many of the plasma proteins (p. 18) are glycoproteins as are many peptide hormones such as follicle stimulating hormone (FSH) and thyroid stimulating hormone (TSH). Some of the proteins of bone matrix are glycoproteins; collagen itself has sugar units attached to it during its maturation (Chap. 17). During the life of a red cell its haemoglobin has increasing numbers of glucose units attached to it; the proportion of such 'glycosylated' haemoglobin provides a measure of the blood sugar level over the previous two months in diabetics (p. 205).

Further reading

Lennarz, W. J. (ed) (1980) *The Biochemistry of Glycoproteins and Proteoglycans*. New York: Plenum

Pearson, J. P. (1983) Proteoglycan biochemistry in connective tissue disease. In *Recent Advances in Rheumatology*, ed. Dick, W. C. and Moll, J. M. H., pp. 49–72. Edinburgh: Churchill Livingstone

Rosenberg, L. C. (1980) Proteoglycans. In *Scientific Foundations of Orthopaedics and Traumatology*, ed. Owen, R., Goodfellow, J. and Bullough, P., pp. 36–42. London: Heinemann

Sharon, N. and Lis, H. (1982) Glycoproteins. In *The Proteins*, 3rd edn, Vol. 5, ed. Neurath, H. and Hill, R. L. New York: Academic Press

Silbert, J. E. (1983) Proteoglycans and glycosaminoglycans. In *Biochemistry and Physiology of the Skin*, ed. Goldsmith, L. A., pp. 448–461. Oxford: University Press

Stasse-Wolthuis, M. (1981) Influence of dietary fibre on cholesterol metabolism and colonic function in healthy subjects. *World Review of Nutrition and Dietetics* **36**, 100–140

Vahouny, G. V. and Kritchevsky, D. (eds) (1982) *Dietary Fibre in Health and Disease*. New York: Plenum

Whelan, W. J. (ed.) (1975) *Biochemistry of Carbohydrates*. London: Butterworths

CRP

4 Lipids

THE lipids are a group of substances which share the property of being relatively insoluble in water and readily soluble in organic solvents such as ether or chloroform. They have several distinct functions in the body: (1) storage of the body's reserve of energy, (2) as components of membranes, (3) as components of some enzyme systems, (4) as hormones (the steroids), (5) heat insulation (in the subcutaneous adipose tissue) and (6) electrical insulation (notably in nerve fibres).

Triglycerides

Energy is stored in the body mainly as triglycerides in which three fatty acids are joined by ester linkages to the trihydric alcohol glycerol. Triglyceride molecules are relatively inert with no charge and no reactive groupings.

The fatty acids have the general formula $CH_3(CH_2)_nCOOH$. The shortest fatty acid found in human fat is butyric acid with four carbon atoms, but the most abundant fatty acids are those with 16 or 18 carbon atoms. The $CH_3(CH_2)_nCO-$ chain is called an *acyl* radical. All common fatty acids have a long hydrophobic chain with a highly reactive hydrophilic carboxyl group at one end.

In the triglyceride *tripalmitin* three molecules of the 16-carbon fatty acid palmitic acid are esterified to glycerol:

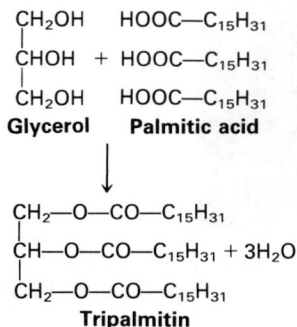

$$
\begin{array}{ll}
CH_2OH & HOOC-C_{15}H_{31} \\
CHOH\ + & HOOC-C_{15}H_{31} \\
CH_2OH & HOOC-C_{15}H_{31} \\
\textbf{Glycerol} & \textbf{Palmitic acid}
\end{array}
$$

$$\downarrow$$

$$
\begin{array}{l}
CH_2-O-CO-C_{15}H_{31} \\
CH-O-CO-C_{15}H_{31}\ +\ 3H_2O \\
CH_2-O-CO-C_{15}H_{31} \\
\qquad\textbf{Tripalmitin}
\end{array}
$$

Similarly *tristearin* is formed from glycerol and three molecules of the 18-carbon fatty acid stearic acid. Both triglycerides occur in large amounts in human fat. When the three fatty acids in a triglyceride are identical, the triglyceride is described as a simple triglyceride, but when the three fatty acids are not identical, the triglyceride is called a mixed triglyceride. The general formula for a triglyceride is

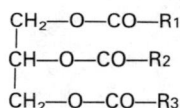

$$
\begin{array}{l}
CH_2-O-CO-R_1 \\
CH-O-CO-R_2 \\
CH_2-O-CO-R_3
\end{array}
$$

In many mixed triglycerides, R_2 is unsaturated. Diglycerides contain two fatty acids and monoglycerides contain one fatty acid esterified to glycerol.

Polyunsaturated fatty acids occur in the body in small amounts in triglycerides and phospholipids. Examples include linoleic acid, linolenic acid and arachidonic acid (fig. 4.1). These are known as *essential fatty acids* since they cannot be synthesised in the body and must be provided in the diet from vegetable sources. However, arachidonic acid can be formed from dietary linoleic acid. Deficiency of essential fatty acids has been recognised for several years (fig. 4.2). The main, and possibly the only function of the essential fatty acids is as precursors of the prostanoids (p. 46).

Energy storage. Triglycerides, unlike the glycogen of liver or muscle, can be laid down in virtually unlimited amounts in adipose tissue; the reserves are only slowly exhausted in fasting. Such storage is economical in both weight and space. Oxidation of one gram of stearic acid produces 40 kJ whereas one gram of glycogen produces 16 kJ. In other words, the amount of energy obtainable from a given weight of triglyceride is more than twice that from the same weight of carbohydrate. It is also economical in terms of bulk since fatty acids have flexible hydrocarbon

$$CH_3.(CH_2)_4.CH{=}CH.CH_2.CH{=}CH.(CH_2)_7.COOH$$

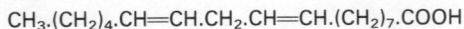

Linoleic acid

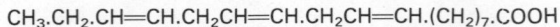

$$CH_3.CH_2.CH{=}CH.CH_2CH{=}CH.CH_2CH{=}CH.(CH_2)_7.COOH$$

Linolenic acid

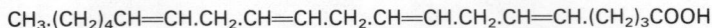

$$CH_3.(CH_2)_4CH{=}CH.CH_2.CH{=}CH.CH_2.CH{=}CH.CH_2.CH{=}CH.(CH_2)_3COOH$$

Arachidonic acid

Fig. 4.1 The principal polyunsaturated fatty acids (essential fatty acids)

chains and can be stored in a more compact way than the rigid molecules of glycogen. Much of the adipose tissue of the body is subcutaneous and, because fat is a poor conductor of heat, it provides excellent insulation. This means that in cold conditions the fat store provides both an insulating blanket and an energy source. The factors which control the deposi-tion and mobilisation of triglycerides in adipose tissue are discussed in Chapter 12.

The structural lipids

The principal lipids found in cell membranes and in the membranes of intracellular organelles are the

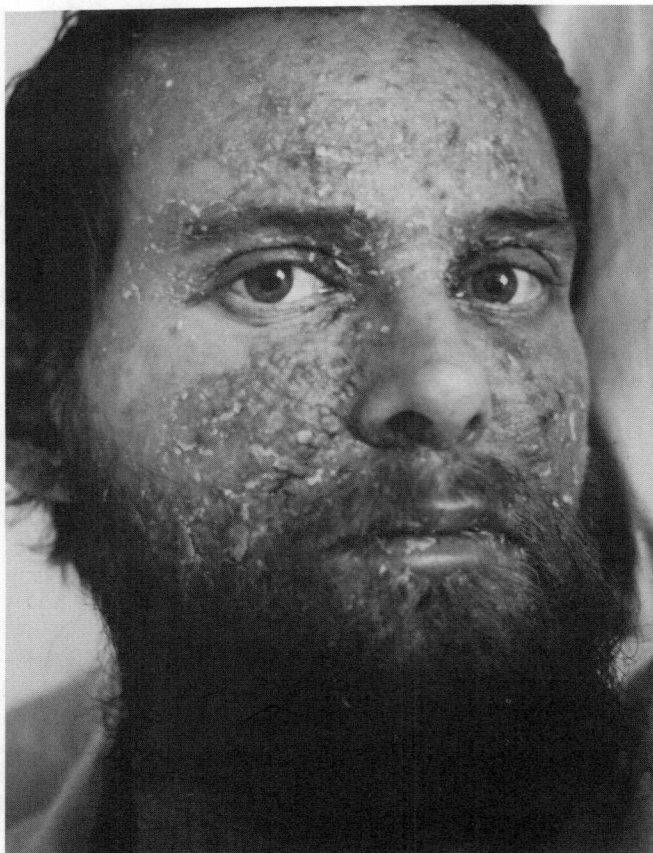

Fig. 4.2 Skin changes of essential fatty acid deficiency. This patient had received total parenteral nutrition for several months. Since, at that time, fat emulsions were not available, the patient received no polyunsaturated fatty acids. (From Hodges, R. E. (1980) *Nutrition in Medical Practice*. Philadelphia: Saunders, by courtesy of author and publisher)

polar or complex lipids which are classified in three main groups: the *glycerophospholipids* (phosphoglycerides), the *sphingophospholipids* and the *glycolipids*. The first two groups are together known as *phospholipids*.

The role of lipids in membranes is discussed later (p. 61). Phospholipids are essential for the activity of many enzymes such as cytochrome oxidase, succinate dehydrogenase and glucose-6-phosphatase.

The glycerophospholipids (phosphatides)

These have a glycerol 'backbone' to which is esterified a phosphoric acid and, with the exception of the plasmalogens, two fatty acids. The parent compound is phosphatidic acid.

$$CH_2-O-CO-R_1$$
$$CH-O-CO-R_2$$
$$CH_2-O-\overset{O^-}{\underset{O}{P}}-OH$$

Phosphatidic acid

These compounds differ from one another in the nature of the alcohol which is esterified to the phosphoric acid and in the nature of the fatty acids. Thus phosphatidic acid is not a single compound but a class of compounds. There are several hundred possible phosphatidic acids which differ from each other only in the nature of the fatty acids attached to the glycerol moiety. Although almost any combination of fatty acids is possible, it is usual for that at position 1 to be saturated while that at position 2 is unsaturated, often one of the essential fatty acids.

In common with many other important intermediate compounds in synthetic pathways, phosphatidic acid is found only in very small amounts in tissues but has a rapid turnover.

The remaining glycerophospholipids have the general formula

$$CH_2-O-CO-R_1$$
$$CH-O-CO-R_2$$
$$CH_2-O-\overset{O^-}{\underset{O}{P}}-O-X$$

where X is an alcohol. The compound formed by esterifying X to a phosphatidic acid is termed phosphatidyl X.

Phosphatidyl choline (lecithin) is the most abundant phospholipid in animal tissues, constituting 30 to 50 per cent of the total phospholipid. Egg yolk is a particularly rich source. In addition to its role in cell membranes phosphatidyl choline is important as a constituent of plasma lipoproteins, particularly chylomicrons.

$$CH_2-O-CO-R_1$$
$$CH-O-CO-R_2$$
$$CH_2-O-\overset{O^-}{\underset{O}{P}}-O-CH_2-CH_2-\overset{+}{N}(CH_3)_3$$

Phosphatidyl choline (lecithin)

Dipalmitoyl lecithin is important as the principal 'surfactant' of the lungs. It reduces surface tension and its lack in premature infants causes the *respiratory distress syndrome* which is often fatal. This disorder can be anticipated, and delivery postponed if possible, by finding a low lecithin level (or a low lecithin/sphingomyelin ratio) in samples of amniotic fluid.

Phosphatidyl ethanolamine (cephalin) is another phospholipid which is widely distributed in animal cells. It is thought to be the platelet factor involved in blood clotting.

$$CH_2-O-CO-R_1$$
$$CH-O-CO-R_2$$
$$CH_2-O-\overset{O^-}{\underset{O}{P}}-O-CH_2CH_2\overset{+}{N}H_3$$

Phosphatidyl ethanolamine (cephalin)

In *phosphatidyl serine* phosphatidic acid is esterified to the amino acid serine (p. 26). It is found in all tissues but in far smaller amounts than either of the previous classes of phospholipids.

Phosphatidyl inositol and derivatives with one or two additional phosphate groups are particularly important in nervous tissue. Little inositol can be synthesised in the body; most is derived from the diet.

Phosphatidyl glycerol is important only as a precursor of *cardiolipin* which has three glycerol moieties and four fatty acid chains. Cardiolipin occurs only in mitochondria, predominantly in the inner membrane (p. 68).

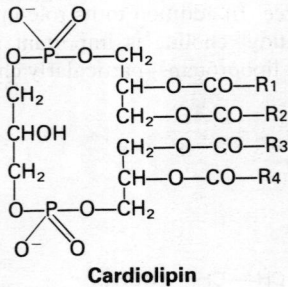

Cardiolipin

Plasmalogens differ from the other glycerophospholipids in having an ether-linked vinyl side-chain instead of an ester-linked fatty acyl chain in position 1.

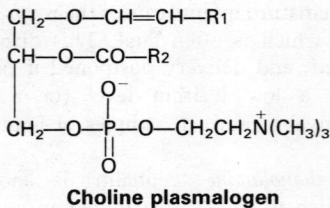

Choline plasmalogen

The plasmalogens, principally choline and ethanolamine plasmalogens, are an important fraction of the phospholipids of nervous tissue.

The sphingophospholipids

Sphingomyelin is the only lipid of this type found in animals. It resembles phosphatidyl choline except that the glycerol moiety is replaced by sphingosine, to which the one fatty acid is linked by an amide bond.

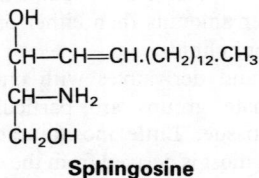

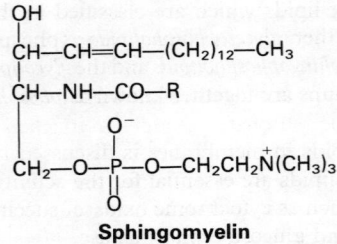

Sphingosine

Sphingomyelin

Sphingomyelin is found in most tissues but is particularly abundant in nervous tissue where, with the glycolipids, it is an important constituent of the myelin sheath.

Glycolipids

In animal cells all the important glycolipids are based on sphingosine. A precursor of the glycolipids is *ceramide* in which a fatty acid is linked by an amide bond to sphingosine.

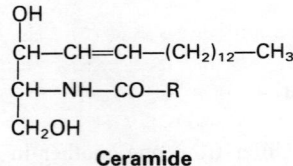

Ceramide

The glycolipids are formed by the addition of one or more monosaccharide residues by a glycoside linkage. Most *cerebrosides* contain galactose, and are known as galactocerebrosides, but glucose may also be found.

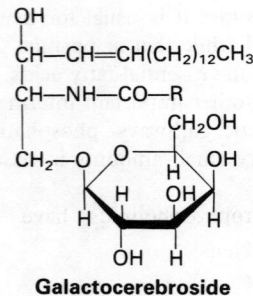

Galactocerebroside

The cerebrosides are major constituents of the myelin sheath. They are characterised by a high content of unusual fatty acid substituents, not found in the phospholipids, including behenic acid and lignoceric acid (saturated fatty acids with 22 and 24 carbon atoms respectively), nervonic acid (24 carbons and one double bond) and cerebronic acid (24 carbons and an α-hydroxy group).

The *sulphatides* are sulphuric acid esters of the cerebrosides. The sulphate group is attached at position 3 on the galactose.

Ceramides with two, three or four monosaccharide units linked together by glycosidic bonds are known as *ceramide oligosaccharides*. Ceramides with four monosaccharides (tetraglycosyl ceramides) are antigenically active and usually have *N*-acetyl galactosamine (p. 30) as one of the monosaccharides. These compounds are precursors of the most complex of the glycolipids, the *gangliosides*. These are tetraglycosyl ceramides to which are attached one, two or three residues of sialic acid (*N*-acetyl neuraminic acid, p. 30). An example is

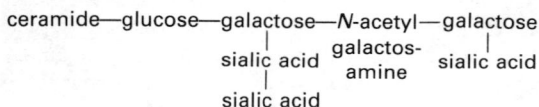

```
ceramide—glucose—galactose—N-acetyl—galactose
              |          galactos-   |
          sialic acid    amine    sialic acid
              |
          sialic acid
```

The importance of sphingomyelin and the glycolipids in the function of the nervous system is illustrated by the rare neurological diseases which result from inherited defects in the enzymes responsible for their breakdown. In each case there is a massive accumulation of the lipid: sphingomyelin in *Niemann–Pick disease*, gangliosides in *Tay–Sachs disease*, cerebrosides in *Gaucher's disease* and triglycosyl ceramides in *Fabry's disease* (fig. 6.14, p. 68). In Gaucher's disease cerebrosides are deposited not only in the nervous system but also in bone and liver (figs. 4.3, 4.4 and 4.5).

Cells of malignant tumours have unusual gangliosides on their cell membranes.

Fig. 4.3 Gaucher's disease. Autopsy specimen of the femora to show deposits of lipid in the marrow and alteration of the normal bone shape. (From Beighton, P. (1978) *Inherited Disorders of the Skeleton*. Edinburgh: Churchill Livingstone, by courtesy of author and publisher)

Steroids

The steroids include cholesterol and other sterols (steroid alcohols), the bile acids, the sex hormones and the hormones of the adrenal cortex. They all contain the same basic nucleus:

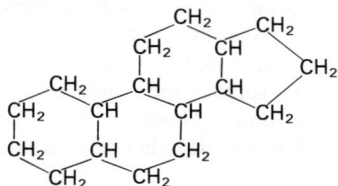

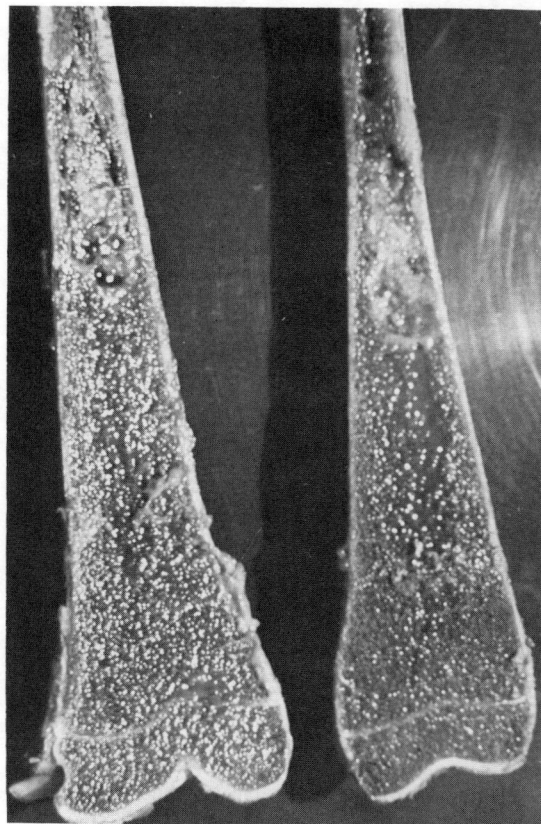

The steroid nucleus is usually written in a 'skeleton' formula. The rings are designated A, B, C, D and the carbon atoms are conventionally numbered as follows:

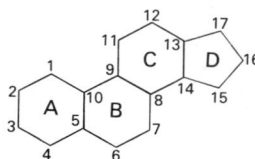

Steroids are classified according to the number of

41

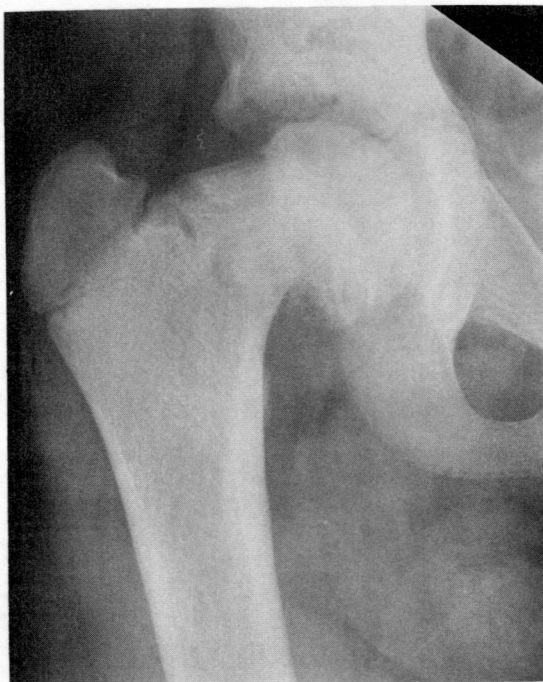

Fig. 4.4 Gaucher's disease in a ten-year-old girl. Fracture through the femoral neck due to replacement of bone by lipid deposits. (From Ozonoff, M. B. (1979) *Pediatric Orthopaedic Radiology*. Philadelphia: Saunders, by courtesy of author and publisher)

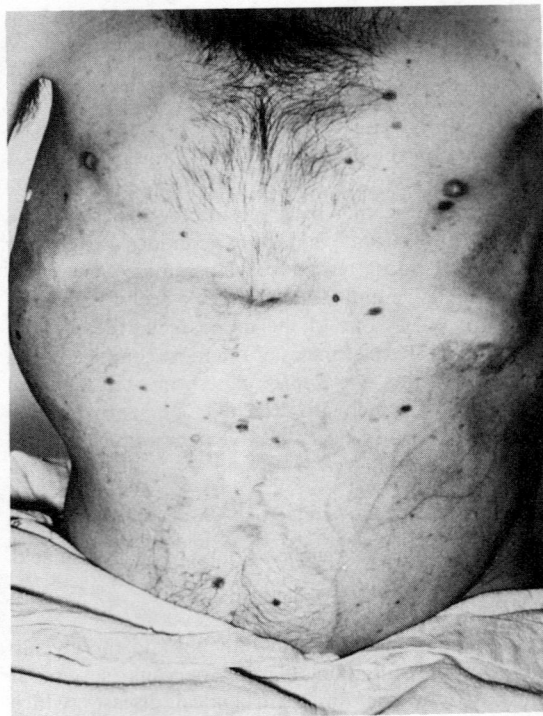

Fig. 4.5 Gaucher's disease. A middle aged male with gross enlargement of the liver due to lipid deposition. (From Beighton, P. (1978) *Inherited Disorders of the Skeleton*. Edinburgh: Churchill Livingstone, by courtesy of author and publisher)

carbon atoms in the ring structures and the side-chains (Table 4.1). X-ray diffraction shows that the methyl groups at C-10 and at C-13 project above the plane of the rings. Substituents projecting above the ring are described as having a β configuration and indicated by a solid line. Substituent groups projecting below the plane are designated α and indicated by a dotted line. Substituents whose projections are unknown are designated ε and indicated by a wavy line.

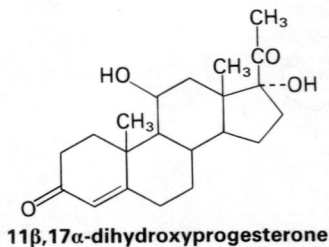

11β,17α-dihydroxyprogesterone

The rings which make up the steroid nucleus are not planar but take up a 'chair' conformation.

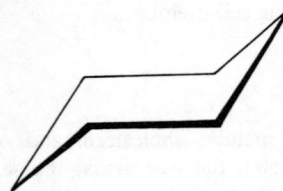

Isomers of steroids occur not only because of differences in the position of substituent groups but also because the relationship between the rings themselves may vary. In naturally occurring steroids the junctions between rings B and C and between C and D are always trans. However the junction between rings A and B may be either cis or trans (fig. 4.6).

Table 4.1   The principal groups of steroids

No. of carbon atoms	Parent hydrocarbon	Biologically important derivatives
18	Oestrane	Oestrogens
19	Androstane	Androgens
21	Pregnane	Progesterone Cortisol Aldosterone
24	Cholane	Bile acids
27	Cholestane	Cholesterol Vitamin D

* In conventional structural formulae a methyl (CH₃₋) side-chain is frequently indicated in this way.

Steroids of the oestrane and androstane series are predominantly of the trans form (giving a 5α-hydrogen) while those of the pregnane series are predominantly cis giving a 5β-hydrogen. When the A ring of the oestrane series is converted to the benzene-like ring as in oestradiol-17β the ring becomes planar and the bonding electrons become involved in the ring structure. This removes the possibility of isomerism in that ring.

The sterols

Cholesterol is widely distributed in tissues and is an essential component of membranes as well as a

Cis A–B junction (5β)

Trans A–B junction (5α)

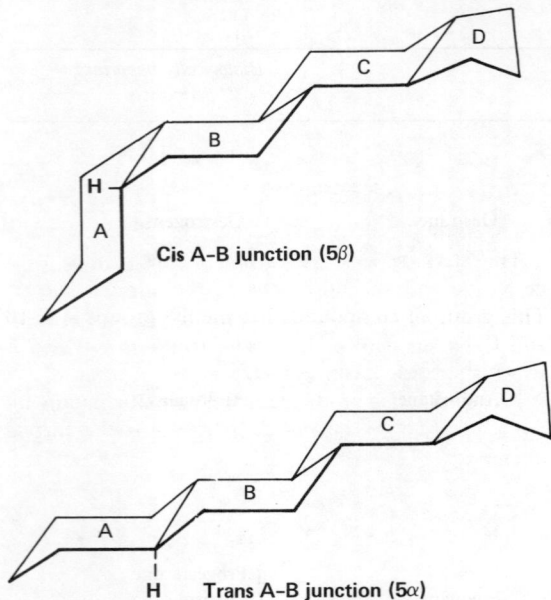

Fig. 4.6 Two conformations found in steroid hormones

metabolic precursor of the bile acids and the steroid hormones. It is abundant in brain, nervous tissue, adrenal glands and skin; it is also found in egg yolk, in gall-stones and in arterial walls affected by the common pathological process called atheroma. Cholesterol has the structure:

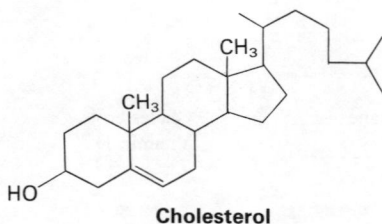

Cholesterol

All sterols are alcohols and can form esters with long-chain fatty acids. In the tissues cholesterol exists partly in the free state and partly as esters. About two-thirds of the cholesterol in plasma lipoproteins is esterified.

Other important sterols include precursors of the various forms of vitamin D (Chap. 21). The vitamin D of animal tissues is vitamin D_3 (cholecalciferol) which is produced in the skin by the action of ultraviolet radiation on 7-dehydrocholesterol.

The bile acids

The bile contains a number of steroid acids which are metabolites of cholesterol. They include cholic acid, deoxycholic acid, chenodeoxycholic acid and lithocholic acid, all of which have a five-membered side-chain at position 17 and hydroxyl groups attached at other positions on the nucleus (fig. 4.7).

These acids are secreted by the liver into the bile in combination with the amino acids glycine and taurine; such compounds are known as *bile salts*. In the intestine they act as biological emulsifiers, emulsifying fats to form *micelles* (p. 3) which present a large surface area for the action of lipolytic

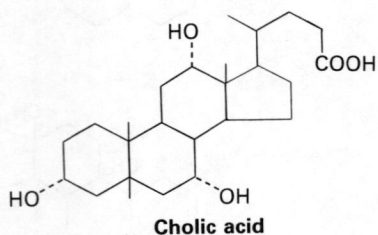

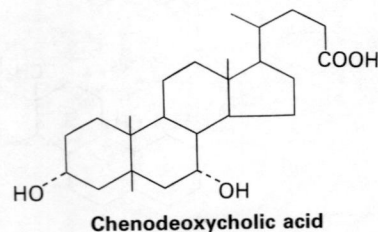

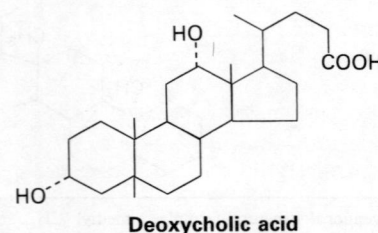

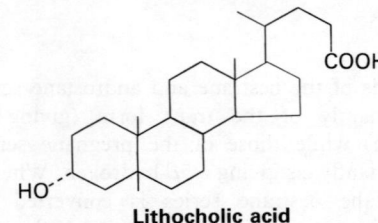

Cholic acid

Chenodeoxycholic acid

Deoxycholic acid

Lithocholic acid

Fig. 4.7 The principal bile acids

enzymes. Their three-dimensional structure (fig. 1.5, p. 2) shows that the hydroxyl groups with an affinity for water are all on the one side of the molecule while the other side is hydrophobic.

The steroid hormones

The steroid hormones can be divided into groups according to the number of carbon atoms they contain. The compounds with 19 carbon atoms include the androgens (male sex hormones) such as testosterone and its metabolite, androsterone.

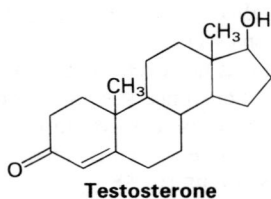

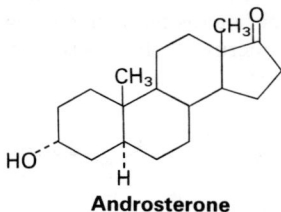

Testosterone

Androsterone

These compounds have methyl groups at C-10 and C-13, no side-chains at C-17 and oxo- or hydroxy-groups at C-3 and C-17.

The steroids with 21 carbon atoms include progesterone and the hormones of the adrenal cortex. The androgens and oestrogens have no side-chain at C-17 but instead an oxo- or hydroxy- group. They have no methyl group at C-10. The oestrogens are unique in that ring A is benzenoid, and thus, because of the hydroxyl group at C-3, they are acidic in nature.

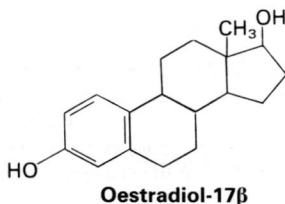

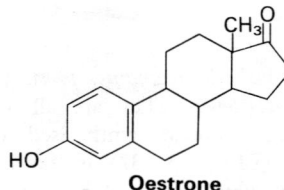

Oestradiol-17β

Oestrone

The steriods with 21 carbon atoms include progesterone and the hormones of the adrenal cortex. This group of compounds has methyl groups at C-10 and C-13, an oxo- or hydroxy- group at C-3 and a two-carbon side-chain at C-17.

Progesterone is produced in very small amounts by the adrenal cortex but its main source is the corpus luteum.

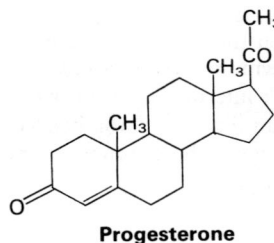

Progesterone

The principal hormones of the adrenal cortex are cortisol and aldosterone.

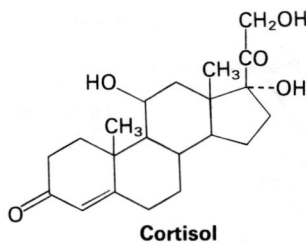

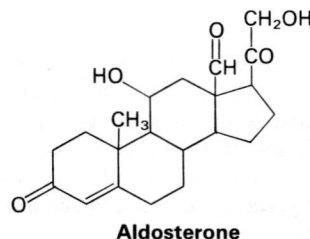

Cortisol

Aldosterone

Prostanoids

This term includes the *prostaglandins, prostacyclin, thromboxanes* and *leucotrienes*. They are all acidic lipids with 20 carbon atoms, synthesised from essential fatty acids (EFA) (p. 37) in particular arachidonic acid. The sole function of EFA may be as prostanoid precursors.

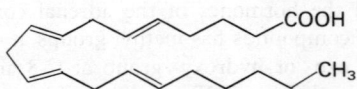

The intracellular source of these EFA is the membrane phospholipids from which they are released by the action of the membrane-bound enzyme, phospholipase A_2 (p. 127). Depending on the cell type, the EFA are then the substrates for two other membrane-bound enzymes: *cyclo-oxygenase*, leading to the production of an intermediate from which the prostaglandins, prostacyclin and thromboxanes are formed, and *lipoxygenase*, whose products include the leucotrienes (fig. 4.8).

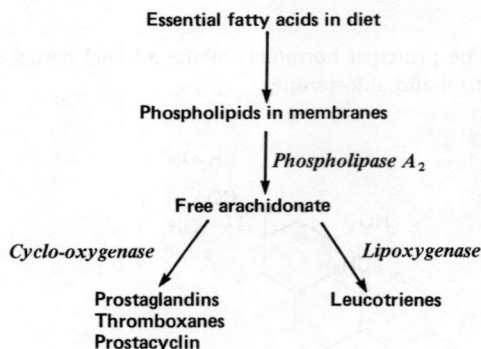

Fig. 4.8 The pathways for the formation of the prostanoids

Most prostanoids are synthesised at, or very near to, their site of action and have very short half-lives; highly active degradative enzymes are present in all tissues. The lung is the most important site for the breakdown of circulating prostanoids, almost all being completely degraded after one passage through the pulmonary bed.

Prostaglandins

As early as the 1930s it was recognised that an acidic lipid from seminal fluid was a potent stimulator of smooth muscle contraction but it was not until some 30 years later that the active compounds, the prostaglandins, were crystallised and identified. Primate seminal fluid is the richest known source of prostaglandins (1 mmol/l) but all other tissues contain them in smaller amounts (15 to 30 μmol/l).

The structures of the major classical prostaglandins are shown in figure 4.9. All are characterised by a cyclopentane ring, one or two oxygen atoms and at least one double bond.

Prostaglandins are highly active; tissues respond to nanomolar concentrations. The physiological function of the prostaglandins is however obscure in most cases. Many of the functions once suggested for the prostaglandins in the cardiovascular system can now be attributed to the more recently discovered thromboxanes and prostacyclin (p. 48).

Prostaglandins have a wide variety of effects on smooth muscle. PGE_1 and PGE_2 inhibit motility and tone of the non-pregnant uterus but increase contraction of the pregnant uterus. $PGF_{2\alpha}$ causes uterine

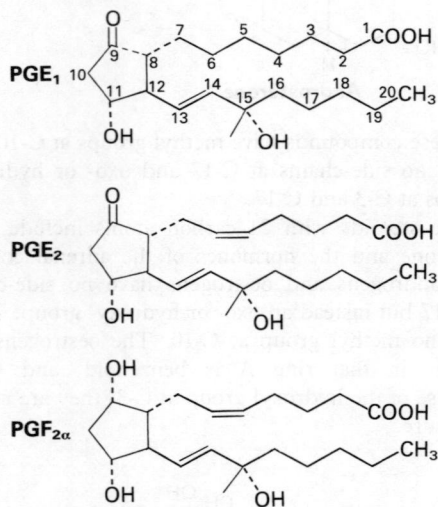

Fig. 4.9 Structures of some of the 'classical' prostaglandins. The E series are those with an oxo- and a hydroxyl group in the ring while the F series are those with two hydroxyl groups. The numerical subscript indicates the number of double bonds while the α refers to the configuration of the hydroxyl group at position 11

contraction and may have a physiological role in parturition and menstruation. Both $PGF_{2\alpha}$ and PGE_2 are used as alternatives to oxytocin for the induction of labour. PGEs and PGAs have a direct action on arterial smooth muscle to cause vasodilatation and hypotension; the PGFs have opposite effects. The PGEs and PGFs also have opposite actions in the lung: PGEs cause bronchodilatation whereas PGFs cause bronchoconstriction.

At one time it was suggested that prostaglandins were hormones and neurotransmitters. It now seems more likely that they act by modifying the response by tissues to hormones, particularly those which act through adenylate cyclase. In most tissues, prostaglandins enhance the hormonal stimulation of adeny-

late cyclase; they may be essential for the interaction of the hormone with its receptor on the cell surface and for the coupling of the hormone–receptor complex with the adenylate cyclase. In a few tissues prostaglandins inhibit both the basal activity of adenylate cyclase and also the response to hormones. For example PGE_1 inhibits the stimulation of lipolysis in adipose tissue by hormones such as adrenaline and glucagon, the gastrin-induced HCl secretion in the stomach and the ADH-stimulated water reabsorption in the kidney.

Prostaglandins also modify the function of neurotransmitters. In the sympathetic nervous system, PGEs are released within synapses after nerve stimulation and inhibit the release of the transmitter,

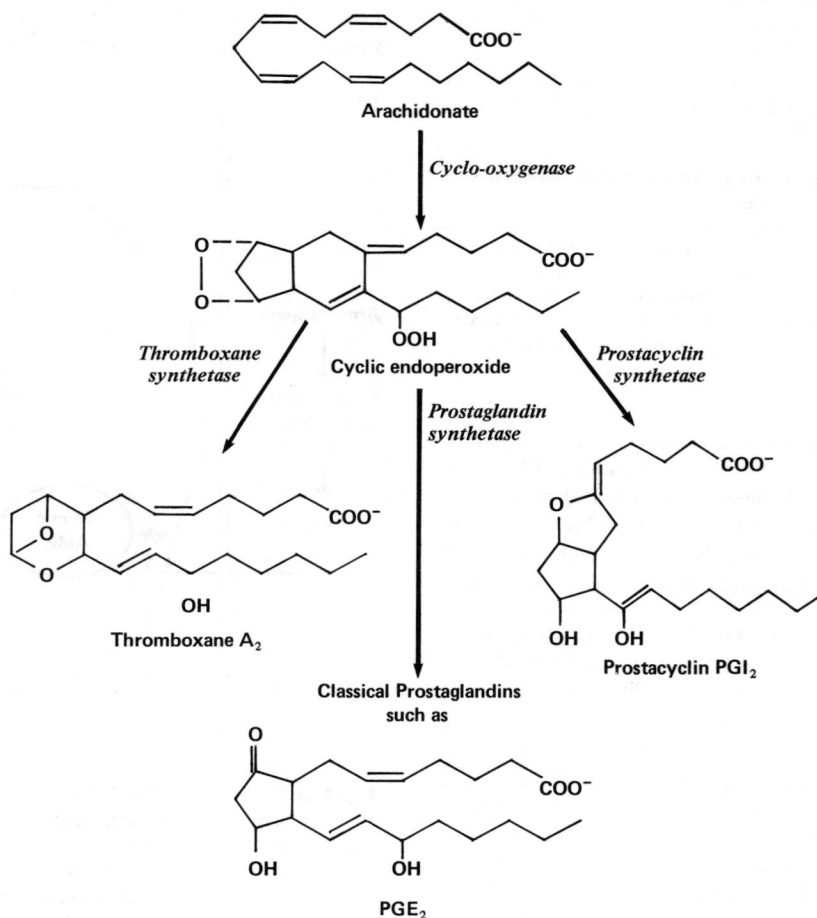

Fig. 4.10 Structures and metabolic origins of prostacyclin and the thromboxanes compared with the classical prostaglandins

noradrenaline. It is suggested that as with the effect on hormone action the release of prostaglandins serves as a negative feedback control to damp down the response.

Prostacyclin, thromboxanes and blood vessels

Prostacyclin (PGI_2) and thromboxane (TXA_2) were discovered in the mid-1970s. PGI_2 is produced from arachidonic acid by the endothelial cells of blood vessels. TXA_2 is the major product formed from arachidonic acid in platelets. These compounds have structures which differ from those of the classical prostaglandins (fig. 4.10). Both are extremely unstable, PGI_2 having a half-life of 3 minutes whilst TXA_2 is even more labile (half-life, 30 sec). PGI_2 is broken down enzymatically and TXA_2 non-enzymically to inactive products.

PGI_2 and TXA_2 have important and antagonistic effects on the cardiovascular system (Table 4.2).

Table 4.2 Effects of prostacyclins and thromboxanes on the cardiovascular system

	Prostacyclin	Thromboxanes
Platelet aggregation	Inhibited	Stimulated
Coronary arteries	Relaxed	Constricted
Blood pressure	Lowered	Raised

PGI_2 lowers the blood pressure by causing vasodilatation and prevents platelet aggregation. TXA_2 has opposite actions. Platelet aggregation appears to be controlled by the levels of cyclic AMP within the platelet; both compounds exert their effect by increasing (PGI_2) or depressing (TXA_2) cyclic AMP concentrations.

The suggested physiological role of prostacyclin and thromboxane in the control of platelet function is illustrated in figure 4.11. Platelets synthesise TXA_2 when stimulated, for example, by thrombin, collagen or adrenaline. In turn this results in aggregation. Platelet aggregation (and possibly occlusion of a blood vessel) could occur within a vessel as a result of mechanical damage to platelets resulting from turbulent flow through small (and especially atheromatous) blood vessels. A mechanism to prevent such inappropriate thrombosis has been suspected for some time and it now seems that the continuous produc-

tion of PGI_2 by the endothelium may be such a mechanism.

PGI_2 interacts with a specific receptor on the platelet membrane and stimulates adenylate cyclase. The resulting increase in cyclic AMP inhibits platelet aggregation and opposes the aggregatory effects of mechanical damage. An interesting finding in keeping with its local anti-aggregatory role is that the capacity for PGI_2 synthesis decreases from the luminal surface to the outside of the blood vessel. Conversely, pro-aggregatory elements such as collagen increase from the lumen to the outside. This explains how platelets can plug a punctured blood vessel without the thrombus extending into the lumen.

A balance between prostacyclin and thromboxane levels thus seems to be important in ensuring that platelet thrombi normally form only in response to vascular damage. A disturbance of that balance

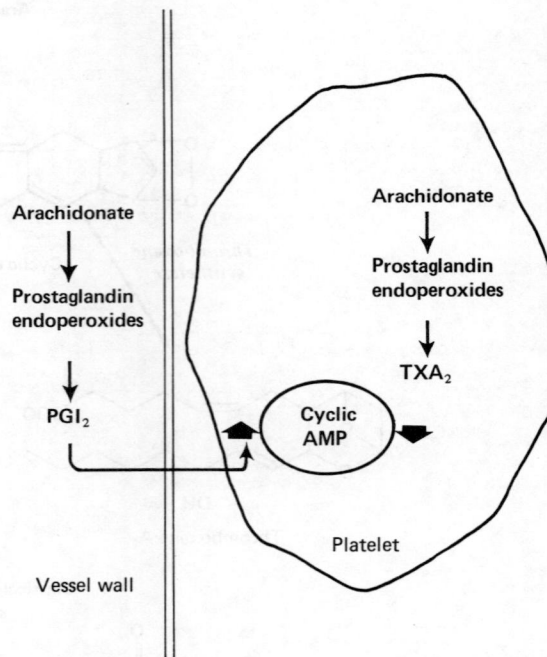

Fig. 4.11 Interaction of platelets with a vessel wall. Stimulation of platelets to aggregate leads to the production of thromboxane (TXA_2) which prevents stimulation of adenylate cyclase. Prostacyclin (PGI_2) is synthesised by the vessel wall and stimulates adenylate cyclase of platelets so inhibiting platelet aggregation. (After Moncada, S. and Vane, J. R. (1978) *British Medical Bulletin* **34**, 129)

through overproduction of TXA_2 or deficiency of PGI_2 has been reported in coronary thrombosis and in other disorders, such as diabetes mellitus and the hyperlipidaemias, in which thrombosis frequently occurs.

Modification of platelet behaviour. The reduction of the TXA_2/PGI_2 ratio is clearly an attractive possibility in the treatment and prevention of vascular disorders. Aspirin, for example, inhibits cyclo-oxygenase, and may be effective in reducing the incidence of coronary artery thrombosis, since it appears to inhibit the enzyme more markedly in platelets than in vascular endothelium. Platelet concentrates from blood donations whose donors had taken aspirin within a week are of greatly reduced value to a recipient with a bleeding tendency due to platelet lack.

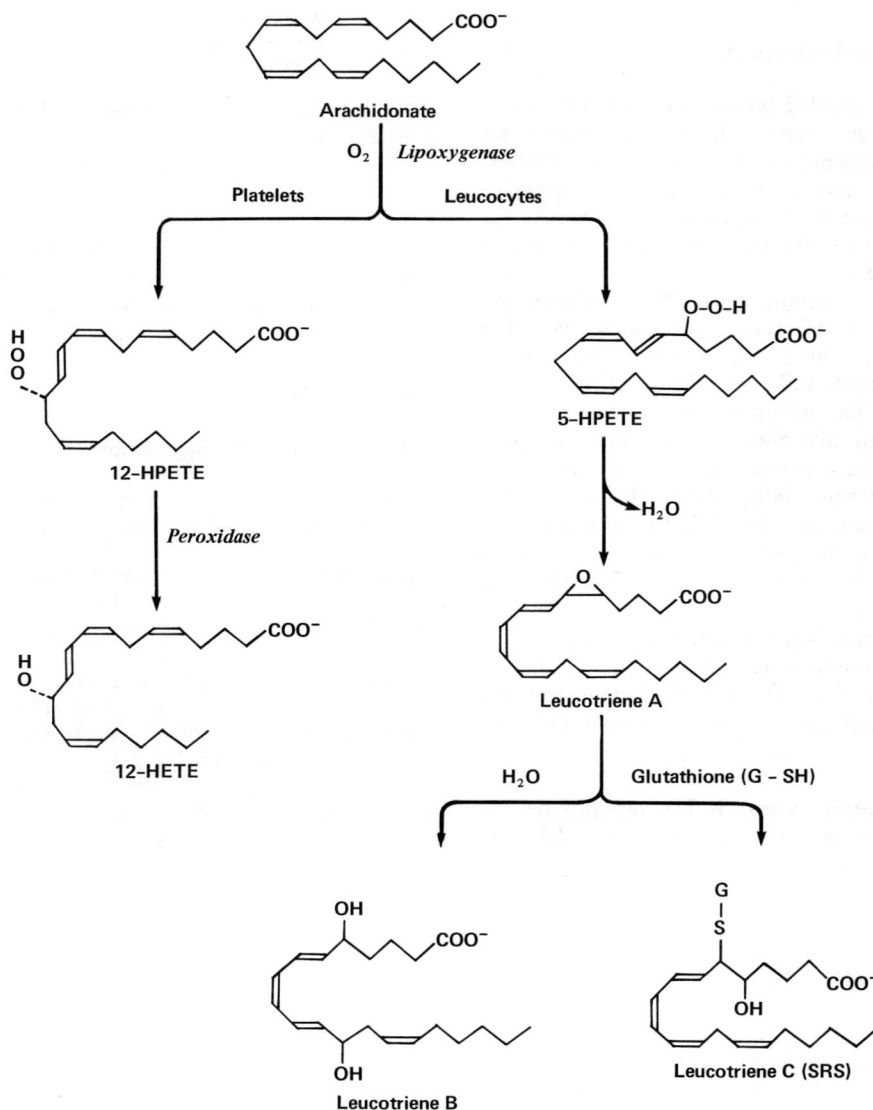

Fig. 4.12 The principal products of the lipoxygenase pathway of arachidonic acid metabolism

Eskimos, whose diet is rich in oily fish such as mackerel, have a reduced tendency for platelet aggregation and prolonged bleeding times. The fish contain an analogue of arachidonic acid, eicosapentaenoic acid, which is metabolised into both a prostacyclin analogue and a thromboxane analogue. The prostacyclin is an effective anti-aggregation compound but the thromboxane is inactive in promoting platelet aggregation.

Leucotrienes and inflammation

Leucotrienes (so-called because they are synthesised by leucocytes and possess three double bonds) are formed from arachidonic acid by the lipoxygenase pathway. The lipoxygenase products are a complex mixture of non-cyclic hydroxy- and hydroperoxy-C_{20} acids (HETEs and HPETEs) as well as the leucotrienes (fig. 4.12).

Inflammatory stimuli cause the synthesis and release from cells of classical prostaglandins and of lipoxygenase products. The prostaglandins cause local vasodilatation and potentiate the effects of other mediators of the inflammatory response such as bradykinin and histamine. These cause the pain, increased capillary permeability and oedema which are features of acute inflammation. Products of the lipoxygenase pathway, notably HPETE and HETE, are released from leucocytes and are strongly chemotactic to cells of the immune system. This results in the rapid invasion of an inflamed or wounded area by such cells. Leucotriene C has now been shown to be identical with the slow reacting substance of anaphylaxis. This is a major factor in the hypotension and bronchial constriction of allergic reactions including asthma.

Anti-inflammatory drugs. It has recently become clear that aspirin and other non-steroidal anti-

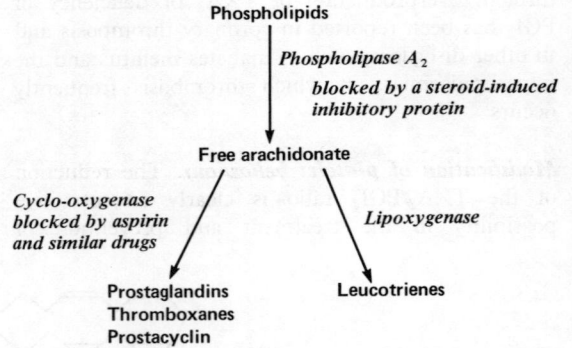

Fig. 4.13 Sites of action of drugs which affect prostanoid metabolism

inflammatory drugs act by inhibiting cyclo-oxygenase, while steroids inhibit the phospholipase A_2-mediated release of arachidonic acid (fig. 4.13).

Further reading

Brady, R. O. (1978) Sphingolipidoses. *Annual Review of Biochemistry* **47,** 687–713

De Gaetano, G. (1981) Platelets, prostaglandins and thrombotic disorders. *Clinics in Haematology* **10,** 297–326

Gurr, M. I. and James, A. T. (1980) *Lipid Biochemistry*, 3rd edn. London: Chapman and Hall

Hammarström, S. (1983) Leukotrienes. *Annual Review of Biochemistry* **52,** 355–377

Moncada, S. (ed.) (1983) Prostacyclin, Thromboxane and leukotrienes. *British Medical Bulletin* **39,** 209–295

Robertson, R. P. (ed.) (1981) Prostaglandins. *Medical Clinics of North America* **65,** 711–938

Robertson, R. P. (ed.) (1981) Prostaglandins. *Medical Clinics of North America* **65,** 711–938

MISH DT MCKB CRP

5 Nucleotides & nucleic acids

GENETIC information is stored in the nuclei of cells as the sequence of nucleotides which make up deoxyribonucleic acid (DNA). The information on the DNA is applied to protein synthesis by a number of ribonucleic acids (RNA). This chapter describes DNA, RNA and the nucleotides which make up both macromolecules.

Nucleotides

A nucleic acid consists of a chain of pentose and phosphate units; to the pentose units are attached the bases which are either purines or pyrimidines.

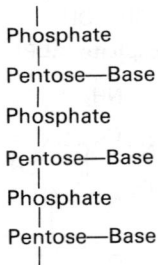

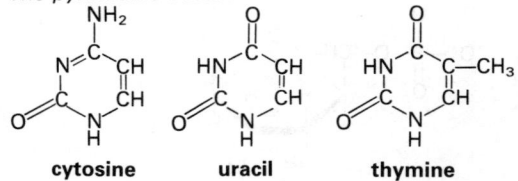

```
   |
Phosphate
   |
Pentose—Base
   |
Phosphate
   |
Pentose—Base
   |
Phosphate
   |
Pentose—Base
   |
```

In DNA the pentose is deoxyribose and in RNA it is ribose.

The phosphate groups are attached to the 3′ and 5′ carbon atoms of the pentose and the base is attached to the 1′ carbon (fig. 5.1). The repeating unit of one pentose, one phosphate and one base is known as a *nucleotide*.

Bases. In DNA there are four commonly occurring bases. Two, adenine and guanine, are purines and two, cytosine and thymine, are pyrimidines. In RNA the four principal bases are adenine, guanine, cytosine and uracil (fig. 5.2). In some forms of RNA, particularly transfer RNA, smaller amounts of bases such as thymine, 5-methyl-cytosine and 6-methyl-

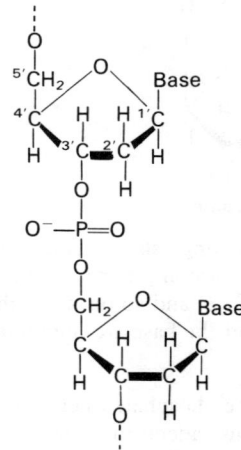

Fig. 5.1 The way in which the deoxyribose units of DNA are linked to each other and to the base. In RNA the pentose is ribose with a hydroxyl group at position 2

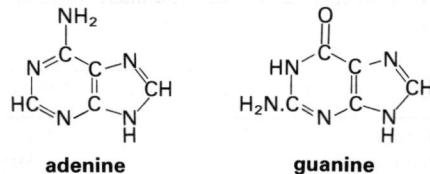

Fig. 5.2 The principal pyrimidine and purine bases found in DNA and RNA

adenine are found. Very small amounts of methylated bases are also found in DNA.

Nucleosides. Compounds of bases and pentoses are known as nucleosides. The condensation product of adenine and ribose is adenosine:

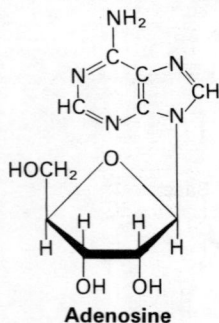

Adenosine

In order to distinguish the atoms in a particular compound the carbon atoms of the pentose are numbered 1′, 2′, 3′ and so on while the carbon and nitrogen atoms of the base are numbered 1, 2, 3 and so on.

Nucleotides. The phosphate ester of a nucleoside is a nucleotide. Thus adenosine forms adenosine 5′-monophosphate (AMP) sometimes known as adenylate.

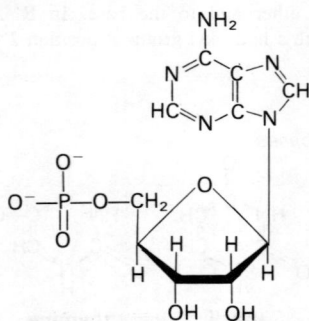

Adenosine 5′-monophosphate (AMP)

The names of the principal nucleosides and nucleotides which make up DNA and RNA are given in Table 5.1.

Nucleotide coenzymes

In addition to their role in the structure of nucleic acids, nucleotides have many important functions as coenzymes (Chap. 8).

Adenosine diphosphate (ADP) and adenosine triphosphate (ATP). These play a central role in metabolism as the molecules which link energy-yielding reactions to those which require energy (fig. 7.1 p. 75).

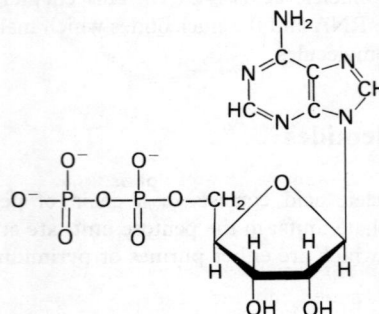

Adenosine 5′-diphosphate (ADP)

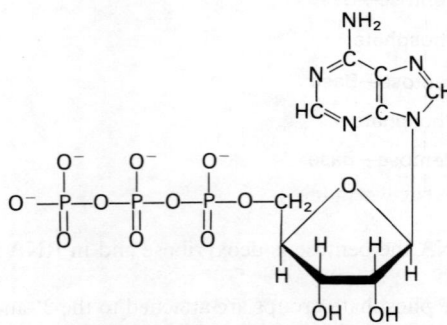

Adenosine 5′-triphosphate (ATP)

Table 5.1 Principal nucleosides and nucleotides of RNA and DNA

	Adenine (A)	Guanine (G)	Thymine (T)	Uracil (U)	Cytosine (C)
Ribonucleoside	Adenosine	Guanosine	—	Uridine	Cytidine
Ribonucleotide	AMP	GMP	—	UMP	CMP
Deoxyribonucleoside	Deoxyadenosine	Deoxyguanosine	Deoxythymidine	—	Deoxycytidine
Deoxyribonucleotide	dAMP	dGMP	dTMP	—	dCMP

(Base)

Cyclic AMP. Cyclic 3′,5′-adenosine monophosphate plays an essential part in hormone action (Chaps. 12 and 22). Its production within the cell, in response to the arrival of a hormone molecule at a receptor on the cell membrane, leads to the activation of intracellular enzymes (for example in fig. 12.4, p. 133).

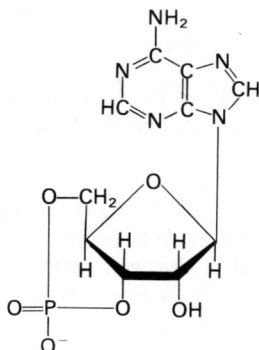

Cyclic 3′,5′-adenosine monophosphate (cyclic AMP)

Nicotinamide nucleotides. These contain two bases, nicotinamide and adenine. Nicotinamide whose structure is

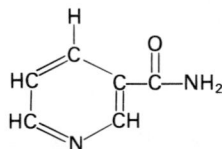

Nicotinamide

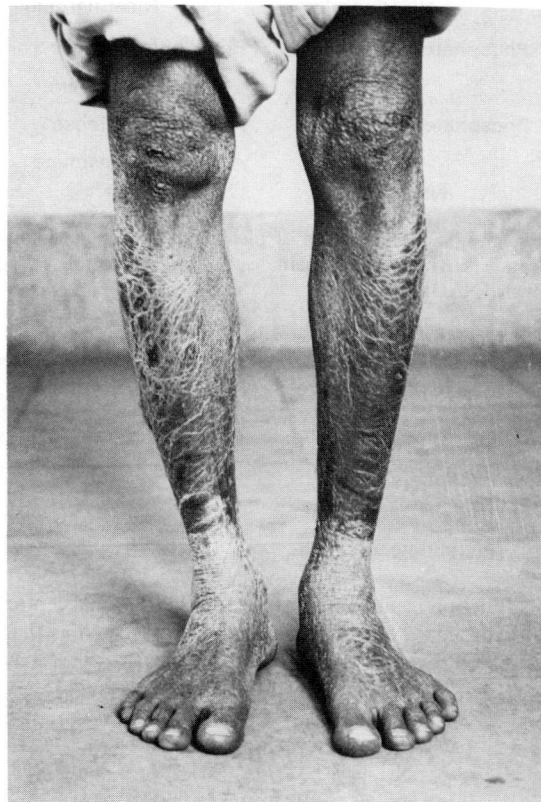

Fig. 5.3 Skin changes in pellagra due to lack of nicotinamide. (Courtesy of M. Mohan Ram)

and oxidised as described in Chapter 10. The reactive site in each case is in the nicotinamide component.

can be synthesised in the body from dietary tryptophan (p. 193). Pellagra occurs if the diet is deficient in both tryptophan and nicotinamide (fig. 5.3).

The principal nicotinamide nucleotides are *nicotinamide adenine dinucleotide* (NAD) which has an essential role in many energy-yielding reactions such as glycolysis (Chap. 9) and nicotinamide adenine dinucleotide phosphate (NADP) which is essential for fatty acid synthesis (Chap. 11) and the reduction of met-haemoglobin in red cells (Chap. 16). Their structures are given in figure 5.4.

These compounds are important as carriers of electrons and hydrogen ions; they are readily reduced

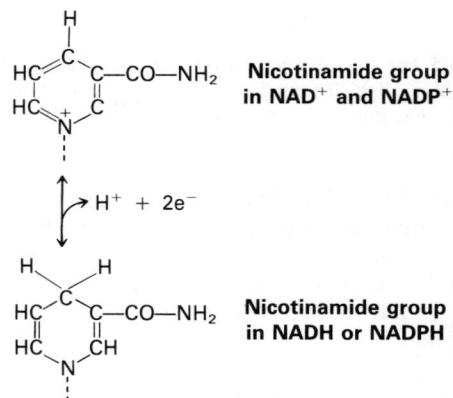

Nicotinamide group in NAD$^+$ and NADP$^+$

$H^+ + 2e^-$

Nicotinamide group in NADH or NADPH

53

```
        Nicotinamide                    Nicotinamide
             |                               |
  Phosphate—Ribose                Phosphate—Ribose
             |                               |
          Adenine                         Adenine
             |                               |
  Phosphate—Ribose                Phosphate—Ribose
                                            |
                                        Phosphate

        NAD                             NADP
```

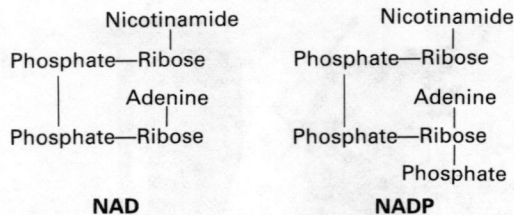

Fig. 5.4 The components of nicotinamide adenine dinucleotide (NAD) and nicotinamide adenine dinucleotide phosphate (NADP)

Flavin nucleotides. Flavin adenine dinucleotide (FAD) has the structure

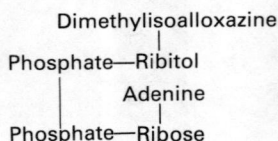

```
      Dimethylisoalloxazine
             |
  Phosphate—Ribitol
             |
          Adenine
             |
  Phosphate—Ribose
```

Its function is also similar in that it acts as an acceptor of hydrogen atoms in oxidative reactions such as that catalysed by succinate dehydrogenase (p. 98). The active component is shown in figure 5.

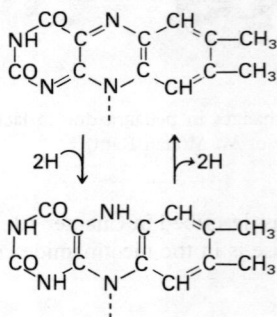

Fig. 5.5 The role of flavin adenine dinucleotide (FAD) in accepting and giving up hydrogen atoms. The dimethyl-isoalloxazine group only is shown

The vitamin *riboflavin* consists of a dimethylisoalloxazine group and ribitol. Pure riboflavin deficiency does not occur naturally but lack of riboflavin probably contributes to the symptoms of patients with general malnutrition particularly in chronic alcoholics.

Coenzyme A. This plays an essential role in the metabolism of carbohydrates and fats (Chaps. 9 and 11). Its structure is

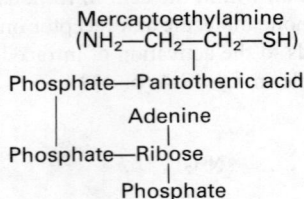

```
      Mercaptoethylamine
      (NH₂—CH₂—CH₂—SH)
             |
  Phosphate—Pantothenic acid
             |
          Adenine
             |
  Phosphate—Ribose
             |
          Phosphate
```

Pantothenic acid can be shown to be a vitamin in some animals but a deficiency disease has not been described in man.

Uridine coenzymes. The most important uridine coenzyme is *uridine diphosphate glucose* (UDP glucose) which is involved in glycogen synthesis (p. 103) and in the conversion of galactose to glucose (p. 103). Its structure is

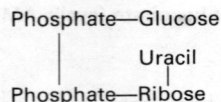

```
  Phosphate—Glucose
             |
          Uracil
             |
  Phosphate—Ribose
```

Cytidine coenzymes. Derivatives of cytidine diphosphate such as CDP choline are involved in the synthesis of phospholipids (Chap. 11).

DNA

It has long been recognised that a gene, the fundamental unit of heredity, must have two basic properties: it has to have a special function within a cell to direct a specific activity of the cell and it has to be able to replicate itself accurately. In addition it was recognised that genes were subject to an occasional sudden change or *mutation* which was then passed on to subsequent generations.

The model of DNA put forward by Watson and Crick in 1953 provided a structural explanation for these properties. They proposed that DNA is made up of two very long polynucleotide chains coiled round each other to form a double helix. The backbone of each chain consists of phosphate and deoxyribose; the interior of the helix is occupied by the purine and pyrimidine bases. The structure of Watson and Crick is a right-handed helix with a pitch of 3.4 nm or 10 base pairs per turn (fig. 5.6). Studies

54

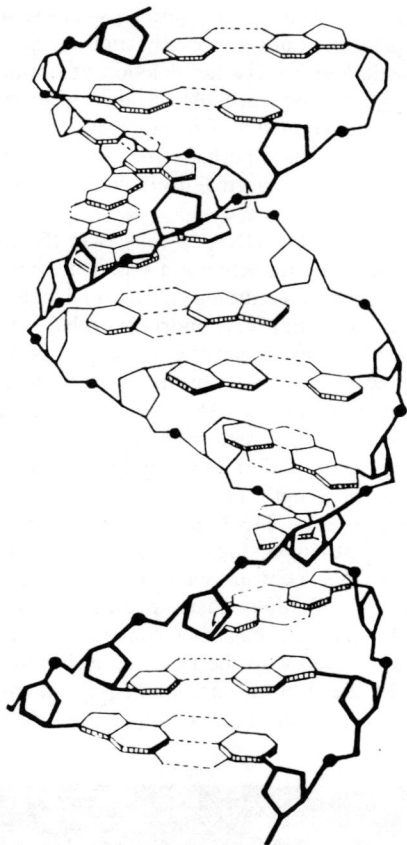

Fig. 5.6 The double helix of DNA. The two strands are linked by hydrogen bonds (dotted lines) between purine and pyrimidine residues as shown in fig. 5.7. (After Davidson, J. N. (1972) *The Biochemistry of the Nucleic Acids*. 7th edn. London: Methuen)

on fibres of pure DNA have shown however that the characteristics of the helix vary with the hydration of the DNA and the ionic composition of the environment. The possibility that some DNA exists in a largely non-helical form with the chains side by side has not been excluded.

The two chains of DNA are held together by hydrogen bonds between the bases projecting at the same level in each chain. A pair of bases must consist of one purine and one pyrimidine and, of the possible combinations, only two can occur: adenine with thymine and guanine with cytosine (fig. 5.7).

The individual hydrogen bonds are weak but the very large number in a double helix gives the whole

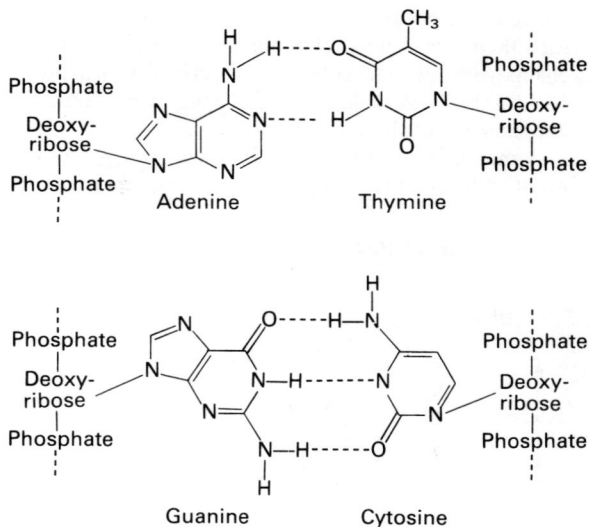

Fig. 5.7 The normal base-pairing in DNA

molecule great stability. In addition the partial overlapping of the hydrophobic parts of bases of successive nucleotides (fig. 5.6) provides additional forces to maintain the stability of the structure.

A molecule of DNA is very large; its mass ranges from 2 to 3 million in small viruses to a hundred thousand million in some animal cells. A gene can be regarded as a length of DNA typically containing several thousand base pairs. The base pairs can occur in any sequence and an enormously large number of permutations is therefore possible. Each gene has its own unique structure from which is derived its special function. The way in which the precise information in the sequence of bases is used to direct the synthesis of a protein is described in Chapter 14.

The nature of one base in a pair determines the nature of the other. Thus the two chains which make up a molecule of DNA, though different, are exactly complementary (fig. 5.8). For replication of the DNA

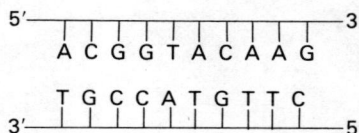

Fig. 5.8 The pairing of bases in part of a double strand of DNA. Note that the two strands are running in opposite directions (*antiparallel*)

molecule to occur the chains unwind and separate; each then acts as a template for the formation of new chains with free nucleotides as described in Chapter 13. In this way from one molecule two identical daughter molecules are formed (fig. 13.14).

A mutation can be regarded as the result of some event which alters the base-pair sequence of a

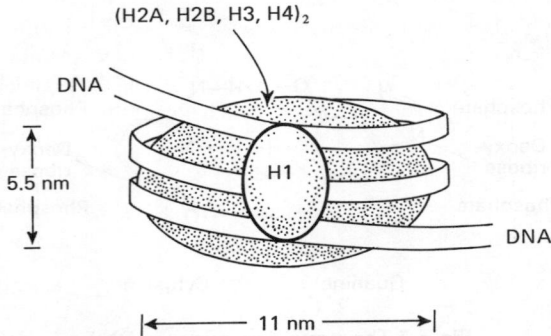

Fig. 5.9 Structure of a nucleosome. A length of the DNA double helix (equivalent to 170 to 200 base pairs) is wrapped round a complex of 8 histone proteins (H2A, H2B, H3, H4)$_2$. Histone H1 binds to DNA but is not associated with the other histones. (After Kornberg, A. (1980) *DNA Replication*. San Francisco: Freeman)

particular gene. Many such alterations consist simply of the replacement of one base by another. Some are more drastic and involve the deletion or duplication of a part of the sequence. Alterations in genes give rise to the inborn errors of metabolism described in several chapters and, perhaps most graphically, to the great variety of abnormal haemoglobins described in Chapter 16.

While most of the DNA in cells is in the nuclei, some is found in mitochondria where it may have unusual structures, including circular molecules. The presence of DNA in mitochondria is probably related to the fact that these organelles are self-replicating.

It is now clear that a chromosome contains one continuous molecule of DNA. The largest human chromosome, if fully extended as a double helix, would be some 7.3 cm in length. However at metaphase the length of the chromosome is only about 10 μm. The DNA in a chromosome must therefore exist in a highly condensed state. In addition to DNA a chromosome consists of a number of proteins including basic proteins known as *histones*. Human chromosomes contain five different histones associated with the DNA in *nucleosomes* (fig. 5.9). Under the electron microscope isolated chromosomes are seen as strings of such particles (fig. 5.10).

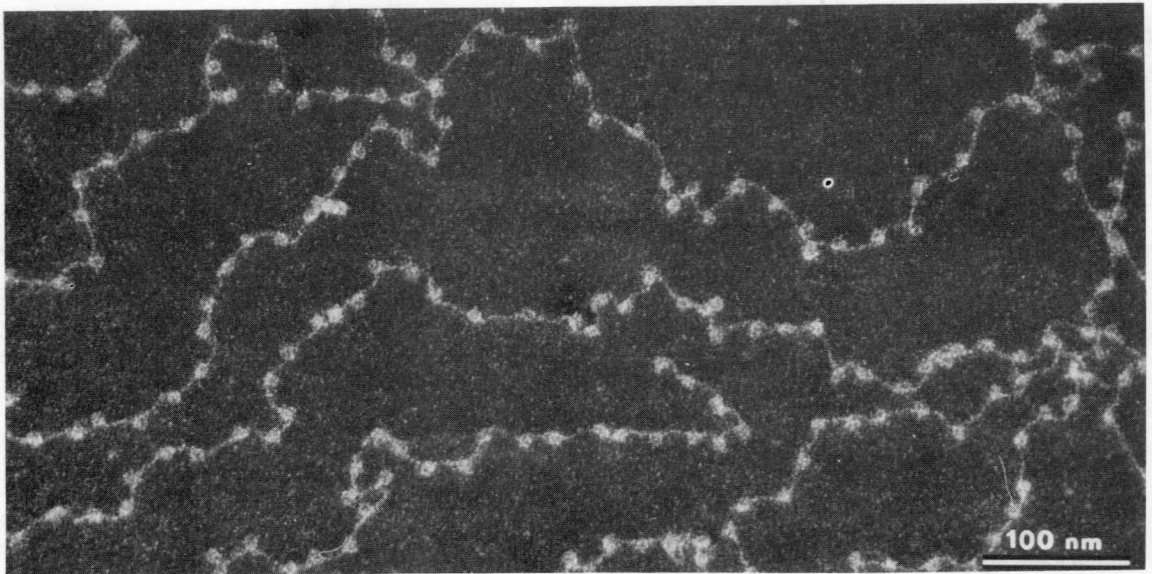

Fig. 5.10 Electron micrograph of chromatin from chicken erythrocytes to show nucleosome 'beads' on the strands of DNA. (Courtesy of Ada L. Olins and D. E. Olins)

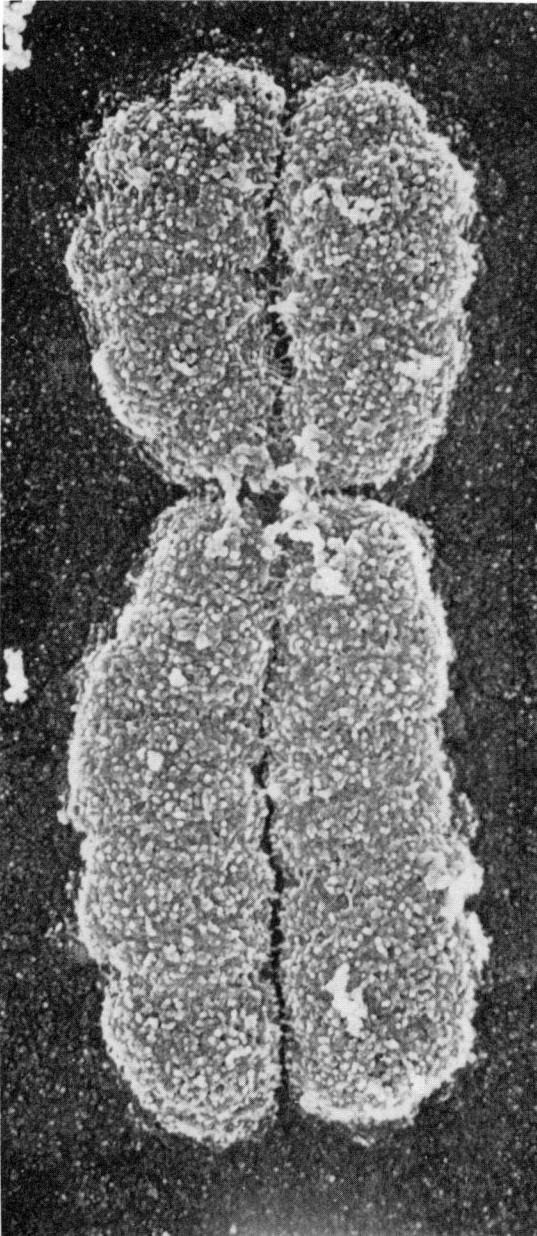

Fig. 5.11 Scanning electron micrograph of human chromosome 2, ×30 000 (Courtesy of Christine Harrison; the technique for the production of this figure is described by C. Harrison *et al.* (1981) *Experimental Cell Research* **134**, 141)

In the intact chromosome (Fig. 5.11) the nucleosomes themselves are thought to exist in highly folded structures which involve proteins other than histones.

RNA

Like DNA, RNA is a long polynucleotide chain but with ribose instead of deoxyribose. A further difference is that uracil is a major base instead of thymine. Unlike DNA, RNA is largely single stranded although there are loops which form short double-helical segments in which hydrogen bonds maintain the structure as in DNA (fig. 5.12).

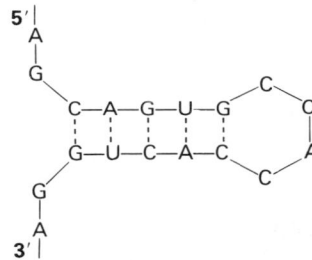

Fig. 5.12 Part of a strand of RNA to show a double-helical segment with base-pairing. The principal pairs are between guanine (G) and cytosine (C), and between adenine (A) and uracil (U). Weaker pairing can occur between G and U

Some RNA conveys genetic information from DNA to the sites of protein synthesis (messenger RNA). Other forms of RNA provide the mechanism of protein synthesis (ribosomal RNA and transfer RNA).

Messenger RNA

Messenger RNA (mRNA) is formed in the nucleus by *transcription* from DNA in such a way that the information on a particular section of DNA (a gene) is reflected in the sequence of bases in the mRNA (Chap. 13). After transcription the mRNA molecule is modified in two ways. At the 5′ end a 7-methylguanosine residue is attached by a 5′–5′ pyrophosphate link; the next two nucleotides may also be methylated. This modified end is known as the *methyl cap*. At the 3′ end a chain of up to 200 adenylic acid residues, the 'polyA tail,' is added.

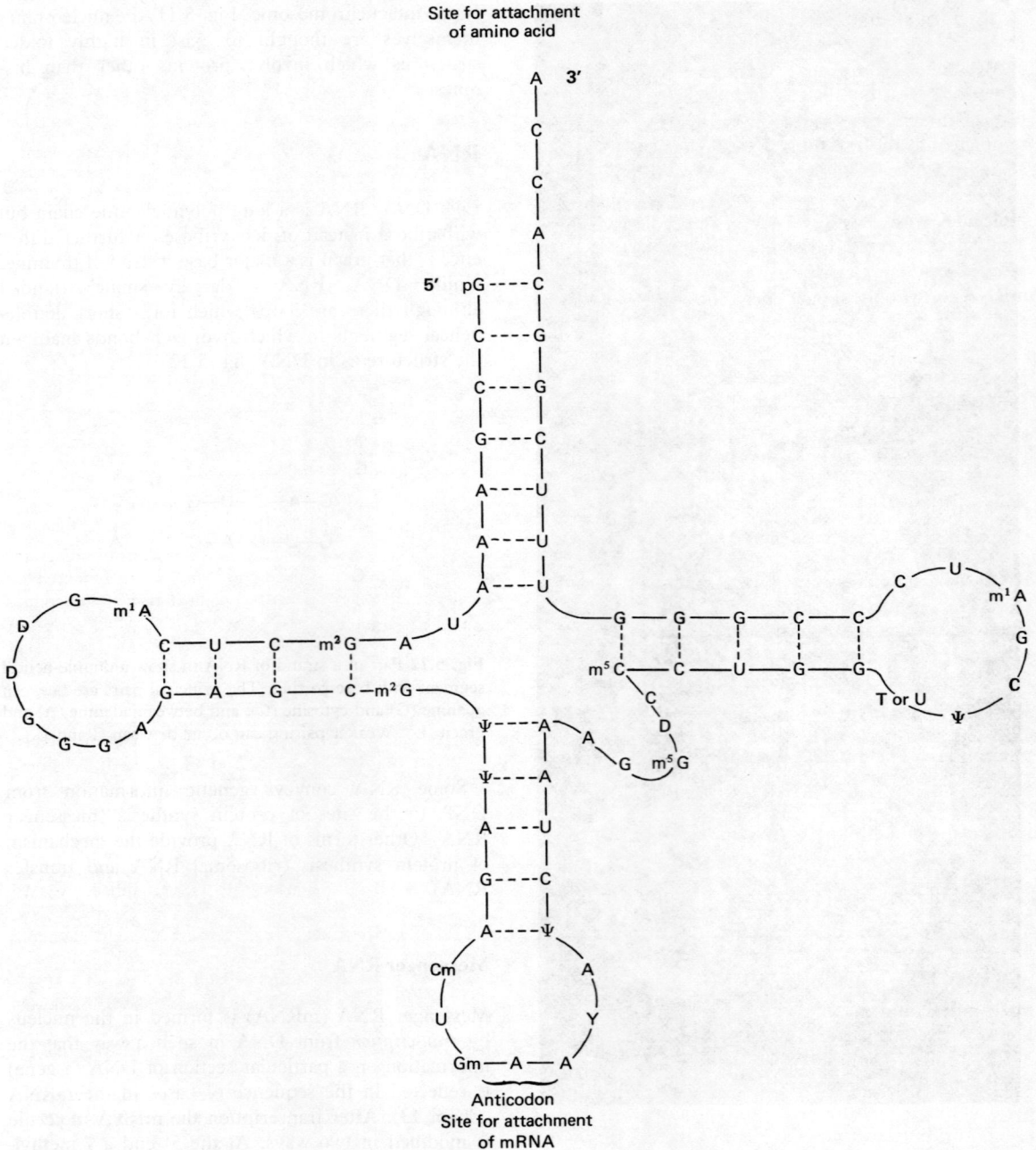

Fig. 5.13 Conventional ('clover leaf') representation of the sequence of bases in transfer RNA for phenylalanine in human placenta. The modified nucleosides are as follows: m^1A = 1-methyl-adenosine, m^2G = 2-methyl-guanosine, m^5G = 5-methyl-guanosine, m^5C = 5-methyl-cytidine, Cm = 2'-O-methyl-cytidine, Gm = 2'-O-methyl-guanosine, Ψ = pseudouridine (5-ribosyl-uracil), D = 5,6-dihydrouridine, Y = unidentified pyrimidine. (After Roe, B. A. *et al.* (1975) *Biochemical and Biophysical Research Communications* **66**, 1097)

These two modifications (fig. 14.4) are thought to make the mRNA more stable.

Messenger RNA in combination with certain proteins leaves the nucleus and migrates to the sites of protein synthesis, the *ribosomes*, where it acts as a template for the assembly of a sequence of amino acids by a process known as *translation* (Chap. 14).

Messenger RNA is a very heterogeneous group of molecules with a wide variety of molecular weights. They have a rapid turnover with half-lives in the range 7 to 24 hours in mammals. At any one time mRNA forms only about 5 per cent of the RNA in a cell.

Ribosomal RNA

Ribosomal RNA (rRNA) comprises about 60 per cent of the mass of a ribosome, the remainder being proteins. The role of ribosomes in protein synthesis is described in Chapter 14.

Ribosomal RNA forms about 80 per cent of the RNA of mammalian cells and includes several distinct molecules (p. 165) which are usually classified according to their sedimentation rate in an ultracentrifuge. The results of such analyses are expressed in Svedberg units (S); a small (5 S) component of rRNA contains about 120 nucleotides while a larger molecule (28 S) has about 5500 nucleotides. The full sequence of bases in several rRNAs has now been determined, including one (from *E.coli*) of 2904 residues. Some of the nucleotides have methyl groups attached either to the base or to the ribose.

Transfer RNA

A transfer RNA (tRNA) molecule contains 75 to 80 nucleotides and has a mass of about 25 000. Transfer RNAs serve as carriers of activated amino acids during protein synthesis (p. 164).

Transfer RNA contains unusual bases including hypoxanthine (inosine, I) dihydrouracil (UH_2), pseudo-uridine, methyl-guanine and methyl-inosine.

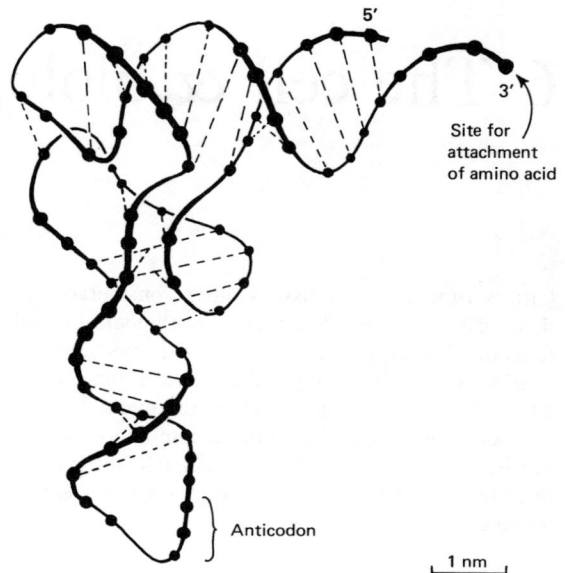

Fig. 5.14 Three-dimensional model of transfer RNA for phenylalanine in yeast. The nucleotides are indicated by 'beads' and the bonds between complementary nucleotides in base pairs are indicated by dotted lines. The structure is maintained further by additional hydrogen bonds (not shown) between bases and between ribose and phosphate components. (After Kim, S. H. (1975) *Nature* **256**, 680)

The sequence of nucleosides in a typical tRNA is given in figure 5.13 and a three-dimensional model of tRNA is shown in figure 5.14.

Further reading

Adams, R. L. P., Burdon, R. H., Campbell, A. M., Leader, D. P. and Smellie, R. M. S. (1981) *The Biochemistry of the Nucleic Acids*, 9th edn. London: Chapman and Hall

Bradbury, E. M., Maclean, N. and Matthews, H. (1981) *DNA, Chromatin and Chromosomes*. Oxford: Blackwell

Mainwaring, W. I. P., Parish, J. H. and Pickering, J. D. (1982) *Nucleic Acid Biochemistry and Molecular Biology*. Oxford: Blackwell

Zimmerman, S. B. (1982) The three-dimensional structure of DNA. *Annual Review of Biochemistry* **51**, 395–427

C R P   G A J G

59

6 The cell & biological membranes

CELLS from different tissues differ considerably in their size and shape. Many are morphologically and functionally highly specialised and their biochemistry is also specialised. A 'typical' cell has a diameter of 10 to 20 μm. Large cells, such as the ovum and the adipocyte, owe their size to the accumulation of vast nutrient reserves. The liver cell (hepatocyte) is often regarded as typical but it too is highly specialised for its functions.

Intracellular compartments

A cell is not simply a bag containing a soup of uniformly distributed enzymes, substrates and cofactors; it is divided into a number of distinct compartments, such as mitochondria, lysosomes, nucleus and endoplasmic reticulum. Each is bounded by its own membrane, which has specific permeability and enzymic properties and which encloses an environment with a unique composition. In addition, each membrane can be regarded as a compartment in its own right since many biochemical reactions occur within the thickness of a membrane. The division of cellular contents into compartments ensures that the enzymes and cofactors of a specific metabolic process are restricted to a single subcellular site. This confers several advantages:

(1) Just as the plasma membrane surrounding a cell regulates the overall metabolism of the cell by controlling the rate of entry of important fuels, such as glucose or amino acids, so the membrane surrounding subcellular organelles acts as an additional control by regulating the inflow of substrates (and the out-flow of products) of the enzymes within that compartment. For example the respiration rate of cells can be increased by the action of thyroid hormones which make the mitochondrial membrane more permeable to substrates for the enzymes of the tricarboxylic acid cycle (p. 96).

(2) The existence of several membrane-enclosed micro-environments allows the simultaneous operation of several metabolic processes which require different optimal conditions. For example the enzymes of fatty acid synthesis (p. 123), which require high concentrations of NADPH and ATP, are located in the cytosol where these occur; fatty acid breakdown (p. 121) takes place within the mitochondria where the optimum conditions (low ATP and NADH) obtain. Thus both breakdown and synthesis can occur simultaneously if necessary. The micro-environment within a lysosome has a low pH (about 4) compared with the cytoplasm (about 7). The acid hydrolases of a lysosome therefore operate at their optimum pH and only those materials which are taken into the lysosomes are digested.

(3) The immobilisation of certain enzymes within a membrane allows their organisation into multi-enzyme complexes. Consecutive reactions in a metabolic pathway proceed efficiently if the product of one enzyme has only a short distance to diffuse to reach the active site of the next (fig. 6.1).

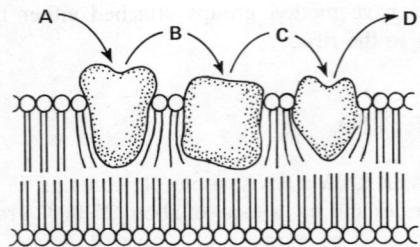

Fig. 6.1 The role of a series of membrane-bound enzymes in catalysing a sequence of reactions. Details of the structure of a membrane are given in figure 6.3

Biological membranes

Membranes have two main components, lipids and proteins, with a small proportion of carbohydrate

Table 6.1 Protein/lipid ratios in different membranes

Membrane	Protein/lipid ratio
Myelin	0.3
Red cell plasma membrane	1.1
Liver cell plasma membrane	1.4
Mitochondrial inner membrane	3.2

Organisation. The fundamental structure on which all membranes are based is the *lipid bilayer*. All polar lipid molecules possess both hydrophilic regions (sugar residues of glycolipids, phosphate and organic alcohols of phospholipids) and hydrophobic regions (fatty acyl or sphingosine side-chains) (figs. 1.4, 1.5 and 1.6, p. 2). In an aqueous environment the most stable arrangement for these polar lipids is either as a *micelle* (p. 3) or a *bilayer* (fig. 6.2) in which the hydrophobic side-chains associate and exclude water. In living cells, it is the bilayer that forms spontaneously.

Lipids are not distributed uniformly in membranes. First, in plasma membranes at least, there is asymmetry across the membrane; all the glycolipids and most of the choline-containing phospholipids are in the extracellular side whereas phosphatidyl ethanolamine and phosphatidyl serine predominate in the cytoplasmic side (Table 6.3). Secondly, there is evidence that molecules of a similar class cluster together within a particular half of a bilayer under certain circumstances, for example in response to certain hormonal stimuli.

associated with them as either *glycolipids* or *glycoproteins*. It is largely the proteins which confer specific properties on a particular membrane, whereas general characteristics, such as electrical resistance and passive permeability, are features of the lipid components. The proportions of lipid to protein vary widely from membrane to membrane (Table 6.1). Generally a membrane with many functions, such as the inner mitochondrial membrane, has a high protein content whereas a membrane with few functions, such as myelin, has much less protein.

Lipids

The principal lipids in membranes are phospholipids, glycolipids and cholesterol (Chap. 4). The lipid composition of a particular membrane is remarkably constant even in the face of dietary changes and is characteristic of the species, of the tissue and of the particular subcellular organelle it surrounds (Table 6.2). This is true not only of the lipid class, as shown in the table, but also of the fatty acid composition within each class.

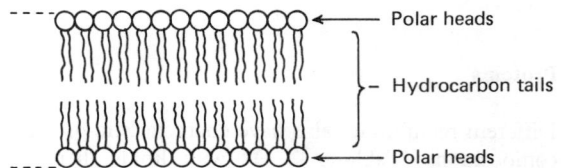

Fig. 6.2 Essential features of a lipid bilayer separating two aqueous compartments. A typical bilayer is approximately 4.5 nm in thickness

Table 6.2 Lipid composition (per cent of total lipid) of different membranes from human tissues

Lipid class	Red cell plasma membrane	Myelin	Liver cell Plasma membrane	Liver cell Endoplasmic reticulum	Liver cell Mitochondria
Cholesterol	25	28	20	5	3
Phosphatidyl choline	23	11	19	48	38
Phosphatidyl ethanolamine	20	17	12	19	29
Phosphatidyl serine	8	6	7	4	Trace
Phosphatidyl inositol	3	1	3	8	3
Sphingomyelin	18	7	12	5	—
Cardiolipin	—	—	—	—	14
Glycolipid	Trace	30	Trace	—	—

Table 6.3 Distribution of phospholipids in the erythrocyte membrane

	Per cent of total phospholipid	
	Outside	Inside
Phosphatidyl serine	0	12
Sphingomyelin	24	5
Phosphatidyl choline	21	10
Phosphatidyl ethanolamine	4	24

After Marchesi, V. T. *et al.* (1976) *Annual Review of Biochemistry* **45**, 667.

Table 6.5 Distribution of some proteins within the plasma membrane

Protein	Localisation
Acetylcholinesterase	Extracellular side
5'-Nucleotidase	Extracellular side
Na^+/K^+ ATPase	Transmembrane
Anion channel	Transmembrane
Adenylate cyclase	Cytosol side

Fluidity. A vital feature of the lipid bilayer in biological membranes is its fluidity. Thermal energy causes a continuous random motion of the hydrocarbon chains in the centre of the bilayer which therefore behaves as a viscous liquid. Individual lipid molecules are able to move relative to one another within the plane of the bilayer. This fluidity is known to be essential for many functions of membranes, such as permeability and enzyme activity; if a membrane is cooled below the temperature at which the lipids become solid, these functions are abolished or greatly impaired.

Proteins

Different membranes also have characteristic protein compositions (Table 6.4). As with lipids there is considerable asymmetry in protein distribution between the two sides of a membrane. Table 6.5 shows the position of some proteins in the plasma membrane; a similar asymmetry is found in the membranes of intracellular organelles. Proteins cluster together in specific areas of a membrane to form multi-enzyme (p. 61) or respiratory complexes (p. 113).

Table 6.4 Some characteristic proteins in particular membranes

Plasma membrane	5'-Nucleotidase, Na^+/K^+ ATPase
Endoplasmic reticulum	Glucose-6-phosphatase, NADPH oxidase
Inner mitochondrial membrane	Succinate dehydrogenase, cytochrome oxidase
Golgi complex	Glycosyl transferases
Lysosomes	Acid phosphatase, aryl sulphatase, β-glucuronidase

The currently accepted model of membrane structure is the *lipid-protein mosaic* model proposed by Singer and Nicolson in 1972 (fig. 6.3). The proteins are classified as either *integral* (intrinsic) or *peripheral* (extrinsic). Most membrane proteins are integral, deeply inserted into the fluid lipid bilayers. The protein is folded so that there is extensive interaction between the non-polar side-chains of amino acid and the non-polar hydrocarbon chains of the lipids. Peripheral proteins are associated only with the membrane surface, being attached by electrostatic and hydrogen bonds to integral proteins or to polar lipid head groups (fig. 6.4).

Integral proteins include cytochrome b_5, the HLA antigens and the Na^+/K^+-dependent ATPase; examples of peripheral proteins are cytochrome *c*, erythrocyte actin and spectrin. An important sub-group of integral membrane proteins are the *transmembrane* or *spanning* proteins (for example protein A in fig. 6.3) which span the lipid bilayer and have regions in contact with the aqueous compartments on both sides of the membrane. The Na^+/K^+-dependent ATPase and the anion transporting protein of red cells, as well as the Ca^{2+}-dependent ATPase of sarcoplasmic reticulum are examples. It is probable that all transport or carrier proteins will prove to be of this type; a likely feature of the mechanism by which they transfer solutes is the formation of hydrophilic pores, either through the protein or between the proteins in an aggregate (fig. 6.5).

The role of lipids in membrane protein function

One consequence of the mobility of the lipid molecules is the freedom of movement of the integral proteins within the bilayer. The lipid-protein mosaic has been likened to a sea of lipid in which icebergs of integral protein float.

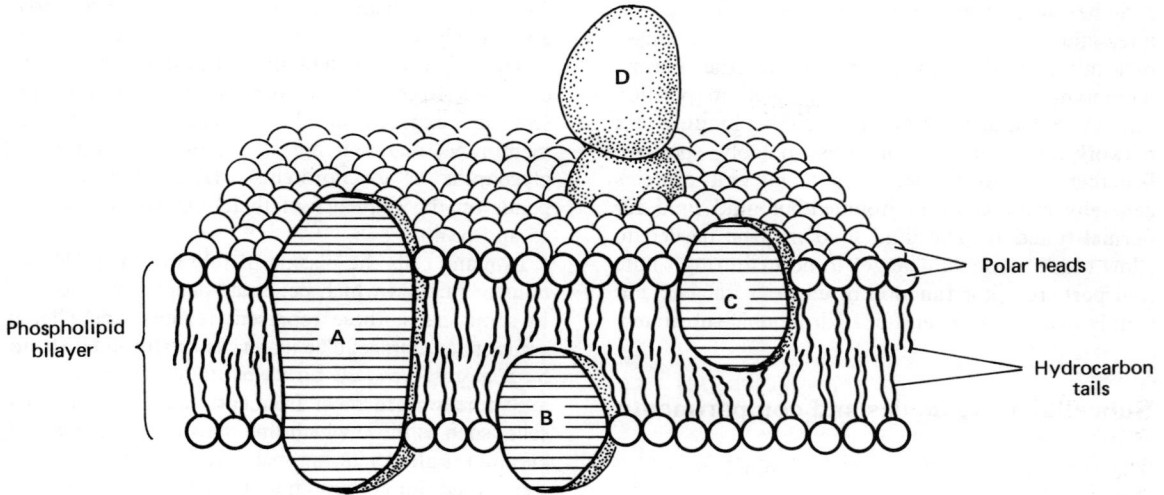

Fig. 6.3 The principal components of a membrane. The lipid bilayer consists of phospholipids (Chap. 4) and four types of protein are seen. A, B and C are integral proteins; these can be on one side or the other, or traverse the membrane (A). Protein D is a peripheral protein

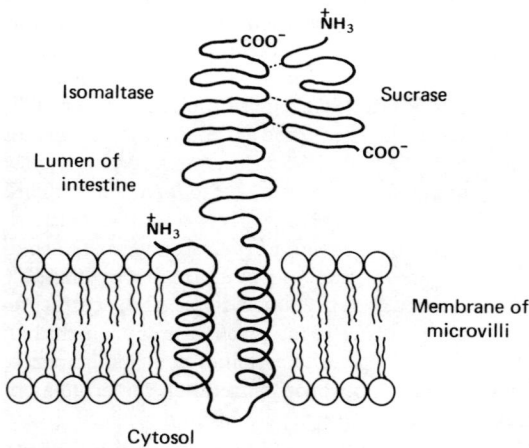

Fig. 6.4 An example of a peripheral protein. Sucrase is situated on the luminal surface of the plasma membrane of intestinal cells. It is attached by hydrogen bonds (dotted lines) to isomaltase which is an integral protein. The function of these enzymes is described in Chapter 18. (After Brunner, J. *et al.* (1979) *Journal of Biological Chemistry* **254**, 1821, and Semenza, G. (1981) *Clinics in Gastroenterology* **10**, 691)

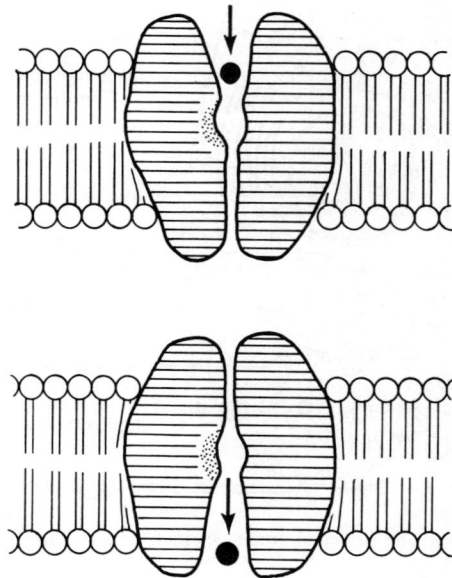

Fig. 6.5 Suggested role of a protein involved in the active transport of a solute through a membrane. A molecule binds to an active site in the protein (dotted) and this causes conformational changes in the protein. In turn this 'squeezes' the molecule through the membrane. (After Singer, S. J. (1975) In *Cell Membranes: Biochemistry, Cell Biology and Pathology*, ed. Weissmann, G. and Claiborne, R. San Francisco: Freeman)

Some proteins are fixed in the bilayer because of interactions with peripheral proteins. For example molecules of the anion channel of the plasma membrane of erthyrocytes are tethered in position relative to one another because of their binding to a network of peripheral proteins, notably spectrin. Whether free to diffuse or not, integral proteins generally require an environment of fluid lipid for normal function. The fluid lipid may be needed to allow conformational changes to occur during solute transport, receptor function or enzymic catalysis, or simply to act as a solvent for hydrophobic substrates.

Subcellular organelles and compartments

The major features of a typical mammalian cell are shown in figure 6.6.

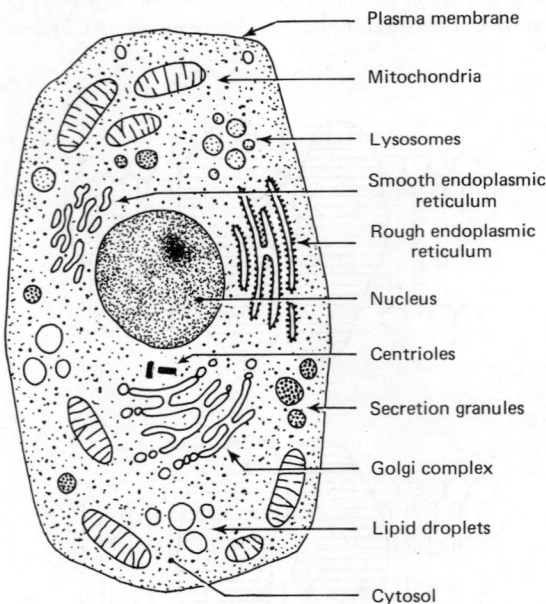

Fig. 6.6 The principal organelles in a 'typical' mammalian cell

Plasma membrane
Mitochondria
Lysosomes
Smooth endoplasmic reticulum
Rough endoplasmic reticulum
Nucleus
Centrioles
Secretion granules
Golgi complex
Lipid droplets
Cytosol

Plasma membrane

This membrane is essential for the homeostasis of the cell since its unique permeability properties determine the composition of the intracellular fluid. The

fuels of metabolism (such as glucose and amino acids) are actively taken up by special transport proteins (Chap. 17) whilst potentially harmful substances are excluded. Receptors for hormones or neurotransmitters are exposed at the surface of the plasma membrane; many, if not all, of these receptors are glycolipids or glycoproteins whose specificity depends on the sequence of their oligosaccharides (p. 35 and p. 40).

In many cells the plasma membrane has invaginations or furrows which penetrate deeply into the cell. In some cases, these represent regions where large amounts of extracellular materials are being taken up by *pinocytosis (endocytosis)* (Chap. 18). These pinocytotic furrows are most frequently seen in epithelial cells, such as those which line the small intestine and kidney tubules. The luminal surface of these cells is convoluted into many fine filamentous projections called *microvilli* which greatly increase the surface area for absorption (fig. 6.7). In muscle, these channels form the elaborate *T-system* of tubules which conduct the electrical impulse into all parts of the muscle cells. In some cells, infoldings of the plasma membrane connect with the *perinuclear space* (p. 69) and provide direct communication between it and the extracellular fluid. The width of the plasma membrane is approximately 8 nm; the extracellular surface is frequently overlaid by a thick, amorphous

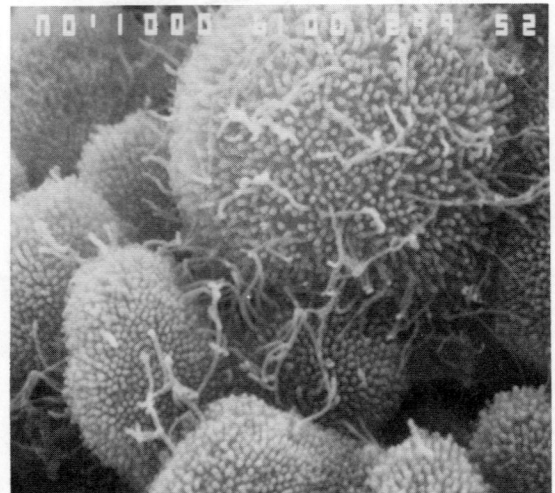

Fig. 6.7 Scanning electron micrograph to show villi and microvilli from human gall bladder (×10 000). (Courtesy of G. Milne)

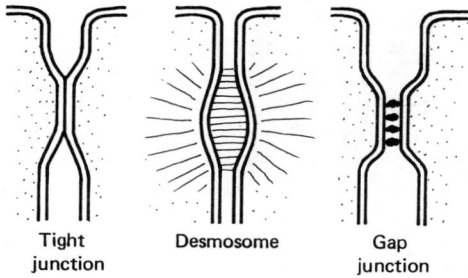

Tight   Desmosome   Gap
junction   junction

Fig. 6.8 The principal types of intercellular junction described by electron microscopists. The lines radiating into each cell from the desmosome represent microfilaments

cell coat or *glycocalyx*, consisting mainly of glycoproteins.

In a tissue such as muscle or liver the plasma membrane of adjacent cells is frequently modified to provide connections between them known as *gap junctions* and *tight junctions* or *occlusion zones* (fig. 6.8). At gap junctions, the two plasma membranes are linked by polygonal projections so that they adhere. These junctions have decreased electrical resistance and are probably involved in intercellular communication. Such junctions in myocardial cells allow electrical communication and coordinated contraction of each part of the myocardium. A *desmosome* (fig. 6.9) is a wide gap between the adjacent membranes, frequently containing dense material. The cytoplasm immediately opposite the desmosome is dense and many *microfilaments* (p. 70) are attached to the membrane. Both kinds of junction are important in cementing adjacent cells together to form a tissue.

Tight junctions occur particularly between epithelial and endothelial cells and constitute a complete ring around each cell forming a continuous barrier between the lumen lined by the cells and the extracellular fluid. This means that solutes absorbed from the lumen must pass through the cells and cannot pass between them. Tight junctions between capillary endothelial cells restrict the passage of solutes between the blood and the brain cells so that it is the cells themselves which constitute the selectively permeable *blood brain barrier*. Figure 6.8 shows that at tight junctions the outer layers of adjacent plasma membranes are fused together.

One elaborate modification of the plasma membrane is seen in the *myelin sheath* (fig. 6.10) which surrounds the axons of most neurones. It is formed

Fig. 6.9 Electron micrograph to show a desmosome with associated microfilaments (×72 000). (From Tucker, J. B. (1981) *Journal of Embryology and Experimental Morphology* **65**, 1, by courtesy of author and publisher)

by the rotation of supporting *Schwann cells* around the axon during development; the cells, in the process, lose their contents and leave behind a 'swiss roll' arrangement of plasma membrane that acts as an electrical insulator.

Endoplasmic reticulum

The endoplasmic reticulum is a complex arrangement of membranes which form tubes or flattened discs (*cisternae*) enclosing a cisternal space. Usually the endoplasmic reticulum lies near the nucleus. The

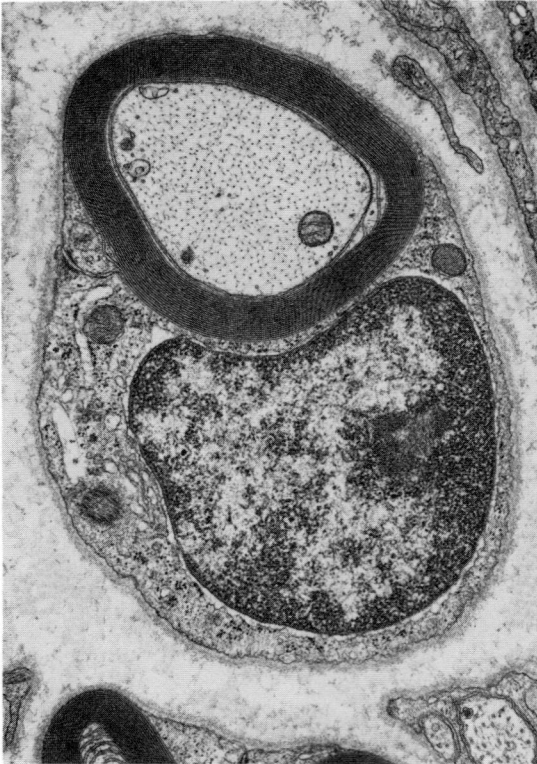

Fig. 6.10 Electron micrograph of an axon with myelin sheath and Schwann cell. (From Schochet, S. S. (1981) In *Introduction to Diagnostic Electron Microscopy*, ed. Mackay, B. New York: Appleton Century Crofts)

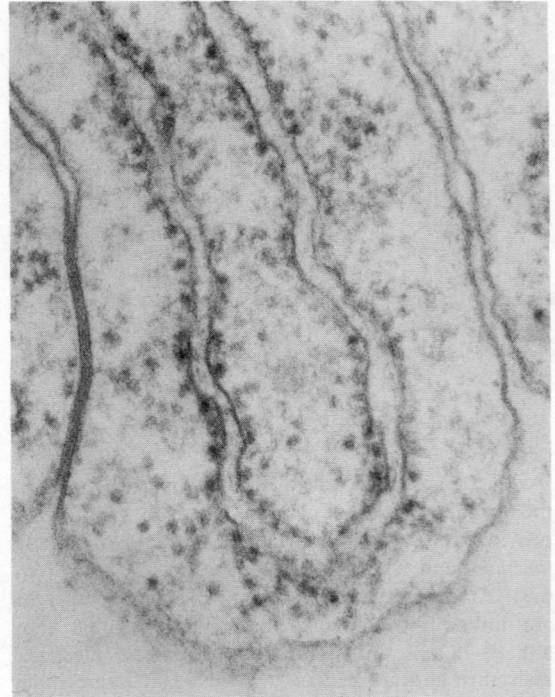

Fig. 6.11 Electron micrograph (×146 000) of rough endoplasmic reticulum. A gap junction is also seen. (Courtesy of J. B. Tucker)

amount of endoplasmic reticulum is variable; secretory cells such as pancreatic acinar cells have a particularly abundant network. The *rough endoplasmic reticulum* has ribosomes attached to the cytoplasmic surface of its membrane (fig. 6.11) whereas the *smooth endoplasmic reticulum* has none.

The rough endoplasmic reticulum is the site of synthesis of secreted proteins such as enzymes, hormones and antibodies and also of some membrane proteins. Protein synthesis proceeds whilst the ribosome is bound to the membrane; the growing polypeptide chain of a secreted protein passes through the membrane and into the cisternal space (fig. 14.8, p. 171) whereas integral membrane proteins remain embedded in the membrane.

The smooth endoplasmic reticulum possesses several important membrane-bound enzymes (Table 6.4) including those concerned with glycosylation of glycoproteins and glycolipids. It also has many of the enzymes of lipid metabolism (most of the cell's polar and neutral lipids are made here), of steroid synthesis and of detoxication. A specialised form of the smooth endoplasmic reticulum is the *sarcoplasmic reticulum* of muscle which is concerned with the rapid and reversible uptake and release of calcium ions.

Golgi complex

This organelle consists of stacks of curved membranous cisternae which lie close to, and may be continuous with, the smooth endoplasmic reticulum. At the surfaces near the plasma membrane, the Golgi complex is pinched off into vesicles. This is particularly obvious in secretory cells in which proteins extruded into the cisternal space of the endoplasmic reticulum are passed to and through the

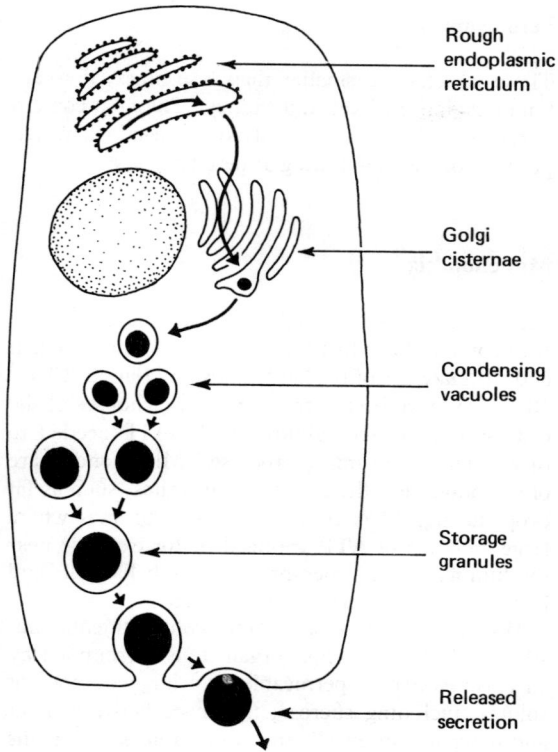

Fig. 6.12 Integrated function of rough endoplasmic reticulum, the Golgi complex and vacuoles in the production, modification, storage and release of proteins, such as hormones. (After Finean, J. B. *et al.* (1974) *Membranes and their Cellular Functions.* Oxford: Blackwell)

Golgi complex by a continuous process of vesicle production and fusion. Finally, secretory vesicles are formed which migrate to, and fuse with, the plasma membrane, releasing their contents (fig. 6.12).

The Golgi complex also possesses enzymes concerned with the further glycosylation and sulphation of proteins and glycoproteins destined for use outside the cell. The endoplasmic reticulum and the Golgi complex thus form part of a well organised production, packaging and export system which results in the flow not only of products but also of membrane from the endoplasmic reticulum to the plasma membrane. This process would lead to the steady increase in the amount of plasma membrane were it not for the endocytosis and recycling of plasma membrane vesicles.

Lysosomes

These vesicles, also derived from the Golgi complex, contain a wide range of acid hydrolases, enzymes whose pH optimum is about 4. These hydrolases, produced in the endoplasmic reticulum, include cathepsins, lipases, oligosaccharidases, RNAase and DNAase; these enzymes allow the lysosomes to digest material of biological origin, such as bacteria ingested from outside the cell, as well as aged cellular components such as mitochondria. Primary lysosomes, containing their battery of degradative enzymes can fuse with vesicles containing ingested extracellular material or can engulf intracellular debris to become secondary lysosomes in which digestion occurs (fig. 6.13). The products of digestion may be released either within or outside the cell.

Lysosomes play an important role in the breakdown of cellular components as part of normal

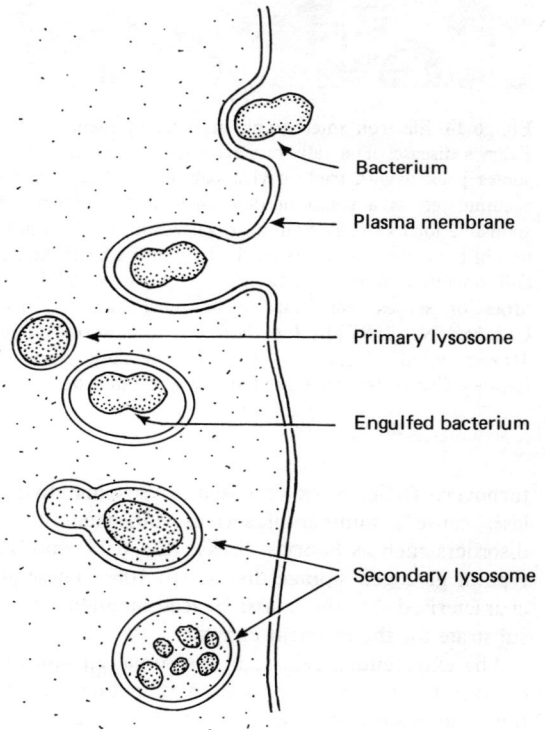

Fig. 6.13 Function of lysosomes in destroying a foreign particle (in this case a pneumococcus) taken up by endocytosis

67

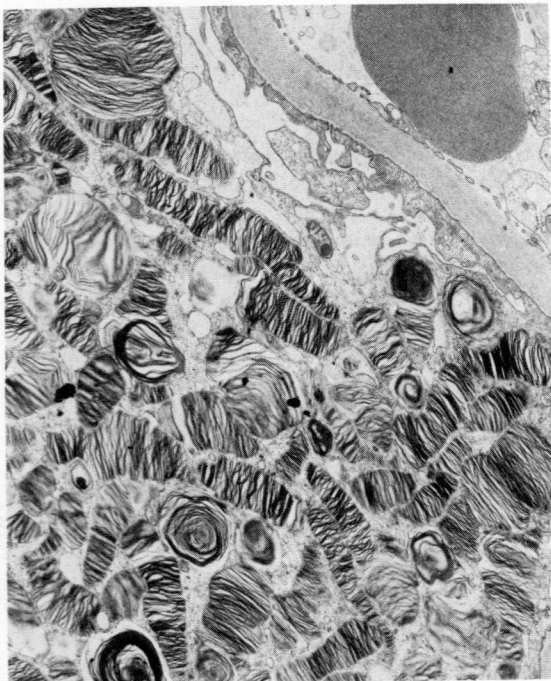

Fig. 6.14 Electron micrograph of renal epithelial cell in Fabry's disease. The cell contains a large number of lysosomes packed with triglycosyl ceramide (p. 40). This has accumulated as a result of deficiency of the enzyme responsible for its breakdown, an α-galactosidase. Accumulation of the ceramide occurs particularly in the heart, kidney and smooth muscle of blood vessels; death occurs in the fifties or sixties from cardiac or renal failure. (From Ordoñez, N. (1981) In *Introduction to Diagnostic Electron Microscopy*, ed. Mackay, B., p. 120. New York: Appleton Century Crofts, by courtesy of author and publisher)

turnover. Deficiencies of specific lysosomal hydrolases cause a number of serious, inherited human disorders such as Fabry's disease (fig. 6.14) and one type of glycogen storage disease (p. 108). These are characterised by the massive accumulation of the substrate for the deficient enzyme.

The extracellular release of lysosomal enzymes by osteoclasts plays an important part in bone resorption. The release of lysosomal enzymes by cells of the immune system results in the tissue damage which accompanies chronic inflammation. Release of lysosomal enzymes within a cell is responsible for the *autolysis* which occurs after its death.

Peroxisomes

These vesicles are smaller than lysosomes; they are found mainly in liver and kidney. They too contain enzymes, chiefly those involved in reactions which produce or destroy hydrogen peroxide.

Mitochondria

These organelles are vital to all aerobic cells since they contain the cytochromes of the respiratory chain and the enzymes of oxidative phosphorylation (Chap. 10). They have been termed the power houses of the cell as they produce almost all the ATP needed to drive energy-consuming processes. Mitochondria are often found near the potential substrates, such as fat droplets (fig. 10.7, p. 112), or close to sites where large amounts of ATP are needed, for example near the luminal plasma membrane of epithelial cells and between the myofibrils of muscle cells.

Mitochondria consist of four compartments (fig. 10.1, p. 109). The whole organelle is surrounded by an *outer membrane* permeable to a large number of solutes, including sucrose; the space between outer and inner membranes is sometimes referred to as the *sucrose permeable space*. The *inner membrane* is thrown into folds or *cristae* which penetrate deeply into the innermost compartment, the *matrix*. The inner membrane is impermeable to sucrose and most metabolites; specific and elaborate transport systems exist in the membrane to move substrates in and products out (Chap. 10). The matrix surface of the inner membrane is lined with knobs of ATP synthetase whose function is also described in Chapter 10.

Each compartment of a mitochondrion has its characteristic complement of enzymes (Table 6.6) and this subdivision is vital for the overall control of metabolism. In addition to the enzymes listed, the matrix also contains mitochondrial DNA, ribosomes and the remainder of the machinery for protein synthesis (Chap. 14) which allows the mitochondrion to make some of its own proteins.

Nucleus

The nucleus is usually about 5 μm in diameter and is surrounded by a double membrane (fig. 6.15)

Table 6.6 Some proteins found in each mitochondrial compartment

---

Outer membrane
   Monoamine oxidase
   Cytochrome b_5

Sucrose permeable space
   Adenylate kinase

Inner membrane
   Cytochromes a, a_3, b, c, c_1
   NADH dehydrogenase
   Succinate dehydrogenase
   ATPase

Matrix
   Enzymes of the tricarboxylic acid cycle
   Glutamate dehydrogenase
   Enzymes needed for β-oxidation of fatty acids

---

which is permeated by large pores that allow the passage of molecules such as RNA. The membranes of this nuclear envelope enclose the perinuclear space which is variable in width but generally less than 100 nm across. The nucleus contains *chromatin* consisting of 15 per cent DNA, 10 per cent RNA and 75 per cent proteins such as *protamines* and *histones* (p. 56). Chromatin is usually amorphous except during cell division when the individual chromosomes become apparent. The *nucleoli* (one or more per nucleus) are the only morphologically distinct bodies within the nucleus at other times. They are rich in RNA since they are the sites of the synthesis and temporary storage of ribosomal RNA, before it is assembled into ribosomal precursors and exported from the nucleus.

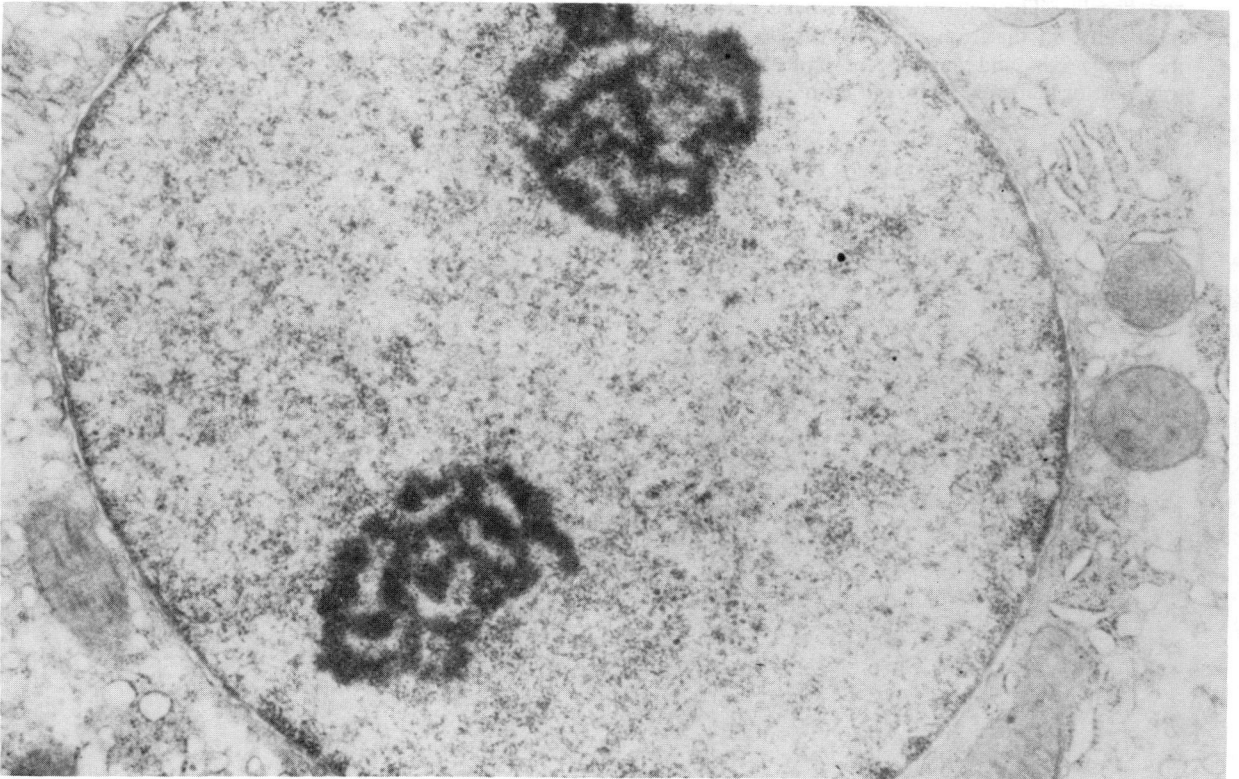

Fig. 6.15 Electron micrograph of a human hepatocyte nucleus to show two nucleoli and the perinuclear space. (Courtesy of G. Milne)

Microfilaments, microtubules, centrioles, cilia and flagella

These structures are considered together since they are non-membranous subcellular inclusions and are structurally related.

Microfilaments. These are fine fibrous structures seen in many cells and associated, for example, with desmosomes, the microvilli of epithelial cells and the plasma membrane of other cells (fig. 6.9). They provide strength and may determine cell shape. They are probably involved in the movements of cells such as phagocytes.

Microtubules. Microtubules are larger than microfilaments and frequently link other structures in the cell such as centrioles and the basal bodies of cilia and flagella. They also form a skeletal network in most cells. In the axon microtubules lie parallel to the long axis and may be responsible for the transport of materials by *axonal flow.*

Centrioles. Two centrioles are found close to the nucleus where they play a key role in cell division. Each centriole (fig. 6.16) is a small cylinder containing a series of microtubules grouped in a ring of nine.

Cilia and flagella. The cilia, for example in the respiratory tract, and the flagella in spermatozoa also contain microtubules in the characteristic 9 + 2 arrangement shown in figure 6.17. A flagellum is simply a long cilium; each has at its base a *basal body* or *kinetosome* which controls the whipping motion.

The cytosol

The cytosol is the soluble, non-particulate fraction of the cell in which all the subcellular structures are suspended. It contains all the soluble enzymes such as those of glycolysis and fatty acid synthesis. It is now becoming clear that there is a considerable degree of organisation of these enzymes either as multi-enzyme complexes or by loose attachment to membranes.

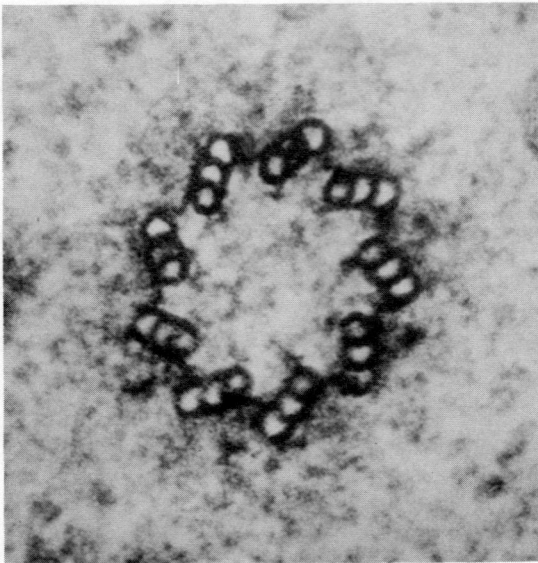

Fig. 6.16 Electron micrograph (×220 000) of a human centriole. (Courtesy of G. Milne)

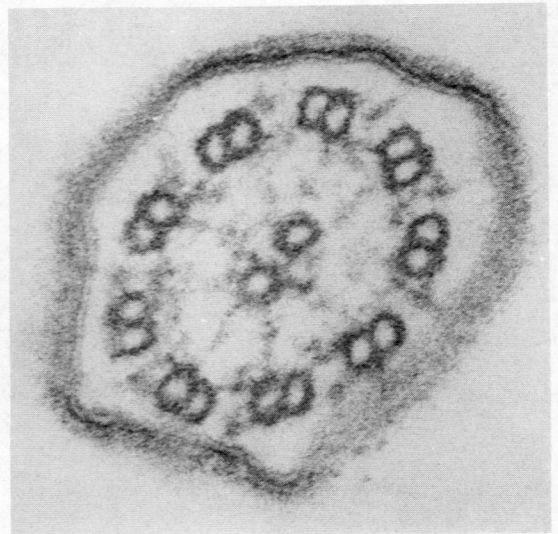

Fig. 6.17 Electron micrograph (×375 000) of newt cilium in cross-section, to show the characteristic arrangement of microtubules. (Courtesy of J. B. Tucker)

Further reading

Alberts, B., Bray, D., Lewis, J., Raff, M., Roberts, K. and Watson, J. D. (1983) *Molecular Biology of the Cell.* New York: Garland

Callaghan, J. W. and Lowden, J. A. (eds) (1981) *Lysosomes and Lysosomal Storage Diseases.* New York: Raven Press

Capaldi, R. A. (1982) Structure of intrinsic membrane proteins. *Trends in Biochemical Sciences* **7,** 292–295

Chapman, D. (1982) *Biomembrane Structure and Function.* London: Macmillan

De Robertis, E. D. and De Robertis, E. M. (1981) *Essentials of Cell and Molecular Biology.* New York: Holt, Rinehart & Winston

Green, D. E. (1983) Mitochondria: structure function and replication. *New England Journal of Medicine* **309,** 182–183

Harrison, R. and Lunt, G. G. (1980) *Biological Membranes: Their Structure and Function.* 2nd edn. Glasgow: Blackie

Neufeld, E. F. (1981) Lessons from genetic disorders of lysosomes. *Harvey Lectures* **75**

Wallach, D. F. H. (1979) *Plasma Membranes and Disease.* New York: Academic Press

MISH

7 Biochemical reactions

EARLIER chapters dealt with the chemical components of tissues and foodstuffs. This chapter and those which follow deal with the way in which these substances react with each other within the body. Since such reactions follow the same laws as the simple chemical reactions that occur in a test tube it is convenient to start with a discussion of chemical reactions in general.

Chemical equilibrium

Some chemical reactions are obviously reversible. For example: when glucose is metabolised in the body it is first phosphorylated to glucose 6-phosphate, which may then be converted to fructose 6-phosphate

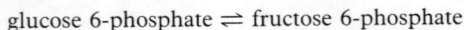

$$\text{glucose 6-phosphate} \rightleftharpoons \text{fructose 6-phosphate}$$

If this conversion is studied in aqueous solution it is found to be reversible. A pure solution of glucose 6-phosphate is eventually partly converted to fructose 6-phosphate. A solution of fructose 6-phosphate is partly converted to glucose 6-phosphate. In either case if the reaction is allowed to proceed long enough, the ratio of the concentration of fructose 6-phosphate to that of glucose 6-phosphate always becomes 31:69. At this point the reaction is said to have reached equilibrium. The equilibrium is a dynamic one; glucose 6-phosphate continues to be converted to fructose 6-phosphate and vice versa. But because the two processes take place at the same rate they balance one another so that there is no *net* conversion. The ratio of the participants when this equilibrium is attained is called the equilibrium constant (K) and is a characteristic of the reaction.

The situation is more complicated if there is more than one compound on each side of the equation. In this case the *product* of the concentrations of all the participants on one side must be compared with the corresponding *product* on the other side. This can be illustrated by the cleavage of fructose 1,6-bisphosphate to give glyceraldehyde 3-phosphate and dihydroxyacetone phosphate

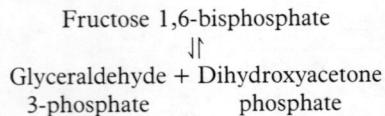

$$\begin{array}{c} \text{Fructose 1,6-bisphosphate} \\ \rightleftharpoons \\ \text{Glyceraldehyde} + \text{Dihydroxyacetone} \\ \text{3-phosphate} \qquad \text{phosphate} \end{array}$$

If this reaction is studied in isolation it is found that, whether the solution contains initially only fructose bisphosphate or a mixture of the cleavage products, the reaction proceeds to an equilibrium

$$\frac{[\text{glyceraldehyde 3-phosphate}] \times [\text{dihydroxyacetone phosphate}]}{[\text{fructose 1,6-bisphosphate}]}$$
$$= 10^{-4} \text{ mol/l}$$

It should be noted that in reactions like this in which the number of products differs from the number of reactants, the equilibrium constant has to be expressed in terms of concentration.

The equilibrium constant of a reaction is important because the probable direction of a reversible reaction can be predicted if the initial concentrations of the participants are known. The reaction by which a glycogen chain can be elongated or shortened by one glucose unit may be used as an example:

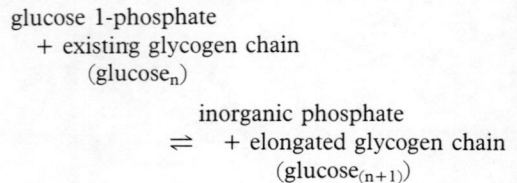

$$\begin{array}{c} \text{glucose 1-phosphate} \\ + \text{ existing glycogen chain} \\ (\text{glucose}_n) \\ \\ \text{inorganic phosphate} \\ \rightleftharpoons + \text{ elongated glycogen chain} \\ (\text{glucose}_{(n+1)}) \end{array}$$

Since the increase or decrease in chain length does not alter the number of chains, the equilibrium constant is given by the ratio at equilibrium of inorganic phosphate to glucose 1-phosphate. At equilibrium

$$\frac{[\text{inorganic phosphate}]}{[\text{glucose 1-phosphate}]} = 3$$

Clearly, therefore, in situations where the actual ratio of inorganic phosphate to glucose 1-phosphate is less than 3 the reaction proceeds from left to right so that the glycogen chains are lengthened. If the ratio is greater than 3 the reaction proceeds from right to left, so that the glycogen chains are broken down. The concentrations of inorganic phosphate and glucose 1-phosphate in liver are normally about 6 mmol/l and 0.06 mmol/l respectively:

$$\frac{[\text{inorganic phosphate}]}{[\text{glucose 1-phosphate}]} = \frac{6}{0.06} = 100$$

We can, therefore, predict that within the liver cell the reaction will proceed from right to left, that is in the direction of chain breakdown.

Even when the concentrations of the participants in a reaction are not accurately known, knowledge of the equilibrium constant can still be helpful in predicting the likely direction of a reaction. For example the equilibrium constant for the hydrolysis of sucrose is

$$\frac{[\text{glucose}] \times [\text{fructose}]}{[\text{sucrose}]} = 100\ 000 \text{ mol/l}$$

The concentration of glucose found in the body (in blood for instance) is of the order of 0.01 mol/l. The concentration of fructose is unlikely to be much greater than this. The known concentration of glucose and the presumed concentration of fructose can be used to calculate what the concentration of sucrose would be if the reaction were at equilibrium:

$$\frac{0.01 \times 0.01}{[\text{sucrose}]} = 100\ 000 \text{ mol/l}$$

$$[\text{sucrose}] = \frac{0.01 \times 0.01}{100\ 000} = 10^{-9} \text{ mol/l}$$

This is so small that even though we can only guess the concentration of fructose, we can be confident that the reaction described is not used for the synthesis of sucrose in tissues. Knowledge of a chemical equilibrium not only allows one to make a prediction of the likely direction of a reaction but also indicates the energy changes that should accompany it.

Energy changes in chemical reactions

Biochemists are particularly interested in *the capacity of energy to do useful work*. This term is used in the ordinary sense of the work done in, say, lifting a weight or generating electricity. It is a matter of common experience that some forms of energy are not available for performing useful work. For example, in water at room temperature the individual molecules have considerable kinetic energy in the form of thermal agitation, but this energy cannot be harnessed to do work.

Energy capable of doing useful work is called 'free energy'. In general free energy can be obtained from any process which takes place spontaneously. For example, water runs downhill and this process can be used to generate electricity. A compressed spring returns spontaneously to its natural shape; in doing so it can be made to drive a clock. In a steam turbine heat flows spontaneously from the boiler, which is at a high temperature, to the condenser which is at a low temperature; the turbine itself is a device for obtaining the free energy made available by this heat transfer. Conversely, to bring about a process which is not itself spontaneous, free energy must be supplied in a suitable form. Water flows uphill only if it is supplied with the necessary energy by some sort of pump. A spring can be compressed only by doing mechanical work on it. Free energy must be supplied to transfer heat from the cold interior of a refrigerator to its warmer surroundings. In short, non-spontaneous processes take place only if they are supplied with free energy.

These principles apply equally to chemical reactions. In any reaction, unless the reactants and the products are initially in equilibrium, the reaction will continue, in one direction or the other until equilibrium is attained. This may occur very slowly but even so it is a spontaneous process in the sense that it takes place of its own accord and can, like any other spontaneous process, provide free energy. Conversely, if we want the reaction to proceed in the direction which takes the reactants and products further from equilibrium, free energy must be supplied. The amount of free energy supplied in the first case, or required in the second, is greater the further the system is from equilibrium.

These theoretical ideas can be illustrated from everyday experience. The engine of a car is started by means of an electric motor (the starter) which is driven by a battery in which the reaction

$$Pb + PbO_2 + 2H_2SO_4 \rightleftharpoons 2PbSO_4 + 2H_2O$$

takes place. The equilibrium of this reaction is far to the right and never reached in normal use. While the reaction proceeds from left to right the system is approaching equilibrium and free energy is made available. The battery is essentially a device for converting this free energy into an electrical form which is converted into mechanical energy by the starter. This is an example of free energy being obtained from a reaction which is proceeding to equilibrium. When the car is running the sequence of events is reversed. The dynamo produces electrical power which is fed into the battery where it provides the energy necessary to drive the reaction from right to left (that is away from equilibrium).

If the free energy made available by a reaction is not used in some way, it appears as heat. In practice, it is not easy to devise mechanisms which make it possible to use directly the free energy produced by chemical reactions. Electric batteries are the best examples of such mechanisms. In relation to the amount of power they provide, they are heavy and clumsy, and they require expensive materials. The main source of chemical energy available to meet the needs of industrial societies is the oxidation of fossil fuels. Although the free energy produced is very large, no means has yet been devised by which it can be converted directly into mechanical or electrical energy. It has to be liberated as heat which is used to drive, say, a steam turbine or a diesel engine.

Energy sources for the cell

One remarkable property of living cells is that they can convert the free energy of chemical reactions directly into other forms of energy without the clumsy and wasteful expedient of converting it first into heat. The most obvious example is muscle. A muscle fibre alternately contracting and relaxing does mechanical work just as a steam engine does but it derives the necessary energy directly from chemical reactions. Another important example of such direct utilisation of chemical energy is the ability of many cells to bring about the transfer of solutes. For instance, the parietal cells of the gastric mucosa can be regarded as pumping H^+ ions into the lumen of the gastric glands; the cells of the kidney tubule pump Na^+ and other ions. These transfers require energy which is obtained directly from the free energy of chemical reactions. It is remarkable that all the energy-requiring activities of cells which have so far been investigated have been shown to be driven by the same reaction, namely, the hydrolysis of ATP:

$$ATP + H_2O \rightleftharpoons ADP + Pi \quad K = 250\,000$$

The concentrations of these substances within cells are known for many tissues. For example in rat brain the values remain remarkably constant at:

ATP	0.002 mol/l
ADP	0.0005 mol/l
Pi	0.005 mol/l

Conventionally in biochemistry the concentration of water is ignored, that is regarded as 1 mol/l. The ratio of products to reactants is therefore

$$\frac{(ADP)(Pi)}{(ATP)} = \frac{0.0005 \times 0.005}{0.002} = 0.00125 \text{ mol/l}$$

This is very different from the ratio (250 000 mol/l) required for equilibrium; the ATP concentration is too high relative to those of ADP and Pi. Consequently under these conditions, hydrolysis of ATP makes available free energy. The amount of free energy depends on how far the ratio of reactants to products differs from that which would be found at equilibrium just as the amount of energy obtainable from a spring depends on the extent to which it has been compressed. The free energy obtainable from a chemical reaction is given by the relationship:

$$\text{Free energy change } (\Delta G) =$$

$$5.9 \log_{10} \frac{\text{actual ratio}}{\text{equilibrium ratio}} \text{kJ/mol}$$

In the case of the hydrolysis of ATP this is

$$5.9 \log_{10} \frac{0.00125}{250\,000} = -49.0 \text{ kJ/mol}$$

It should be noted that when energy is made available, ΔG is, by convention, negative.

Similar calculations can be carried out for other tissues such as heart muscle, skeletal muscle and liver. The ΔG values obtained are usually between -45 and -50 kJ/mol.

The hydrolysis of ATP provides the power for the cell's activities just as power can be obtained as a compressed spring relaxes. The amount of energy stored in the disequilibrium of the reaction is limited

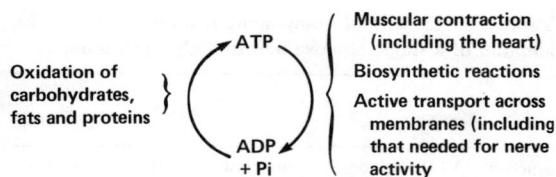

Fig. 7.1 The role of ATP as the principal link between energy-yielding reactions (described in Chapters 9, 10 and 12) and energy-requiring processes

just as is that in a compressed spring. For ATP to be synthesised, as for a spring to be compressed, energy is needed from outside; in the cell this is derived mainly from the oxidation of fats and carbohydrate. The interconversion of ATP and ADP, therefore, provides a link between processes yielding energy and those which require energy (fig. 7.1).

The conversion of potentially useful energy from one form to another is always subject to some loss. For example, not all of the electrical energy generated by a car dynamo is stored in the battery, some is lost as heat. Similarly when the battery operates the windscreen wipers, not all of the chemical energy available is converted into useful mechanical energy. The same is true of the ATP system. When fat and carbohydrate are oxidised in the body, not more than 70 per cent of the available chemical energy is used to form ATP. Similarly when ATP hydrolysis is used to power muscle contraction, less than half the chemical energy of the hydrolysis is converted into mechanical work.

It may be helpful to calculate the amount of ATP needed in practice. A joule (J) is approximately the amount of energy needed to raise 1 kg through a vertical distance of 10 cm. If a 75 kg man goes upstairs one floor he is raising 75 kg about 3 metres. The mechanical energy expended is $75 \times 30 = 2250$ J $= 2.25$ kJ. This energy is provided by the breakdown of ATP in the muscles concerned. We have seen that the concentrations of ATP, ADP and inorganic phosphate in muscle are such that ATP hydrolysis yields about 50 kJ/mol. If skeletal muscle were 100 per cent efficient in converting the energy of ATP hydrolysis into mechanical energy this 2.25 kJ could be obtained by breaking down $2.25 \div 50 = 0.045$ mol ATP. However since muscle is not 100 per cent efficient rather more than 0.045 mol ATP is used up. The muscles used by a 75 kg man in climbing stairs weigh perhaps 11 kg and contain ATP in a concen-

tration of about 0.005 mol/l. Their total content of ATP therefore amounts to about 0.055 mol/l. In other words the modest exertion of climbing a flight of stairs requires as much ATP as the muscles involved contain. But as rapidly as ATP is hydrolysed to provide the energy for muscular contraction, so the ADP and Pi produced are converted back to ATP by the processes of oxidative phosphorylation (Chap. 10) and glycolysis (Chap. 9). The processes of breakdown and resynthesis balance almost exactly and it is only in conditions of very severe exertion, as for example at the end of a sprint, that the ATP concentration in muscle is significantly depressed below the resting level.

The contraction of a skeletal muscle is an obvious example of the use of ATP hydrolysis as a source of energy but it is not the only one. Even at rest a 75 kg man requires energy at a rate of about 6 kJ/min or 100 watts (W). The brain alone requires 20 W, mainly to pump sodium ions out of the nerve cells and to pump potassium ions in. The kidneys use approximately 7 W, mostly for the reabsorption of sodium ions and water from the glomerular filtrate. The beating of the heart requires about 11 W and the maintenance of muscle tone in the skeletal muscle about 30 W. Most of the remaining energy is used for absorption in the gut and for synthetic activities. In every tissue ATP acts as an intermediary between the energy source and the use to which it is put.

Standard free energy change

The exact concentrations of products and reactants are seldom known even for reactions as important as the hydrolysis of ATP. In these circumstances it is convenient to calculate the free energy change when the concentrations of products and reactants are all 1 mol/l. The equation then becomes:

$$\Delta G = 5.9 \log \frac{1}{K} = -5.9 \log K$$

This value of ΔG, called the standard free energy change ($\Delta G^{0\prime}$), represents an artificial situation. Concentrations as high as 1 mol/l are much greater than those found inside the cell. None the less, $\Delta G^{0\prime}$ does represent a useful approximation. For the hydrolysis of ATP:

$$\Delta G^{0\prime} = -5.9 \log 250\,000 = -31.8 \text{ kJ/mol}$$

75

$\Delta G^{0\prime}$ provides an approximate measure of the amount of energy available from a reaction or the amount of energy required to drive it in the reverse direction. But $\Delta G^{0\prime}$ is also useful because it indicates the probable direction of a reaction or set of reactions in living tissues. This arises from the relationship of $\Delta G^{0\prime}$ with K (the equilibrium constant):

$$\Delta G^{0\prime} = -5.9 \log K$$

If one is known the other can be calculated.

Readily reversible reactions where K is close to 1 have a small $\Delta G^{0\prime}$; thus the value of $\Delta G^{0\prime}$ indicates the reversibility (or otherwise) of a reaction in the same way as K, but $\Delta G^{0\prime}$ is easier to apply to a complex series of reactions than is K. This is because the $\Delta G^{0\prime}$ values of two or more reactions in sequence is the sum of the $\Delta G^{0\prime}$ values for each reaction. Thus phosphopyruvate can react with ADP to give pyruvate and ATP.

The ATP produced in this reaction can then react with glucose to give glucose 6-phosphate and ADP.

phosphopyruvate + ADP $\rightleftharpoons$ pyruvate + ATP

$$\Delta G^{0\prime} = -22.6 \text{ kJ/mol}$$

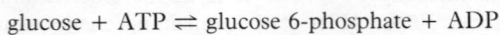

glucose + ATP $\rightleftharpoons$ glucose 6-phosphate + ADP

$$\Delta G^{0\prime} = -18.4 \text{ kJ/mol}$$

When these equations are added the overall result is:

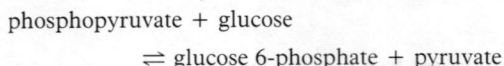

phosphopyruvate + glucose

$\rightleftharpoons$ glucose 6-phosphate + pyruvate

The $\Delta G^{0\prime}$ for this overall process is the sum of the $\Delta G^{0\prime}$ values for the two stages

$$(-22.6 \text{ kJ/mol}) + (-18.4 \text{ kJ/mol}) = -41.0 \text{ kJ/mol}$$

Similarly endergonic reactions, those for which energy is required, proceed only if they are coupled in some way with an energy-yielding reaction, frequently the hydrolysis of ATP. For example, in the synthesis of glycogen (Table 7.1) the addition of each glucose unit involves the breakdown of one molecule of ATP and one of UTP. The synthesis of other complex molecules like lipids, proteins and nucleic acids, is driven in the same way by the breakdown of ATP and other nucleoside triphosphates. For these reactions to continue there must be a continuous supply of nucleoside triphosphates, and especially ATP. The sources of these will be described in later chapters.

Table 7.1 Details of component reactions used in the addition of a single glucose unit to a glycogen molecule

	$\Delta G^{0\prime}$ (kJ/mol)
glucose + ATP $\rightleftharpoons$ glucose 6-phosphate + ADP	−18.4
glucose 6-phosphate $\rightleftharpoons$ glucose 1-phosphate	+6.7
glucose 1-phosphate + UTP $\rightleftharpoons$ UDP glucose + PPi	0.0
PPi + H_2O $\rightleftharpoons$ 2Pi	−31.4
UDP glucose + (glucose)$_n$ $\rightleftharpoons$ (glucose)$_{n+1}$ + UDP	−16.7
Overall reaction	
glucose + (glucose)$_n$ + ATP + UTP + H_2O → (glucose)$_{n+1}$ + ADP + UDP + 2Pi	−59.8

Chemical reactions and free energy changes

What factors determine the $\Delta G^{0\prime}$ of a reaction? The most important of these is the nature of the existing bonds broken and the new bonds formed. A simple example, of considerable biological importance, is the oxidation of fat by molecular oxygen:

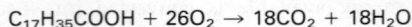

$$C_{17}H_{35}COOH + 26O_2 \rightarrow 18CO_2 + 18H_2O$$

The very large favourable free energy change of this reaction is almost entirely attributable to the fact that the bonds which are broken (C–C, C–H and O=O) are less strong and less stable than those which are formed. This is an example of the general phenomenon that bonds such as C–H, C–C, and O=O, in which the bonding electrons are equally or almost equally attracted by the atoms they link, are less strong and stable than bonds such as H–O and C=O in which the bonding electrons are more strongly attracted to one atom (in this case oxygen) than the other.

A second factor determining the K and hence the $\Delta G^{0\prime}$ of a biochemical reaction is whether the reaction involves the breakdown of a complex, ordered structure or the formation of such a structure from simpler units. If other things are equal, the equilibrium of such a reaction tends to favour the breakdown of the complex structure.

A third factor which influences the $\Delta G^{0\prime}$ of a reaction is whether any of the reactants or products have *resonance*. Figure 7.2 shows an example. A conventional structural formula with single and double bonds may be an incomplete and misleading representation of the structure of a compound.

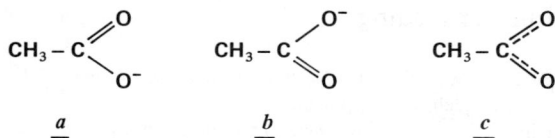

Fig. 7.2 An example of resonance. The acetate ion is not accurately represented by either *a* or *b* but is a hybrid *c* in which the negative charge is shared equally by the two oxygen atoms; the two carbon-to-oxygen bonds are, to an equal degree, hybrids between single bonds and double bonds

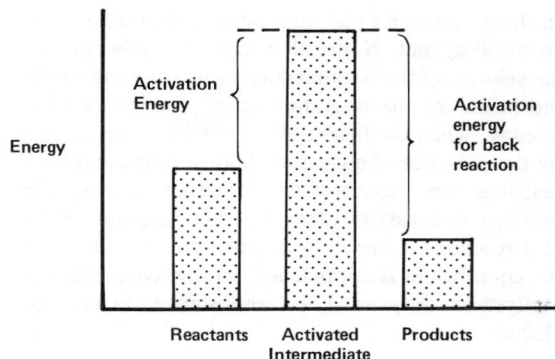

Fig. 7.3 The activation energy concept

A resonance hybrid is more stable than a similar compound which can be accurately represented by a single formula. When a molecule is rearranged in such a way as to make resonance possible considerable amounts of energy may be made available. For example, the pyruvate ion is stabilised by resonance:

A large part of the energy produced by the conversion of glucose to pyruvate (p. 94) can be attributed to this fact.

Activation energy

$\Delta G^{0'}$, then, gives an indication of the probable direction of a reaction in living tissues and gives an estimate of the energy theoretically available if the reaction proceeds in one direction, or of the energy needed to drive it in the opposite direction. However it gives no indication of the rate at which a reaction takes place. Some reactions with a relatively small free energy change take place very rapidly as, for example, the uptake or release of oxygen by haemoglobin. Other reactions capable of releasing enormous amounts of energy, such as the oxidation of mineral oil, take place so slowly at room temperature as to be undetectable.

Several theories have been put forward to explain what determines the rate of a chemical reaction. It is generally agreed, however, that, before a molecule can undergo any reaction, it must possess an amount of energy in excess of a level characteristic of the reaction—the *activation energy* (fig. 7.3). Most biochemical reactions take place in aqueous solution in which the molecules of the reactant are being continuously jostled by collision with solvent molecules. The amount of energy which an individual molecule possesses varies from moment to moment depending on the number and angle of collisions it has recently sustained. Very occasionally a molecule reaches an energy content equal to or greater than the activation energy for a particular reaction. Then, and only then, is it capable of undergoing that reaction. The rate of the reaction, other things being equal, depends on how frequently molecules reach this very high energy content. If the activation energy of a reaction is low, molecules reach it relatively often and the reaction is rapid. If it is high, molecules attain it rarely and the reaction is slow.

In general a reaction can be accelerated in two ways. The first is by raising the temperature of the solution and so increasing the kinetic energy of all the molecules. An increase of 10°, say from 20 °C to 30 °C, brings about only a small (3 per cent) increase in the average energy of the molecules but it may double the frequency with which they reach the activation energy for an average reaction.

The other means of accelerating a reaction is in effect to lower its activation energy. This can be done by using a catalyst. For example the activation energy of the breakdown of hydrogen peroxide is about 75 kJ/mol.

$$2H_2O_2 \rightarrow 2H_2O + O_2$$

In the presence of a platinum catalyst this is reduced to about 50 kJ/mol. Notice that in a reversible reaction the velocity of the back reaction is also determined by the height of the activation energy barrier. Consequently, when the height of the barrier is diminished by the presence of a catalyst, both forward and back reactions are accelerated. In other words, the presence of a catalyst does not alter the equilibrium of a reaction: it merely accelerates the rate at which the equilibrium is approached. In the living cell, the catalysts are *enzymes*. They are discussed in the next chapter.

Further reading

Becker, W. M. (1977) *Energy and the Living Cell.* Philadelphia: Lippincott

Hill, T. L. (1977) *Free Energy Transduction in Biology.* New York: Academic Press

Nicholls, D. G. (1982) *Bioenergetics.* London: Academic Press

Smith, E. B. (1977) *Basic Chemical Thermodynamics.* Oxford: University Press

Wilson, D. F. and Westerhoff, H. V. (1982) Should irreversible thermodynamics be applied to metabolic systems? *Trends in Biochemical Sciences* **7,** 275–279

RYT GCB DGN CRP

8 Enzymes

ENZYMES are biological catalysts which are synthesised within cells. Like a chemical catalyst an enzyme accelerates the rate at which a reaction reaches equilibrium without affecting the equilibrium position; it is unchanged at the end of a reaction and it is effective in very small amounts. Unlike a chemical catalyst an enzyme allows a reaction to proceed rapidly under physiological conditions of temperature and pH and is highly specific for its reactants or *substrates*. The increase in reaction rate achieved by an enzyme is as much as 10^{10} greater than that with ordinary catalysts.

Enzymes are proteins with molecular masses between 10 000 and one million. Under adverse conditions of temperature and pH they rapidly lose activity. The protein nature of enzymes ensures that the catalytic function is highly specific; the protein surface provides specific areas for the binding of a single substrate or a small group of similar substances.

Partly because of this specificity and partly because reactions take place at body temperature, enzyme-catalysed reactions are more likely to give a nearly theoretical yield of the end-product than reactions brought about by other agents. In most organic reactions only a proportion of the starting material is converted to the desired product, the rest being lost in a variety of side-reactions. An organic synthesis usually involves a sequence of reactions, each with some loss. The cumulative effect of such losses may be so great that the final product represents only a small proportion of the starting material. By contrast, the enzymes present in muscle can convert glucose to lactate by a sequence of eleven reactions with an overall efficiency of 100 per cent.

An enzyme possesses a unique region on its surface called the *active site*. When a substrate associates with the active site a number of changes occur both in the substrate and in the enzyme. These are accompanied by a reduction in the activation energy needed for the conversion of substrate to product.

Factors affecting the rate of enzyme-catalysed reactions

Enzymes are very sensitive to various factors in their environment. The rate at which an enzyme-catalysed reaction proceeds depends on:

(1) the concentration of the substrate,
(2) the concentration of the enzyme,
(3) the temperature,
(4) the pH,
(5) the presence of inhibitors and activators

The essential feature of an enzyme-catalysed reaction is that the enzyme first combines with the substrate to give an *enzyme–substrate complex* which then breaks down to yield the products of the reaction and regenerate the free enzyme:

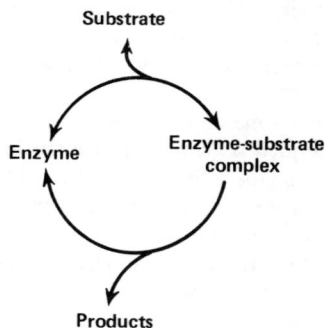

The action of the enzyme acetylcholinesterase in hydrolysing acetylcholine is an example:

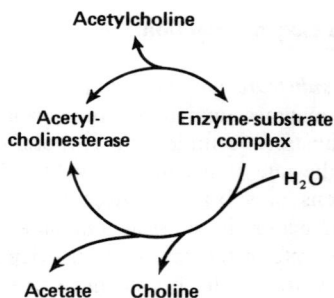

Table 8.1 Amino acid side-chains associated with the binding sites of enzymes

Residues	Structure of side-chain	Function
Arginine } Lysine	- - -CH$_2$—N$^+$H$_3$	Binds negatively charged groups
Glutamate } Aspartate	- - -CH$_2$—COO$^-$	Binds positively charged groups
Hydrophobic amino acids	Alkyl groups	Binds hydrophobic (non-polar) groups

The active site. The substrate is bound to the enzyme at the active site and the specificity of an enzyme depends on the nature of the side-chains of the amino acid residues in this area. These residues can be classified into four groups. The *binding residues* (Table 8.1) bind the substrate and align it so that the bond to be altered is close to the *catalytic residues* which are responsible for the chemical change. The *structural residues* do not participate directly in substrate-binding or catalysis but maintain the correct conformation (tertiary structure) of the enzyme; any disruption of these residues usually leads to deformation of the enzyme and loss of catalytic activity. The *non-essential residues*, usually on the surface, take no part in enzyme catalysis. A schematic diagram of an active site is shown in figure 8.1.

It was at one time thought that the specificity of an enzyme depended on a precise relationship between the substrate and the active site much as a key is related to a lock. It has recently become clear that an enzyme does not exist in a rigid shape complementary to that of substrate but, rather, is induced to change shape much as a glove changes shape when a hand is inserted (fig. 8.2).

Rate of an enzymic reaction

Effect of substrate concentration. On the assumption that enzymes operate by the formation of an enzyme–substrate complex, it is possible to predict how the rate of reaction is likely to be affected by variations in substrate concentration. Since the enzyme can act on the substrate only after combining with it, the rate of the reaction must depend on the frequency with which enzyme molecules encounter

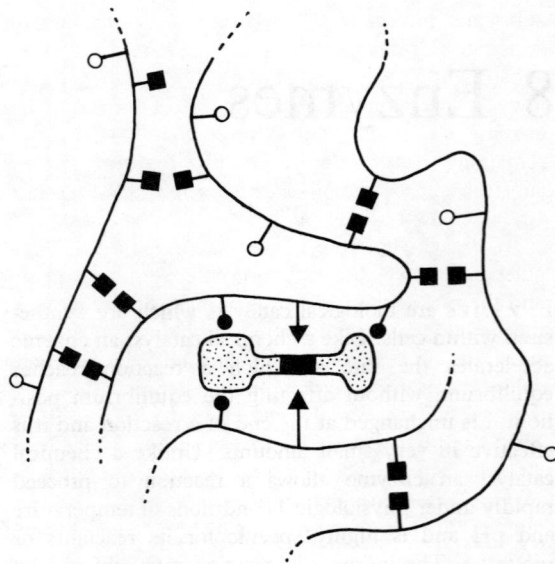

Fig. 8.1 The components of an active site of an enzyme. The amino acid side-chains are shown as ● for binding residues, ▼ for catalytic residues, ■ for structural residues and ○ for non-essential residues. It should be noted that the active site may contain segments of the polypeptide chain which are widely separated in the primary sequence of the protein

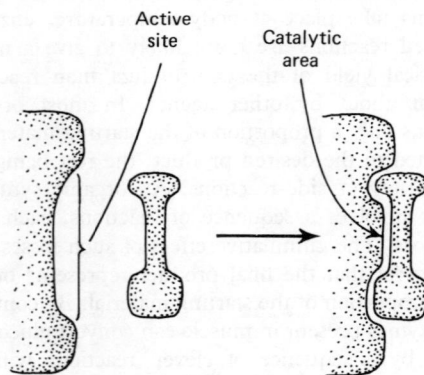

Fig. 8.2 'Induced fit' of a substrate at the active site of an enzyme. The enzyme surface is not at first complementary in shape to the substrate but becomes complementary as the substrate is bound. Once the substrate is bound the catalytic residues can carry out the reaction. When the products leave the active site it returns to its original shape

substrate molecules. When the concentration of substrate is low, enzyme molecules seldom encounter substrate molecules and so the rate of the reaction is low. At any moment only a small proportion of the enzyme molecules are acting on substrate; the remainder are unoccupied. At low concentrations of substrate, the reaction rate is proportional to substrate concentration.

At higher substrate concentrations enzyme molecules encounter substrate molecules more frequently and the rate of the reaction is higher. At any moment most of the enzyme molecules are being used. As the substrate concentration rises further, enzyme molecules encounter substrate molecules so frequently that, as soon as one substrate molecule has been dealt with, another takes its place. At any moment virtually all the active sites are occupied and the enzyme molecules are operating at their maximal rate. A further increase in substrate concentration cannot increase the rate of the reaction.

The theory of enzyme kinetics proposed in 1913 by Michaelis and Menten allows the formulation of an equation for a reaction involving one substrate; its predictions are fulfilled by the experimental results obtained with many simple reactions.

When the initial velocity of a simple enzyme reaction is plotted against substrate concentration a curve (fig. 8.3) is obtained. This has the general equation:

$$v = \frac{V_{max}}{K_m + [S]}$$

where v is the initial rate of the reaction and S is the initial concentration of substrate. The maximum velocity, V_{max}, is reached at high substrate concentration when all the enzyme molecules are occupied. The Michaelis constant, K_m, is defined as the substrate concentration needed to produce one-half the maximum rate ($V_{max}/2$). The value of K_m depends on the type of substrate, the pH of the solution and the temperature. If an enzyme catalyses the reaction of several similar substrates, each substrate has its own K_m.

Effect of enzyme concentration. As with any catalyst, the rate of an enzyme-catalysed reaction depends on the concentration of enzyme when the substrate is present in excess (fig. 8.4).

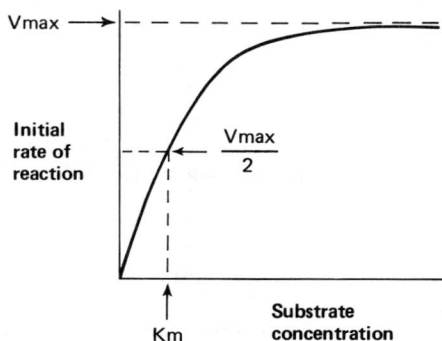

Fig. 8.4 The relationship between enzyme concentration and the initial rate of reaction when the substrate is present in excess

Effect of temperature. With a few exceptions, the rates of chemical reactions are increased as the temperature is raised. An increase in temperature of 10 °C usually doubles the rate of a reaction. Between 0 °C and an *optimal temperature* of about 50 °C, this rule also applies to enzyme-catalysed reactions. This explains why bacterial cultures grow more rapidly at 37 °C than at 20 °C, why food can be preserved in a refrigerator and why, in order to carry out certain operations, a patient's metabolic rate can be lowered by lowering his temperature (*induced hypothermia*).

Above the optimal temperature, which is different for different enzymes, the hydrogen bonds and other forces responsible for the maintenance of the tertiary structure of the enzymes are disrupted. The protein is denatured and its catalytic activity is lost (fig. 8.5).

Effect of pH. Each enzyme has a characteristic pH value for maximal activity, the *pH optimum* (fig. 8.6).

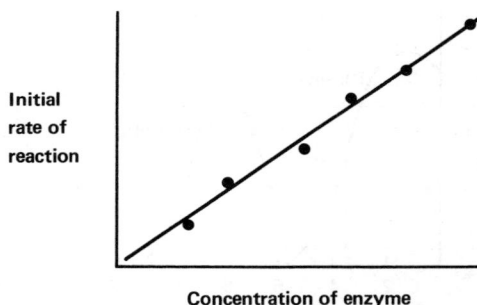

Fig. 8.3 The relationship between substrate concentration and the initial rate of an enzyme-catalysed reaction

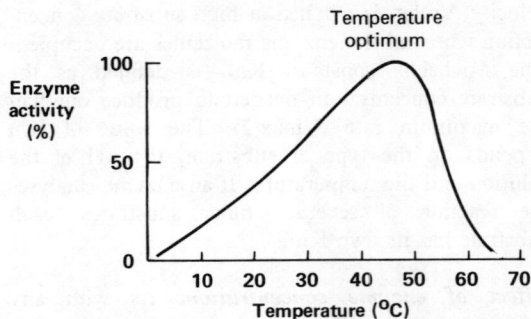

Fig. 8.5 The effect of temperature on the rate of an enzyme-catalysed reaction

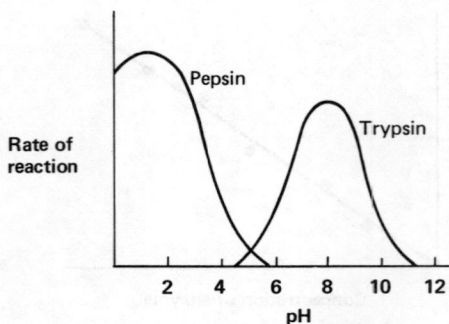

Fig. 8.6 The effect of pH on the activity of two digestive enzymes: pepsin in the stomach and trypsin in the small intestine

For most enzymes this is in the range pH 5 to pH 9. Lower values are found, for example, with the pepsins, proteolytic enzymes in the gastric secretions (pH 2), and with the lysosomal enzymes. Alkaline phosphatase has a pH optimum of 10. The changes in enzyme activity with pH presumably reflect changes in the ionisation of amino acid side-chains, particularly in the regions responsible for the binding of substrate and for catalytic activity.

At extremes of pH enzymes may undergo irreversible inactivation.

Enzyme inhibition

Substances that reduce the activity of an enzyme are termed *enzyme inhibitors*. Two main groups of inhibitors are recognised: reversible and irreversible.

Irreversible inhibition

If an inhibitor causes denaturation of the enzyme protein, catalytic effectiveness is lost permanently. Apart from extremes of pH and temperature, denaturation may result from the action of detergents, proteolytic enzymes or urea. Irreversible inhibition also occurs when essential residues on the enzyme surface are blocked. Such inhibitors are specific for particular enzymes; they bind to the active site in such a way that dissociation from the enzyme is extremely slow. An example of this type of inhibitor is the 'nerve gas', di-isopropyl-fluorophosphate (DIFP). This toxic organophosphate compound was developed during the second world war, first as an insecticide and later as a potential agent for chemical warfare. DIFP irreversibly inactivates enzymes such as trypsin, chymotrypsin and acetylcholinesterase by binding to serine residues in the active site (fig. 8.7). Since acetylcholinesterase

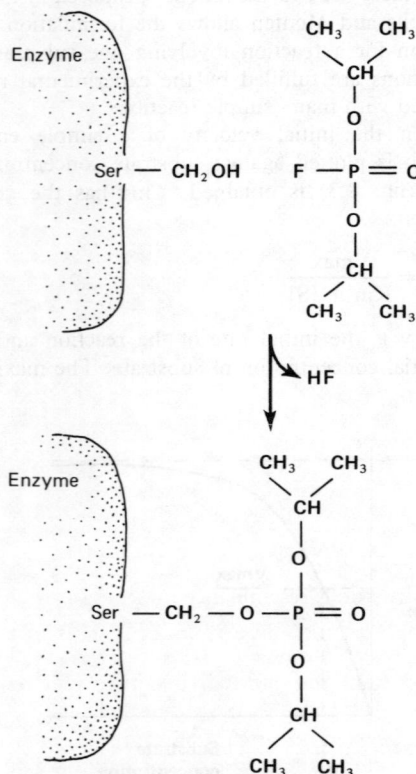

Fig. 8.7 Action of di-isopropyl-fluorophosphate (DIFP) in binding irreversibly to a serine residue in the active site of an enzyme

plays an important part in the transmission of nerve impulses at neuromuscular junctions, poisoning with DIFP causes sustained involuntary contraction of all the skeletal muscle including the muscles of respiration.

The cyanide ion (CN^-) irreversibly inhibits many enzymes which contain copper or iron. The poisonous properties of cyanide are due to its binding to cytochrome a_3 in the respiratory chain in mitochondria (p. 112).

Reversible inhibition

Reversible inhibition is characterised by an equilibrium between enzyme and inhibitor. Four main types of reversible inhibition are recognised: competitive, pure non-competitive, mixed and uncompetitive. Two of these, competitive and pure non-competitive, will be discussed here.

Competitive inhibitors. A competitive inhibitor usually has a structure similar to that of the normal substrate; it binds to the enzyme at, or near, the active site but cannot undergo the catalytic reaction. The substrate and inhibitor compete for the same binding site; the inhibitor occupies active sites that would otherwise be available for the substrate. It should be possible, therefore, to overcome the inhibition by increasing the concentration of substrate and so displacing the inhibitor from the enzyme. If the substrate concentration is increased sufficiently the same maximum velocity (Vmax) should be achieved as in the absence of inhibitor (fig. 8.8).

The classical example of competitive inhibition is the reaction catalysed by the enzyme, succinate dehydrogenase, which converts succinate to fumarate in the tricarboxylic acid cycle. Malonate, malate and oxaloacetate all act as competitive inhibitors of succinate dehydrogenase, presumably because of their structural resemblance to the substrate (fig. 8.9). Malonate binds to the enzyme as succinate does. Whereas succinate is converted to fumarate and released from the enzyme, malonate undergoes no catalytic change; it remains attached to the binding site and prevents the binding of further molecules.

Since a competitive inhibitor usually resembles the substrate, it is specific in its action. It is possible therefore to inhibit an individual enzyme by administering an appropriate competitive inhibitor; the

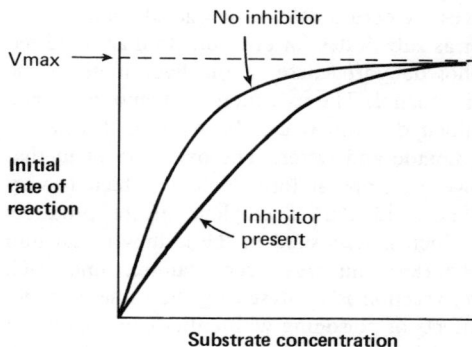

Fig. 8.8 Competitive inhibition. The effect of substrate concentration on the rate of an enzyme-catalysed reaction in the presence and absence of a competitive inhibitor. Note that at low substrate concentration the reaction rate is greatly reduced. As the substrate concentration is increased it displaces inhibitor from the binding sites and the effect of the inhibitor diminishes

action of several important drugs can be explained in this way.

A practical application of competitive inhibition is the use of ethanol in the treatment of poisoning with methanol or ethylene glycol. Methanol is an industrial solvent; ethylene glycol is the main constituent of the antifreeze solutions used in car engines. These

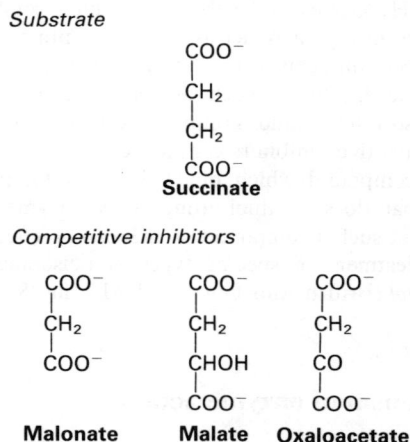

Fig. 8.9 Structures of succinate, the normal substrate for succinate dehydrogenase, and of its competitive inhibitors. Succinate binds to the enzyme by the two carboxyl groups. The competitors also bind in the same way but do not expose comparable parts of the molecule to the catalytic residues

83

substances are occasionally drunk accidentally, suici-
dally, or as substitutes for ethanol. Both are oxidised
by alcohol dehydrogenase in the liver in the same
way as is ethanol. The oxidation of ethanol gives rise
to acetaldehyde, which can be oxidised further to
carbon dioxide and water. The oxidation of methanol, however, stops at formaldehyde which is very
toxic. The oxidation of ethylene glycol produces
oxalate which is deposited in the kidney as calcium
oxalate (p. 186) and causes renal failure. Since each
oxidation reaction is catalysed by the same enzyme,
the ill-effects of poisoning with either compound can
be reduced by the administration of ethanol, which
competes with methanol or ethylene glycol for
binding sites on the alcohol dehydrogenase.

Non-competitive inhibitors. These bind to an enzyme in such a way that binding of the substrate is
not prevented but catalysis is. Typically these
inhibitors do not resemble the substrate; they react
with part of the enzyme destroying its catalytic power
by changing the properties of the active site. The
substrate and inhibitors do not compete for the same
site and inhibition is not reversed by increasing the
amount of substrate. A non-competitive inhibitor
functions by decreasing the effective enzyme concentration and the degree of inhibition is the same
whatever the substrate concentration.

The functioning of many enzymes depends on free
thiol (–SH) groups and substances which combine
with these groups may act as enzyme inhibitors. In
this respect non-competitive inhibitors may resemble
irreversible inhibitors such as lead and mercury
which also bind to thiol groups. The toxic action of
non-competitive inhibitors can, however, be reduced
by any compound which binds the inhibitor more
readily than does the thiol group of the enzyme. An
example of such a compound, which can be effective
in the treatment of specific types of poisoning, is
dimercaprol (British Anti-Lewisite, BAL) (fig. 8.10).

Mechanism of enzyme action

Just as specific amino acid side-chains are involved in
substrate binding, so a specific group of side-chains
are involved in the making and breaking of bonds to
convert substrate(s) to product(s). The mechanism of
action of acetylcholinesterase can be used to illustrate
this principle (fig. 8.11).

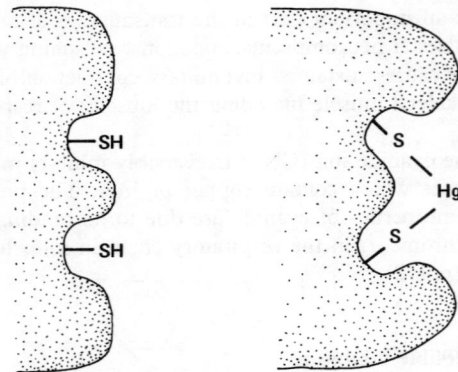

Fig. 8.10 Action of mercury-containing poisons in deform-
ing and inactivating an enzyme by binding to –SH groups.
The inhibition may be relieved by the administration of
substances such as dimercaprol which bind mercury strongly and so remove it from the enzyme

Cholinesterases are enzymes that catalyse the
hydrolysis of choline esters to a greater degree than
other esters. These enzymes can be classified according to their substrate specificity and their sensitivity
to various inhibitors. Acetylcholinesterase is of great
importance; it inactivates acetylcholine released in
autonomic ganglia, in cholinergic postganglionic
synapses and at the neuromuscular junctions of
skeletal muscle.

Acetylcholinesterase is a large molecule with
several subunits. Each molecule has a number of
active sites at which hydrolysis of acetylcholine
occurs. Each active site consists of two separate areas
for binding and for the cleavage of acetylcholine;
these are the *anionic site* and the *esteratic site*
respectively. The anionic site contains a negatively
charged side-chain (COO^-) to which the positively
charged nitrogen of the substrate is bound by
electrostatic forces. The anionic site binds the
substrate and aligns it so that the ester group of
acetylcholine is close to the esteratic site. The
esteratic site consists of serine, tyrosine and histidine
residues.

During the reaction, a proton (H^+) which dissociates from the phenolic group of the tyrosine
combines with the ester oxygen of the substrate. This
generates a positive charge at the carbon of the acetyl
group which is then attracted to the negatively
charged oxygen of the serine residue. The bond
between the acetyl carbon and ester oxygen is broken

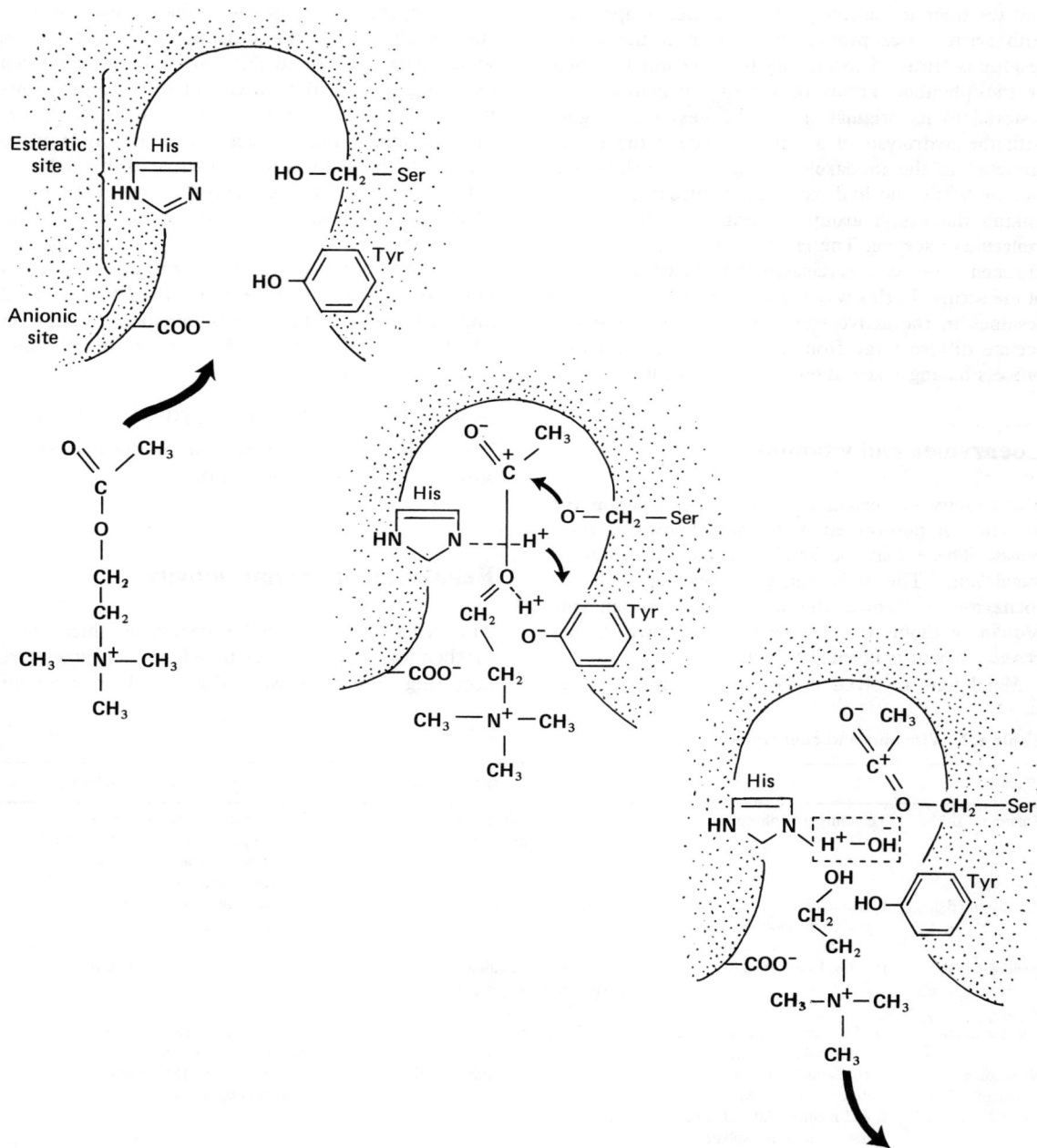

Fig. 8.11 The active site of acetylcholinesterase at three stages of the hydrolysis of acetylcholine. (After Musil, J. *et al.* (1977) *Biochemistry in Schematic Perspective*. Prague: Avicenum)

and the liberated acetyl group combines temporarily with serine. The proton split off from the serine residue is attracted to the negatively charged oxygen of the phenolic group of tyrosine which is thus restored to its original state. The next step begins with the hydrolysis of a water molecule; the H^+ is attracted to the imidazole nitrogen of the histidine residue while the hydroxyl group attacks the bond linking the acetyl group to serine and liberates a molecule of acetate. The H^+ bound transiently to the nitrogen of histidine is released and attaches to the O^- of the serine. In this way the original state of all three residues in the active site is restored. Choline and acetate diffuse away from the active site the entire process having taken about 100 microseconds.

Coenzymes and vitamins

The activity of certain enzymes depends upon the presence of non-protein substances known as *coenzymes*. These may be small organic molecules or metal ions. The active enzyme together with the coenzyme is termed the *holoenzyme;* the enzyme protein without the coenzyme is known as an *apoenzyme* and is normally inactive.

Metal ions involved in enzymic catalysis include iron, copper, zinc, magnesium, manganese, calcium, sodium and potassium. Organic coenzymes have a wide variety of structures; some have already been described (p. 52). Many coenzymes have components which cannot be synthesised so that their precursors must be supplied in the diet as *vitamins*. Most of the water-soluble vitamins have this role (Table 8.2).

One relatively simple coenzyme, pyridoxal phosphate can be used to illustrate the function of coenzymes.

Pyridoxal phosphate is important for many reactions of amino acids (Chap. 15). Figure 8.12 illustrates its role in a transamination reaction in which an amine group is transferred from an amino acid to an oxo acid.

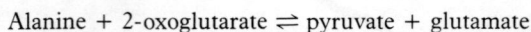

Alanine + 2-oxoglutarate $\rightleftharpoons$ pyruvate + glutamate

Such reactions are important in the disposal of surplus amino acids (Chap. 15).

Regulation of enzyme activity

The living cell is a complex system in which many biochemical reactions catalysed by enzymes are occurring simultaneously. Most of these reactions

Table 8.2 Vitamins and coenzymes

Vitamin	Coenzyme	Coenzyme function	Symptoms of vitamin deficiency
Thiamine (B_1)	Thiamine pyrophosphate	(a) Aerobic decarboxylation (elimination of CO_2) (b) Transfer of aldehyde group	'Dry beri-beri' (muscular weakness, neuropathy loss of weight); 'wet beri-beri' (oedema, due to impaired cardiac function)
Riboflavin (B_2)	Flavin adenine dinucleotide and flavin mononucleotide (p. 54)	Oxidation–reduction	Various skin lesions, corneal vascularisation
Pyridoxine (B_6)	Pyridoxal phosphate	(a) Transamination (b) Decarboxylation (c) Racemisation	Various skin lesions, convulsions
Cobalamin (B_{12})	(a) 5'-deoxyadenosylcobalamin (b) Methylcobalamin	C–C bond rearrangement	Megaloblastic anaemia (Chap. 13) Neurological disorders
Nicotinic acid (niacin)	Nicotinamide adenine dinucleotide (NAD) and nicotinamide adenine dinucleotide phosphate (NADP) (p. 53)	Oxidation–reduction reactions	Pellagra (Chap. 15), diarrhoea, weakness, depression
Pantothenic acid	Coenzyme A (p. 54)	(a) Acyl (R–CO–) reactions (b) Aerobic degradation and synthesis of fatty acids	—
Biotin	Biotin	CO_2 reactions; carboxylation and decarboxylation	—
Folic acid	Tetrahydrofolic acid	Reactions of one-carbon units	Megaloblastic anaemia (Chap. 13)

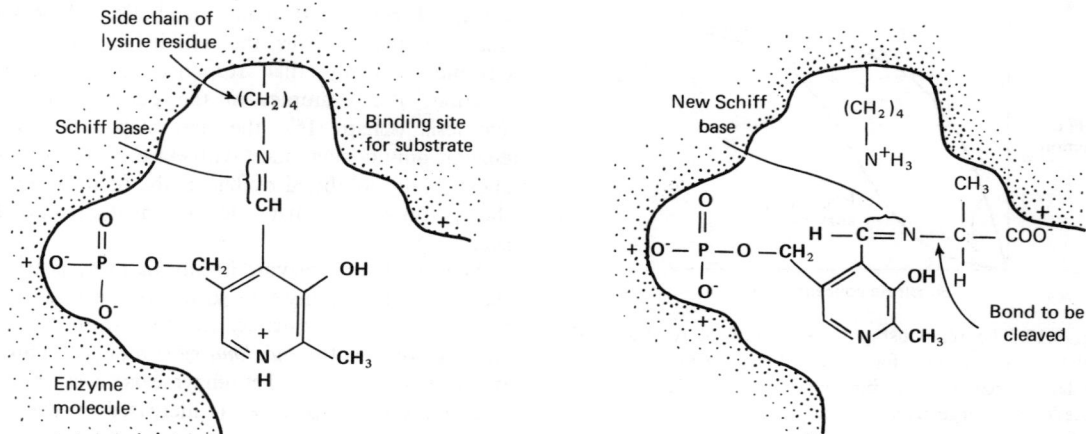

Fig. 8.12 Role of pyridoxal phosphate in a transamination reaction. The first step consists of the binding of the coenzyme at two points on the enzyme surface (left). When an amino acid substrate, in this case alanine, is available the Schiff base (aldimine bond) formed with the lysine residue of the enzyme is broken and a new Schiff base is formed with alanine. In consequence electrons in the alanine are attracted toward the nitrogen of the pyridoxal phosphate and the bonds with the amino acid are weakened. The result is cleavage of the N–C bond at the point shown. The amino group remains attached to the coenzyme while the remainder of the alanine substrate accepts an oxygen atom from water to form pyruvate, which dissociates from the complex. The recipient substrate, 2-oxoglutarate then binds to the same site and accepts the amino group to form glutamate. The coenzyme finally re-forms a Schiff-base linkage with the enzyme thus completing one catalytic cycle

proceed at rates determined by the temperature, the concentrations of substrates, the pH and sometimes by the availability of metal and other ions. In the living cell, the rates of such enzyme-catalysed reactions are subject to additional controls through mechanisms that regulate the location, the amounts and the activity of enzymes.

Synthesis of inactive precursors

Many enzymes are synthesised as inactive precursors (*proenzymes*) which can be rapidly activated when the enzyme is needed. Activation usually consists of the proteolysis of one or more peptide bonds to remove a fragment of the molecule and unmask the active site. Examples include digestive enzymes such as pepsins (secreted as pepsinogens) and trypsin (secreted as trypsinogen). The clotting of blood provides another example of proenzyme activation. A blood clot is formed by a remarkable series of proenzyme modifications involving more than ten different proteins. The activated form of one enzyme is responsible for activating the next enzyme in the series and so producing a 'cascade' of reactions ultimately leading to the formation of cross-linked fibrin.

Covalent modification

Many enzymes are controlled by reversible chemical modifications. The best known examples of this process are provided by the enzymes responsible for the synthesis or breakdown of glycogen (Chap. 12). Glycogen synthase, for example, is activated by the removal of a phosphate group attached to a serine residue.

Inactive glycogen synthase

$$\rightleftharpoons \text{active glycogen synthase} + \text{Pi}$$

Allosteric interactions

Most enzymes possess substrate binding sites which are independent of each other; the binding of one substrate molecule has no effect on the affinity of other sites for substrate. However, in certain enzymes the binding of substrate induces changes in the enzyme that alter the affinity of vacant sites. For such enzymes the relationship between rate of reaction and substrate concentration is usually a sigmoid curve quite unlike that found with ordinary enzymes (fig. 8.13). The name given to this

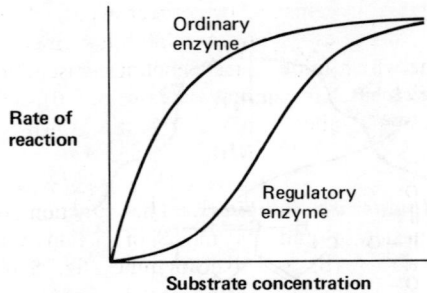

Fig. 8.13 The relationship between rate of reaction and substrate concentration for (a) an ordinary enzyme and (b) a regulatory enzyme. The binding of one molecule of substrate leads to large changes in the affinity of other substrate binding sites. At certain substrate concentrations small changes in concentration lead to large changes in the rate of reaction. In this respect regulatory enzymes resemble haemoglobin in which the binding of one oxygen molecule causes a large change in the affinity of the remaining sites for oxygen (p. 202)

phenomenon is *cooperativity*. Cooperativity may be positive when the affinity of vacant sites increases, or negative when the affinity is decreased.

In some enzymes the affinity of substrate binding sites is altered by the binding of substances other than the normal substrate. Such effects are known as *allosteric interactions* and can consist of either inhibition or activation. Typical allosteric effects are illustrated in figure 8.14; allosteric interactions play an important part in the regulation of metabolic pathways (Chap. 12).

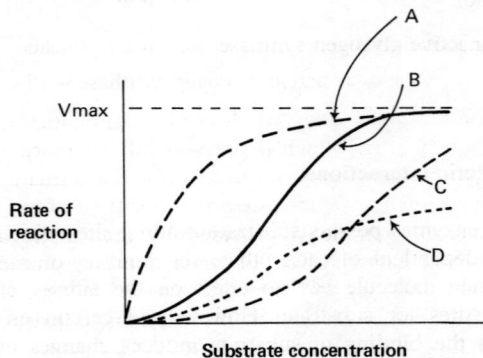

Fig. 8.14 Effect of substances which modify the activity of a regulatory enzyme. A: activator present; B: no modifying substance present; C: inhibitor affecting affinity of enzyme for substrate present; D: inhibitor affecting breakdown of enzyme–substrate complex present

Feedback control. In many metabolic pathways the final product controls the activity of one of the enzymes, usually that responsible for the first reaction. For example, in the pathway of haem synthesis (Chap. 16), the first step involves the enzyme amino-laevulinate synthetase. This enzyme is allosterically inhibited by haem, the final product of the pathway, so that the intermediates do not accumulate.

The regulatory enzyme in any pathway is usually that of lowest activity and so it sets the pace of the whole metabolic sequence. In other words such an enzyme controls the *rate limiting step* of the pathway. If the activity of this key enzyme is increased, the rate of the whole pathway is increased by the same factor.

Isoenzymes

Certain enzymes exist in two or more varieties; although catalysing the same reaction they possess different chemical, immunological and electrophoretic properties. Such variants are known as isoenzymes (or isozymes). One example is lactate dehydrogenase (LD) which catalyses the oxidation of lactate to pyruvate (p. 95). This enzyme in serum can be separated by electrophoresis into five bands (fig. 8.15) designated LD1, LD2, LD3, LD4 and LD5. All these isoenzymes have a similar molecular mass of about 135 000.

The existence of five isoenzymes can be explained in the light of the subunit structure of lactate dehydrogenase. Each enzyme molecule has four subunits which are, by themselves, catalytically inactive. Two types of subunit are recognised, differing slightly in amino acid composition: H, the predominant form in heart muscle, and M, the principal form in skeletal muscle and in liver. Each of the five isoenzymes is made up of a different combination of the subunits as shown in figure 8.15.

The isoenzymes in the heart (mainly H_4) and skeletal muscle (mainly M_4) have distinctly different kinetic properties which are related to the physiological role of the enzyme in each tissue. The LD1 of the heart is active at low pyruvate concentrations but inhibited when the pyruvate level is greater than 1 mmol/l. LD5 of skeletal muscle does not attain its maximal velocity until the pyruvate concentration is 3 mmol/l. This difference means that in muscular

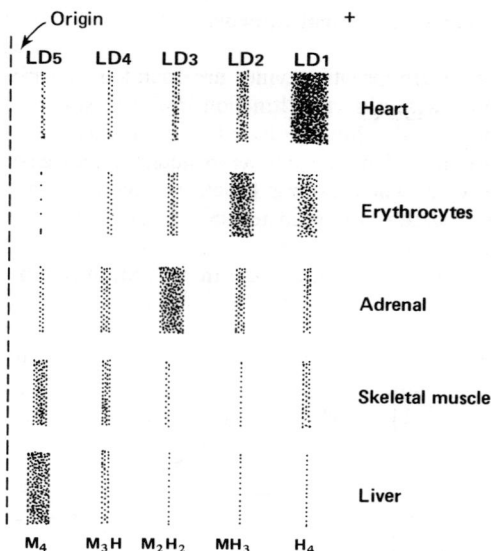

Fig. 8.15 Separation of lactate dehydrogenase isoenzymes from different tissues by electrophoresis. LD1 from the heart moves fastest while LD5 from skeletal muscle and liver moves most slowly. The subunit structure of each isoenzyme is given. (After Wilkinson, J. H. (1970) *Clinical Chemistry* **16,** 882)

exertion, when pyruvate production is greatly increased, the heart obtains its energy from the complete oxidation of glucose through the tricarboxylic acid cycle whereas skeletal muscle converts glucose in large amounts to lactate (Chap. 12).

The separation of isoenzymes and the estimation of their activities in plasma can be useful in diagnosis. In addition to LD, determination of the isoenzyme pattern can be useful in the case of glucose-6-phosphate dehydrogenase, aspartate aminotransferase, alkaline phosphatase and creatine kinase.

The medical significance of enzymes

Applications of enzyme inhibitors

Study of the mode of action of enzymes has provided explanations for the action of certain drugs which are enzyme inhibitors. Once the principles were recognised, deliberate attempts were made to synthesise compounds likely to be specific inhibitors of enzymes in pathogenic organisms or cancer cells. Well recognised examples of enzyme inhibitors include neostigmine which inhibits acetylcholinesterase, allopurinol which inhibits xanthine oxidase (p. 160), and various antimetabolites which inhibit the growth of tumours.

Acetylcholinesterase inhibitors. The function of acetylcholinesterase and its mode of action was described earlier (p. 84). Neostigmine (fig. 8.16) binds at both anionic and esteratic sites in the same way as acetylcholine. In the case of acetylcholine, an acetylated intermediate is formed which is hydrolysed immediately (fig. 8.11). With neostigmine, however, a carbamylated intermediate is formed which undergoes hydrolysis only slowly; the enzyme is meanwhile unable to continue to hydrolyse acetylcholine. Neostigmine and similar compounds are used for the reversal of neuromuscular blockade and in the diagnosis and treatment of myasthenia gravis.

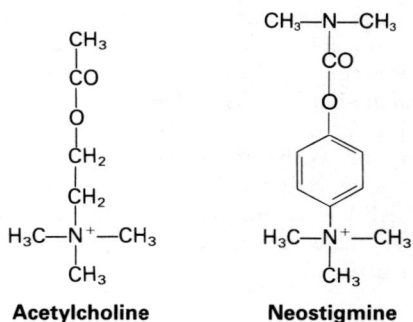

Fig. 8.16 Structures of neostigmine and acetylcholine

Antibiotics and antimetabolites. The antibiotics are a group of drugs which depress or kill microorganisms by blocking enzymes needed for the formation or utilisation of substances essential for their growth. A classical example is provided by the sulphonamides. For bacteria, but not for mammals, p-aminobenzoic acid is an essential growth factor. Its utilisation can be blocked by the sulphonamide drugs which have a similar structure (fig. 8.17).

Antimetabolites are useful in the treatment of malignant disease. Malignant cells proliferate rapidly and therefore need to synthesise large amounts of nucleic acids. For this they require growth factors, such as folate (Chap. 13); certain analogues of folate,

NH₂ structure with COOH
NH₂ structure with SO₂NH₂

p-amino-benzoic acid   Sulphanilimide

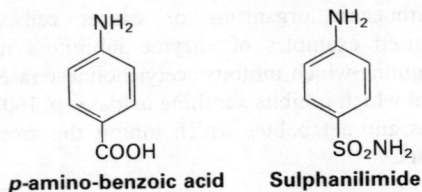

Fig. 8.17 Structures of p-amino-benzoic acid and sulphanilimide

and also analogues of purines and pyrimidines, can block nucleic acid synthesis and so the growth of malignant tumours.

Enzyme-activated inhibitors. These interesting enzyme inhibitors have also been described as *suicide inhibitors* or *k-catalytic inhibitors*. Such an inhibitor has a structural resemblance to the normal substrate of an enzyme and is bound in the normal way. They are then converted by the normal enzymic action to a derivative which irreversibly inactivates the enzyme.

No drugs using this form of inhibition have yet become available for clinical use. Potentially useful substances currently being investigated are analogues of γ-amino-butyric acid (GABA), an inhibitory transmitter in the brain. GABA is broken down by the enzyme GABA-transaminase, which can be inactivated by certain analogues of GABA. Since low brain GABA levels may contribute to some cases of epilepsy, these analogues may prove useful as anticonvulsant drugs.

Enzymes in therapy

In patients with chronic pancreatic insufficiency, digestion of fats, proteins and starch can be greatly improved by the oral administration with meals of *pancreatin,* a mixture of pancreatic enzymes.

Attempts to correct inborn errors of metabolism by the administration of the deficient enzyme have been disappointing because the administered enzyme causes antigenic reactions, or because it is rapidly destroyed in the plasma, or because the enzyme is active only within a particular intracellular compartment. A possible method for avoiding these problems is the use of *immobilised enzymes,* enzymes bound to an inert polymer in such a way that the active sites are unaffected. No immobilised enzymes are yet used in clinical medicine.

Enzymes in clinical diagnosis

Three groups of enzymes are found in the plasma; those with a normal function in the plasma, such as the blood clotting factors, those derived from exocrine glands, such as pancreatic amylase, and enzymes whose normal function is within cells, such as aspartate aminotransferase (AST, p. 177). Enzymes in the last two groups are normally present only in very small amounts but their concentration may

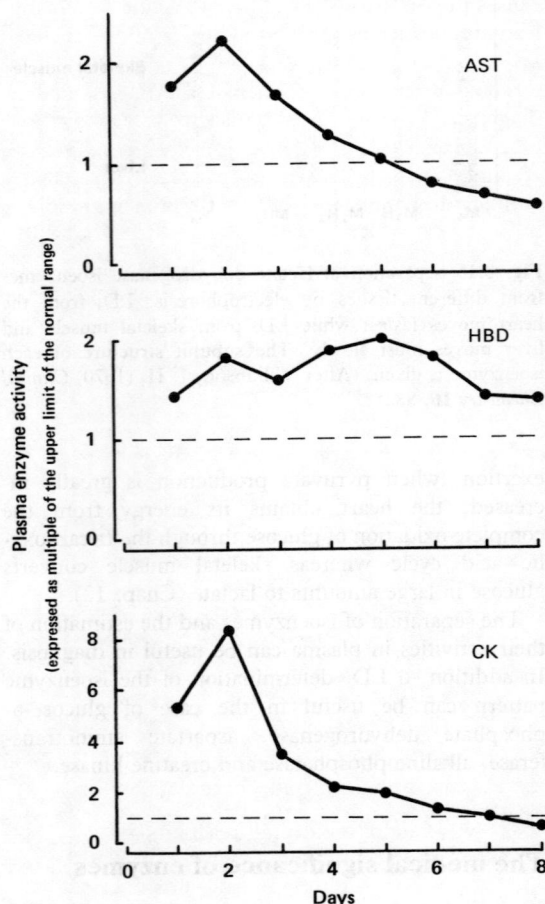

Fig. 8.18 Changes in certain plasma enzymes after myocardial infarction (coronary thrombosis). CK = Creatine kinase (p. 187); AST = aspartate aminotransferase (p. 177); HBD = β-hydroxy-butyrate dehydrogenase (p. 123). The horizontal dotted lines indicate the upper limit of the normal range for the methods used. Mean values for 79 patients. (Courtesy of P. D. Griffiths)

increase greatly in diseases, particularly those involving cell destruction.

One example of this is provided by the enzyme changes after *myocardial infarction* (coronary thrombosis). The necrosis of heart muscle that follows occlusion of a coronary artery leads to the release into the plasma of large amounts of intracellular enzymes (fig. 8.18). Similarly damage to skeletal muscle, for example in *muscular dystrophies*, causes high serum values of creatine kinase (CK) and, to a smaller extent, AST. Liver damage, as in viral hepatitis, causes the release of LD and AST. Acute pancreatitis leads to an increase in plasma levels of amylase and lipase.

The pattern of enzyme changes may give a guide to diagnosis. In some cases the isoenzyme pattern may be helpful (p. 88). For example one isoenzyme of creatine kinase (CK-MB) is particularly associated with cardiac damage whereas the concentration of another isoenzyme (CK-MM) rises in disease of skeletal muscle.

Further reading

Cohen, P. (1983) *Control of Enzyme Activity*, 2nd edn. London: Chapman and Hall

Engel, P. C. (1981) *Enzyme Kinetics. The Steady-state Approach*, 2nd edn. London: Chapman and Hall

Fersht, A. (1984) *Enzyme Structure and Mechanism*, 2nd edn. Oxford: Freeman

Palmer, T. (1981) *Understanding Enzymes*. Chichester: Ellis-Horwood

Price, C. P. (1981) The diagnostic value of isoenzyme measurement. In *Recent Advances in Clinical Biochemistry*, 2nd edn, eds Alberti, K. G. M. M. and Price, C. P., pp. 79–101. Edinburgh: Churchill Livingstone

Price, N. C. and Stevens, L. (1982) *Fundamentals of Enzymology*. Oxford: University Press

APPENDIX—ENZYME KINETICS

Earlier in this chapter it was indicated that a simple enzymic reaction involving a single substrate starts by the formation of an enzyme–substrate complex:

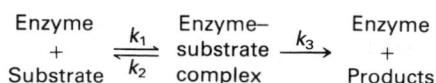

$$\text{Enzyme} + \text{Substrate} \underset{k_2}{\overset{k_1}{\rightleftharpoons}} \text{Enzyme–substrate complex} \xrightarrow{k_3} \text{Enzyme} + \text{Products}$$

where k_1, k_2 and k_3 are the velocity constants for the steps represented by the arrows. It is assumed that the complex is not reformed from the association of enzyme and products. This assumption is likely to be correct in the early stages of a reaction before the product has accumulated.

With this theory it proved possible to predict the way in which the rate of a simple enzymic reaction was related to the substrate concentration.

> v = rate of the overall reaction
> [E] = concentration of enzyme in the free state
> [S] = concentration of free substrate
> [ES] = concentration of enzyme–substrate complex
> [Et] = total enzyme concentration ([E] + [ES])

Since the concentration of enzyme is much lower than that of the substrate, [ES] is extremely small compared with [S] and the total concentration of substrate is virtually identical to that of the free substrate. Further it is assumed that during the period of observation the changes in [S] are negligible and that [ES] remains constant in a *steady state* in which the rate of formation of ES is equal to that of its breakdown.

The rate of formation of ES is therefore $k_1[E][S]$ or $k_1([Et] - [ES])[S]$. The breakdown of ES occurs by two reactions: to products, for which the rate is $k_3[ES]$, and to enzyme and substrate, for which the rate is $k_2[ES]$. The total rate of breakdown is therefore $k_2[ES] + k_3[ES]$. This must be equal to the rate of formation so that:

$$k_1[S]([Et] - [ES]) = k_2[ES] + k_3[ES]$$

or

$$k_1[Et][S] - k_1[ES][S] = [ES](k_2 + k_3)$$

or

$$k_1[Et][S] = [ES](k_1[S] + k_2 + k_3)$$

or

$$[Et][S] = [ES]\left([S] + \frac{k_2 + k_3}{k_1}\right) \qquad (1)$$

Since k_1, k_2 and k_3 are all constants, the ratio $(k_2 + k_3)/k_1$ is also a constant and is known as Km the *Michaelis constant*. Equation 1 above can there-fore be rewritten:

$$[Et][S] = [ES]([S] + Km)$$

or

$$[ES] = \frac{[Et][S]}{Km + [S]} \quad (2)$$

The rate at which the overall reaction takes place (v) is the rate at which the products are formed, $k_3[ES]$. This can be substituted in equation 2 to give:

$$v = k_3[ES] = \frac{k_3[Et][S]}{Km + [S]} \quad (3)$$

This equation describes a hyperbola as illustrated in figure 8.3. When [S] is very large [ES] is virtually identical to [Et] and the rate of the reaction is maximal (Vmax). Under these conditions, $Vmax = k_3[Et]$. This can now be substituted in equation 3 to give

$$v = \frac{Vmax\,[S]}{Km + [S]} \quad (4)$$

This is known as the *Michaelis–Menten equation*. Predictions based on this equation have been fulfilled for many simple enzymic reactions. For more complex reactions, such as those involving two or more substrates, similar calculations can be carried out but a computer is often required. Nevertheless the factors influencing the rates of many enzymic reactions can be analysed in terms of Km and Vmax. For many reactions Km can be regarded as a measure of the enzyme's affinity for substrate and Vmax as a measure of its ability to transform the substrate once it has been bound.

In order to determine the values of Km and Vmax for a particular enzyme, the Michaelis–Menten equation (4) has been transformed in a number of ways to give a straight line. The best known of these

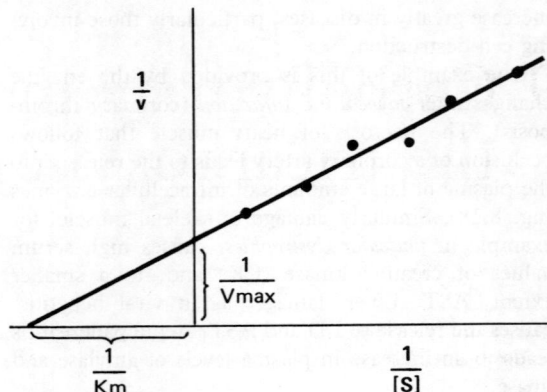

Fig. 8.19 The Lineweaver–Burke plot of the reciprocal of the rate of an enzymic reaction (v) against the reciprocal of the substrate concentration ([s]). The intercepts allow the determination of Km and Vmax

is that devised by Lineweaver and Burke:

$$\frac{1}{v} = \frac{Km}{Vmax}\frac{1}{[S]} + \frac{1}{Vmax}$$

The use of this equation for the determination of Km and Vmax is illustrated in figure 8.19.

Enzyme inhibitors

Calculations similar to those described above can be carried out to predict the effect of different concentrations of reversible inhibitors. It can be shown for example that in competitive inhibition Km is increased but Vmax is unaltered. In pure non-competitive inhibition, which is uncommon, Km is unaltered but Vmax is reduced. In a more frequent form of inhibition, *mixed inhibition*, both Km and Vmax are altered.

RG CRP

9 Carbohydrate metabolism & the tricarboxylic acid cycle

CARBOHYDRATE is the most important source of energy for most of the world's population and in many developing countries it provides as much as 90 per cent of the energy intake. The main carbohydrates in the diet are starch, sucrose and, for infants, lactose. After digestion in the intestine the carbohydrates in a typical western diet yield approximately 250 g glucose, 20 g galactose and 60 g fructose per day.

This chapter is concerned with the ways in which these substances are broken down to yield energy and with the way in which some energy is stored in the body as glycogen. The following chapter will describe the ways in which the energy from these reactions, and from the breakdown of fats and of amino acids, is coupled to the synthesis of ATP.

There are three major pathways to consider: glycolysis, the tricarboxylic acid cycle and the pentose phosphate shunt.

Glycolysis

The first step in the process by which energy is obtained from glucose consists of the conversion of one molecule to two molecules of pyruvate by *glycolysis* or the Embden–Myerhof pathway. This takes place in the cytosol and is summarised in figure 9.1.

Before it can be split the rather asymmetric glucose molecule is converted to the almost symmetrical fructose 1,6-bisphosphate by the transfer of two phosphate groups from ATP. The steps involved are shown in figure 9.2; the ATP expended is recovered later. As described in Chapter 12 this stage is important in the regulation of glycolysis.

Fructose 1,6-bisphosphate is then split by the enzyme *aldolase* (fig. 9.3). The products of this reaction, glyceraldehyde 3-phosphate and dihydroxyacetone phosphate, are interconvertible in the presence of the enzyme *triose-phosphate isomerase*. Since

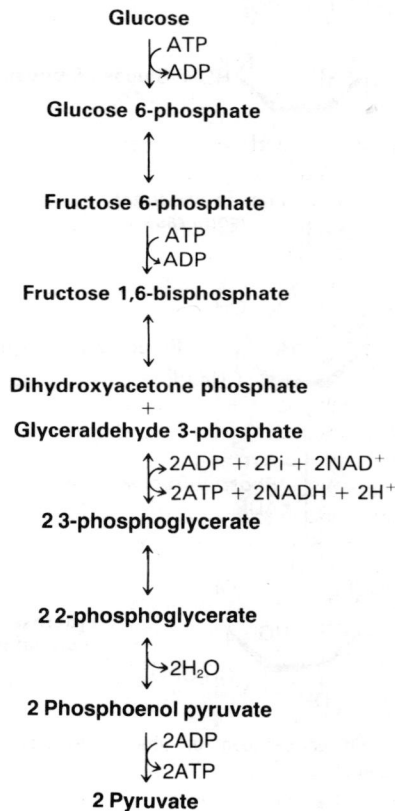

Glucose
ATP
ADP

Glucose 6-phosphate

Fructose 6-phosphate
ATP
ADP

Fructose 1,6-bisphosphate

Dihydroxyacetone phosphate
+
Glyceraldehyde 3-phosphate
$2ADP + 2Pi + 2NAD^+$
$2ATP + 2NADH + 2H^+$

2 3-phosphoglycerate

2 2-phosphoglycerate
$2H_2O$

2 Phosphoenol pyruvate
2ADP
2ATP

2 Pyruvate

Fig. 9.1 Overall sequence of glycolysis

only the aldehyde is used in the next step, the overall effect of this stage is the splitting of the molecule of fructose 1,6-bisphosphate into two molecules of glyceraldehyde 3-phosphate.

In the energy-yielding stage (fig. 9.4), glyceraldehyde 3-phosphate is oxidised by *glyceraldehyde 3-phosphate dehydrogenase*. Reactions in which NAD^+

93

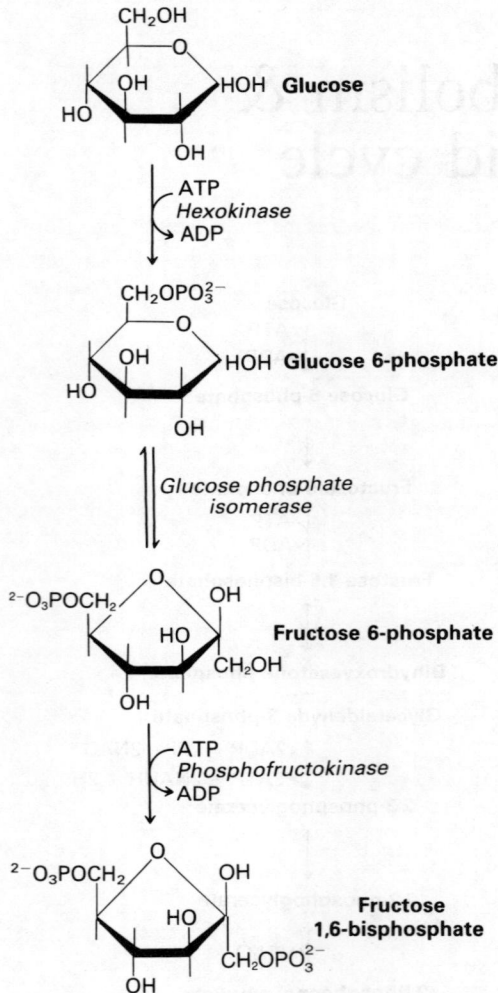

Fig. 9.2 The conversion of glucose to fructose 1,6-bisphosphate

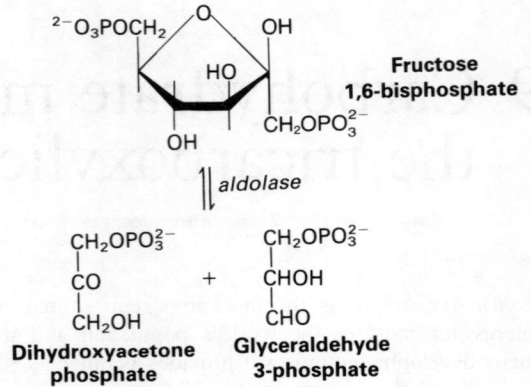

Fig. 9.3 The conversion of fructose 1,6-bisphosphate to two triose phosphates

3-phosphoglycerate still contains the phosphate groups originally derived from ATP. These are now transferred back to ADP as shown in figure 9.5.

Energy yield. It can be seen that breakdown of one glucose molecule to two molecules of pyruvate is

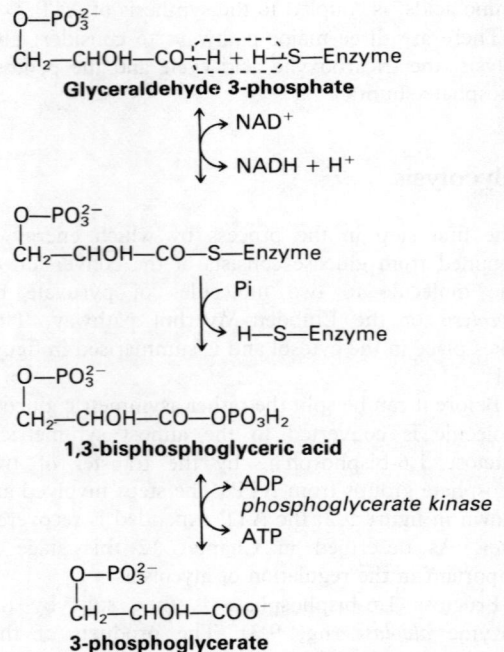

Fig. 9.4 The conversion of glyceraldehyde 3-phosphate to 3-phosphoglycerate

is used to oxidise an aldehyde to an acid are accompanied by the liberation of energy which is used to form ATP. In this reaction a sulphydryl (–SH) group, which is part of the active site of the enzyme, is linked to the aldehyde group by a dehydrogenation reaction in which NAD is reduced to NADH. The complex so formed is then split by addition of inorganic phosphate to give 1,3-bisphosphoglycerate. In turn this loses its newly acquired phosphate to ADP to yield 3-phosphoglycerate.

$$CH_2-O-PO_3^{2-}$$
$$CHOH \qquad \textbf{3-phosphoglycerate}$$
$$COO^-$$

$\parallel$ phosphoglyceromutase

$$CH_2OH$$
$$CH-O-PO_3^{2-} \qquad \textbf{2-phosphoglycerate}$$
$$COO^-$$

H_2O ⇌ H_2O phosphopyruvate hydratase

$$CH_2$$
$$C-O-PO_3^{2-} \qquad \textbf{Phosphoenol pyruvate}$$
$$COO^-$$

ADP
pyruvate kinase
ATP

$$CH_3$$
$$C=O \qquad \textbf{Pyruvate}$$
$$COO^-$$

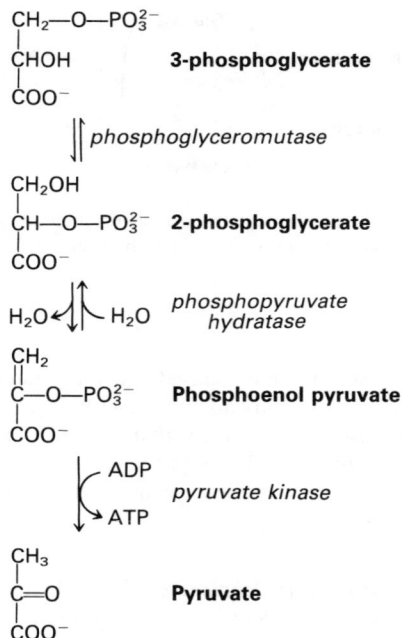

Fig. 9.5 The final stage of the glycolytic pathway leading to the production of pyruvate

accompanied by the net formation of two ATP molecules, and the reduction of two molecules of NAD^+. In aerobic conditions the two NADH molecules can transfer their electrons to molecular oxygen via the electron transport system (p. 110) with the production of four additional ATP molecules.

In anaerobic conditions the NADH reacts with the pyruvate under the influence of *lactate dehydrogenase* to give lactate (fig. 9.6). Since this reaction involves little free energy change there is no uptake or production of ATP. In anaerobic glycolysis therefore one molecule of glucose yields two molecules of lactate and two molecules of ATP. Although this represents only a small amount of energy it is none the less useful in certain circumstances particularly

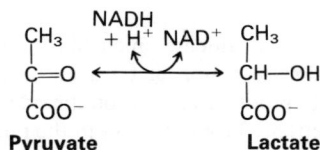

$$CH_3 \qquad \overset{NADH}{\underset{NAD^+}{+ H^+}} \qquad CH_3$$
$$C=O \longleftrightarrow CH-OH$$
$$COO^- \qquad\qquad COO^-$$
$$\textbf{Pyruvate} \qquad \textbf{Lactate}$$

Fig. 9.6 The interconversion of pyruvate and lactate

when tissues are hypoxic. Glycolysis with lactate production takes place even in aerobic conditions in red cells (see below) and in the renal medulla.

Gluconeogenesis

Glucose may be formed from lactate or pyruvate by the use of the reactions of the glycolytic pathway which are reversible with two exceptions: the phosphorylation of glucose and the phosphorylation of fructose 6-phosphate. These reactions can be by-passed by the hydrolytic enzymes *glucose 6-phosphatase* and *fructose 1,6-bisphosphatase*. A third reaction, the conversion of phosphoenol pyruvate to pyruvate, is usually also by-passed when the glycolytic pathway operates in reverse. The by-pass includes the conversion of pyruvate to oxaloacetate with the uptake of CO_2, energy being provided by the hydrolysis of one molecule of ATP. The oxaloacetate then loses CO_2 and takes up a phosphate group from GTP. The whole sequence is shown in figure 9.7.

The overall equation for the aerobic synthesis and breakdown of glucose by the glycolytic pathway is summarised in figure 9.8. It can be seen that while the breakdown of a glucose molecule to lactate yields only 2 ATP molecules its resynthesis requires 6. This illustrates the general rule that the synthesis of a complex molecule requires more energy than is made available by its breakdown.

Disorders of glycolysis

Lactic acidosis. The normal blood lactate level is 1 to 2 mmol/l but higher values are found if the arm from which the blood has been obtained has been exercised. Plasma lactate is high after exertion; a sprinter obtains most of his energy from anaerobic glycolysis. Lactate and H^+ levels also rise rapidly in patients with a failing circulation (shock), or after a cardiac arrest, because of hypoxia of the tissues.

Lactic acidosis occurs in patients with a normal circulation in several disorders such as diabetic ketosis (Chap. 12), renal failure, severe infections, aspirin poisoning and during the use of certain oral antidiabetic drugs including phenformin. The cause of the lactic acidosis in these conditions is not yet clear; in diabetic ketosis pyruvate oxidation is thought to be impaired.

Pyruvate

CO_2 — ATP
pyruvate carboxylase
ADP + Pi

Oxaloacetate

GTP → ADP
CO_2 → GDP → ATP

Phosphoenol pyruvate

→ H_2O

2-phosphoglycerate

3-phosphoglycerate

→ 2ATP + 2NADH + 2H$^+$
→ 2ADP + 2Pi + 2NAD$^+$

Glyceraldehyde 3-phosphate

Fructose 1,6-bisphosphate

→ H_2O
→ Pi

Fructose 6-phosphate

Glucose 6-phosphate ⇌ Glycogen

→ H_2O
→ Pi

Glucose

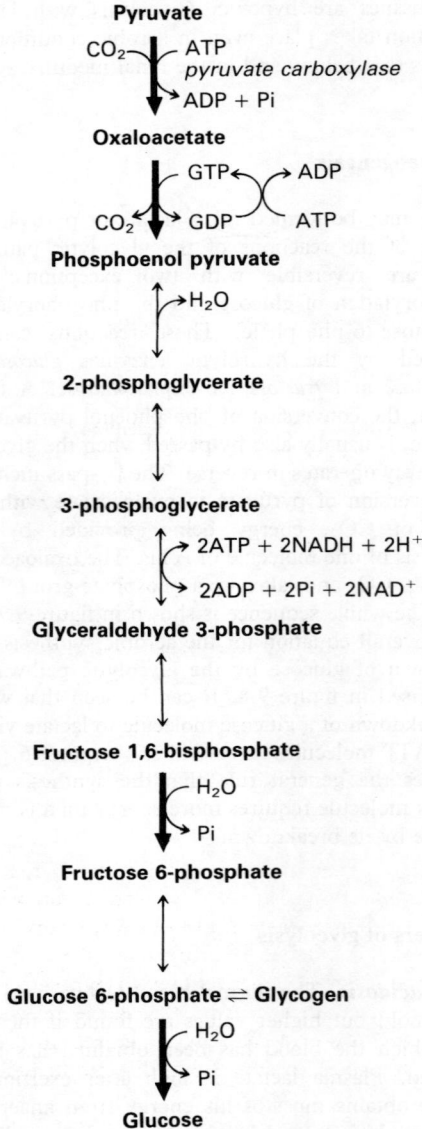

Fig. 9.7 Reverse glycolysis. The steps which are different from those of the forward reaction (fig. 9.1) are shown with heavy arrows

Cells of malignant tumours frequently use glycolysis as their main or only source of energy. This causes a lactic acidosis, the excess lactate being reconverted to glucose in the liver and kidneys. The tumour obtains only two ATP molecules for each

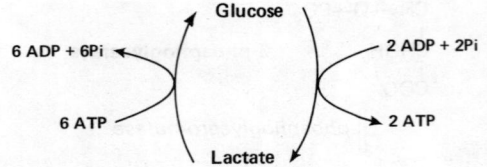

Glucose

6 ADP + 6Pi ←   → 2 ADP + 2Pi
6 ATP   2 ATP

Lactate

Fig. 9.8 A summary of the overall reactions of glycolysis and reverse glycolysis. Together these are known as the Cori cycle

glucose molecule metabolised and the patient's other tissues have to expend six molecules of ATP to resynthesise it. It is likely that this very inefficient use of energy contributes to the rapid loss of weight (cachexia) seen in patients with extensive malignant disease.

Inherited enzyme deficiencies. Since the mature red cell has no nucleus, no mitochondria and no ribosomes, it cannot synthesise proteins or carry out the tricarboxylic acid cycle. The pentose phosphate pathway does operate, but most of the ATP needed by the red cell is derived from glycolysis. This is essential for the maintenance of the cell's normal shape and the electrochemical gradients between plasma and cytosol.

Several different disorders caused by lack of the enzymes of glycolysis have been described but with the exception of pyruvate kinase deficiency they are rare. It has been estimated that the gene for this disorder has a frequency of 1 per cent so that affected cases might be expected in at least 1 in 10 000 births. While the disorder can vary greatly in its severity the most striking abnormality is anaemia due to excessive red cell breakdown (haemolysis). Microscopically the red cells in this, and in other disorders of glycolysis, show characteristic changes of shape (fig. 9.9).

The tricarboxylic acid cycle

In most tissues glucose metabolism in aerobic conditions does not stop at pyruvate but continues until the only products are carbon dioxide and water. The final steps are brought about in the mitochondria by the *tricarboxylic acid cycle* (citric acid cycle) (fig. 9.10) first described by Krebs.

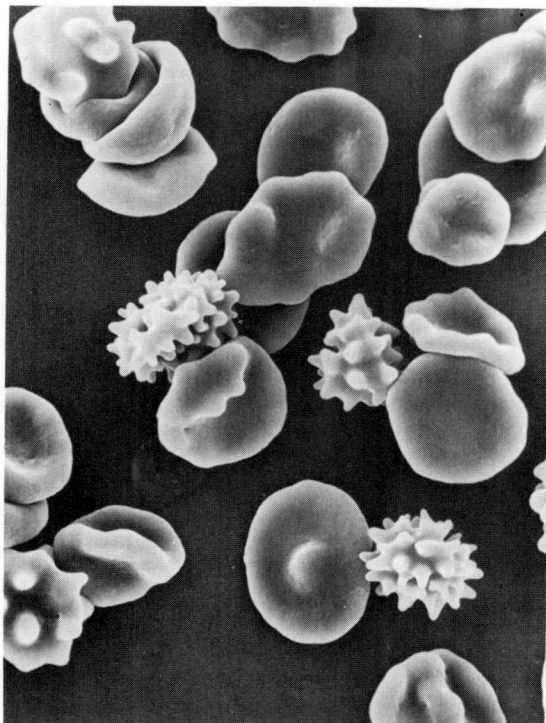

Fig. 9.9 Scanning electron micrograph of red cells in pyruvate kinase deficiency. The patient had had a splenectomy so that these very abnormal cells (echinocytes), instead of being destroyed rapidly, persisted in the circulation. (Courtesy of S. Miwa and N. Matsumoto)

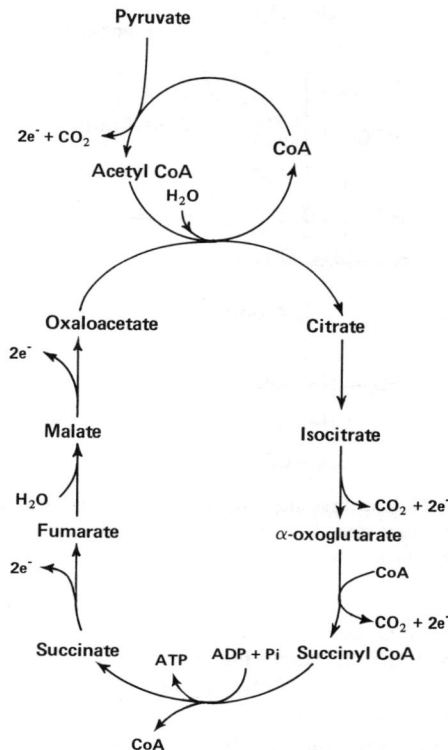

Fig. 9.10 Summary of the reactions of the tricarboxylic acid cycle

Pyruvate produced from glucose in the cytosol passes into the mitochondria and reacts with coenzyme A to form acetyl coenzyme A. This complex reaction, involving thiamine pyrophosphate (TPP) and lipoic acid as cofactors, is summarised in figure 9.11. The acetyl component of the pyruvate is transferred to coenzyme A while the carbon of the carboxyl group is liberated as carbon dioxide. The remaining hydrogen atom is transferred to NAD. Since the overall reaction involves both oxidation and loss of carbon dioxide it is termed an *oxidative decarboxylation*.

The tricarboxylic acid cycle begins with the transfer of an acetyl group from acetyl coenzyme A to oxaloacetate to give citrate. By a series of steps including dehydrogenations and the loss of two molecules of CO_2 the citrate is reconverted to

oxaloacetate. This can then take up another acetyl group from acetyl coenzyme A to form citrate which goes through the cycle again. In this way a very small amount of oxaloacetate can bring about the complete oxidation of a much larger amount of pyruvate. The hydrogen atoms removed in the dehydrogenation steps are passed as described in Chapter 10 to the electron transport system with the production of ATP.

The reactions of the cycle can be divided into three main stages.

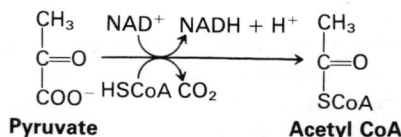

Fig. 9.11 Formation of acetyl CoA from pyruvate. This overall reaction consists of four steps all catalysed by components of the enzyme complex pyruvate dehydrogenase

CO—COO⁻ **Oxaloacetate**
CH₂—COO⁻

H₂O ⟶ CH₃—CO—SCoA **Acetyl CoA**
Citrate synthase
HSCoA

CH₂—COO⁻
HO—C—COO⁻ **Citrate**
CH₂—COO⁻

Aconitase

HO—CH—COO⁻
CH—COO⁻ **Isocitrate**
CH₂—COO⁻

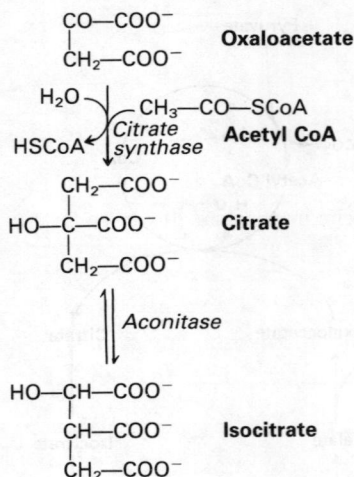

Fig. 9.12 The first stage of the tricarboxylic acid cycle. Acetyl coenzyme A condenses with oxaloacetate to give citrate which is then converted to isocitrate

Stage 1. In this stage (fig. 9.12) the acetyl group of acetyl coenzyme A is transferred to oxaloacetate to form citrate. A molecule of water is needed to hydrolyse the linkage between the acetyl group and the coenzyme A.

In this way the coenzyme A, of which only a minute amount is present in the tissues, is liberated and can react with more pyruvate. The citrate undergoes an internal rearrangement to give isocitrate.

Stage 2. In this stage the 6-carbon isocitrate is converted to the 4-carbon succinate (fig. 9.13). The isocitrate undergoes oxidation and decarboxylation to give α-oxoglutarate. The α-oxoglutarate then undergoes an oxidative decarboxylation to give succinyl coenzyme A.

Succinyl coenzyme A is now split to succinate and free coenzyme A. The energy of this hydrolysis is used to form ATP.

Stage 3. In this sequence of reactions succinate is oxidised to oxaloacetate (fig. 9.14). Succinate is oxidised by the flavoprotein *succinate dehydrogenase* to give fumarate. Fumarate, in the presence of the enzyme *fumarase*, takes up water to give malate which is dehydrogenated by *malate dehydrogenase* and NAD⁺ to give oxaloacetate. The oxaloacetate then

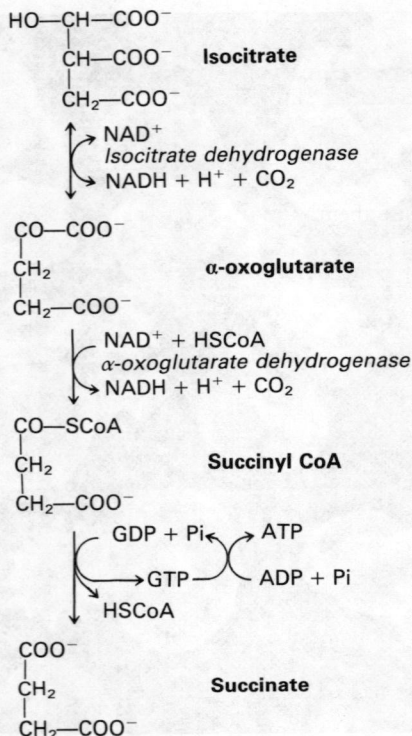

HO—CH—COO⁻
CH—COO⁻ **Isocitrate**
CH₂—COO⁻

NAD⁺
Isocitrate dehydrogenase
NADH + H⁺ + CO₂

CO—COO⁻
CH₂ **α-oxoglutarate**
CH₂—COO⁻

NAD⁺ + HSCoA
α-oxoglutarate dehydrogenase
NADH + H⁺ + CO₂

CO—SCoA
CH₂ **Succinyl CoA**
CH₂—COO⁻

GDP + Pi ⟶ ATP
GTP ⟶ ADP + Pi
HSCoA

COO⁻
CH₂ **Succinate**
CH₂—COO⁻

Fig. 9.13 The production of succinate from isocitrate in the tricarboxylic acid cycle

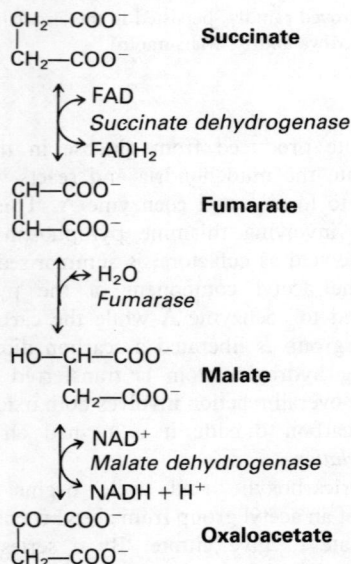

CH₂—COO⁻ **Succinate**
CH₂—COO⁻

FAD
Succinate dehydrogenase
FADH₂

CH—COO⁻ **Fumarate**
CH—COO⁻

H₂O
Fumarase

HO—CH—COO⁻ **Malate**
CH₂—COO⁻

NAD⁺
Malate dehydrogenase
NADH + H⁺

CO—COO⁻ **Oxaloacetate**
CH₂—COO⁻

Fig. 9.14 The final steps of the tricarboxylic acid cycle

combines with a fresh molecule of acetyl coenzyme A and the whole process is repeated.

Two of the reactions of the cycle, the oxidative decarboxylations of pyruvate and α-oxoglutarate, are irreversible. Since these reactions cannot be by-passed the cycle is irreversible; it can be used to degrade pyruvate and acetyl coenzyme A, but not to synthesise them.

Energy yield. The whole cycle causes the oxidation of one molecule of pyruvate to carbon dioxide and water, with the production of one molecule of ATP, the reduction of four molecules of NAD^+ and the reduction of one molecule of FAD. As described in Chapter 10 each molecule of NADH is reoxidised by the electron transfer chain to yield three molecules of ATP; the reoxidation of one molecule of $FADH_2$ yields two molecules of ATP. Thus, for each molecule of pyruvate broken down by the cycle, 15 molecules of ATP are formed.

Central role of the tricarboxylic acid cycle

It is important to note that the tricarboxylic acid cycle is involved in the oxidation of fats and of amino acids, as well as of carbohydrate. Fatty acids are broken down stepwise (Chap. 12) to yield acetyl coenzyme A. This enters the tricarboxylic acid cycle and is dealt with in the same way as acetyl coenzyme A derived from pyruvate. Several amino acids can undergo transamination (Chap. 15) to yield components of the tricarboxylic acid cycle. For example, aspartate yields oxaloacetate and glutamate yields α-oxoglutarate.

The tricarboxylic acid cycle is not only a means of generating ATP; it is also an important source of precursors for synthetic pathways. Oxaloacetate and α-oxoglutarate are precursors for the corresponding amino acids, aspartate and glutamate; succinyl coenzyme A is the starting material for the synthesis of haem (Chap. 16).

When these substances are being drawn off the cycle some means is needed to maintain oxaloacetate levels; acetyl coenzyme A can be oxidised only if oxaloacetate is available to combine with it. To meet this need oxaloacetate is synthesised from pyruvate and CO_2 by the action of pyruvate carboxylase as described earlier (fig. 9.7). Figure 12.1 (p. 131) summarises these important functions of the tricarboxylic acid cycle.

The pentose phosphate pathway

Glycolysis and the tricarboxylic acid cycle together form the principal pathway by which animal tissues oxidise glucose to carbon dioxide and water with the production of useful energy in the form of ATP. One important alternative is the *pentose phosphate pathway*, a cyclic mechanism. It can be described in two phases: the conversion of hexose to pentose and the conversion of pentose to hexose.

Conversion of hexose to pentose. Glucose 6-phosphate is oxidised to 6-phosphogluconolactone by *glucose 6-phosphate dehydrogenase* and NADP. The lactone is hydrolysed by a specific lactonase to give 6-phosphogluconate which is further oxidised and decarboxylated to give the pentose ribulose 5-phosphate. In turn this is isomerised to ribose 5-phosphate (fig. 9.15).

The overall effect of these reactions is the conversion of one molecule of glucose 6-phosphate to one molecule of ribose 6-phosphate and CO_2, together with the reduction of two molecules of $NADP^+$.

Conversion of pentose to hexose. Pentose is converted back to hexose by a complex series of rearrangements involving several pentose molecules. The main types of reaction are transketolation, transaldolation and the aldolase reaction.

Transketolation is a reaction between two sugar phosphate molecules in which a $-CO-CH_2OH$ group is transferred from one to another. For example, two pentose phosphate molecules can react together to give a triose phosphate and a heptose phosphate:

$$C_5 + C_5 \rightleftharpoons C_3 + C_7$$

Transaldolation is a similar reaction in which the unit transferred is $-CHOH-CO-CH_2OH$. This allows a triose phosphate and a heptose phosphate to react together to give a hexose phosphate and a tetrose phosphate:

$$C_3 + C_7 \rightleftharpoons C_6 + C_4$$

Finally the aldolase reaction (p. 94) can be used to combine two triose phosphate molecules to yield one hexose diphosphate:

$$C_3 + C_3 \rightleftharpoons C_6$$

These three reactions are used to convert pentose

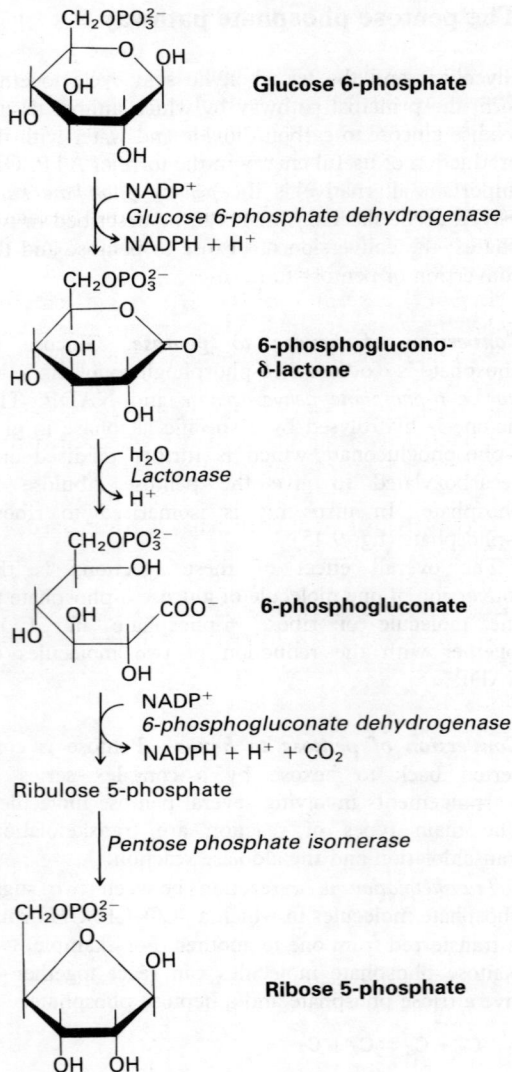

CH$_2$OPO$_3^{2-}$

Glucose 6-phosphate

NADP$^+$
Glucose 6-phosphate dehydrogenase
NADPH + H$^+$

CH$_2$OPO$_3^{2-}$

6-phosphoglucono-δ-lactone

H$_2$O
Lactonase
H$^+$

CH$_2$OPO$_3^{2-}$

6-phosphogluconate

NADP$^+$
6-phosphogluconate dehydrogenase
NADPH + H$^+$ + CO$_2$

Ribulose 5-phosphate

Pentose phosphate isomerase

CH$_2$OPO$_3^{2-}$

Ribose 5-phosphate

Fig. 9.15 The principal steps in the conversion of hexose phosphate to pentose phosphate by the pentose phosphate pathway

quantitatively to hexose. It is convenient to start with six pentose phosphate molecules arranged in pairs as in figure 9.16. Two of these pairs undergo trans-ketolation followed by transaldolation to yield two hexose phosphates and two tetrose phosphates. The latter then undergo transketolation with the remaining pair of pentose phosphates to give two more

hexose phosphates and two triose phosphates. Under the influence of aldolase the two triose phosphates combine to give fructose 1,6-bisphosphate which is hydrolysed by *fructose 1,6-bisphosphatase* to give another hexose monophosphate. The whole sequence leads to the conversion of six molecules of pentose phosphate to five of hexose phosphate:

$$6C_5 \rightarrow 5C_6$$

Overall reaction. In the pentose phosphate pathway the two reaction sequences are combined. Six hexose monophosphates are oxidised to give six pentose monophosphates and six molecules of CO$_2$. The six

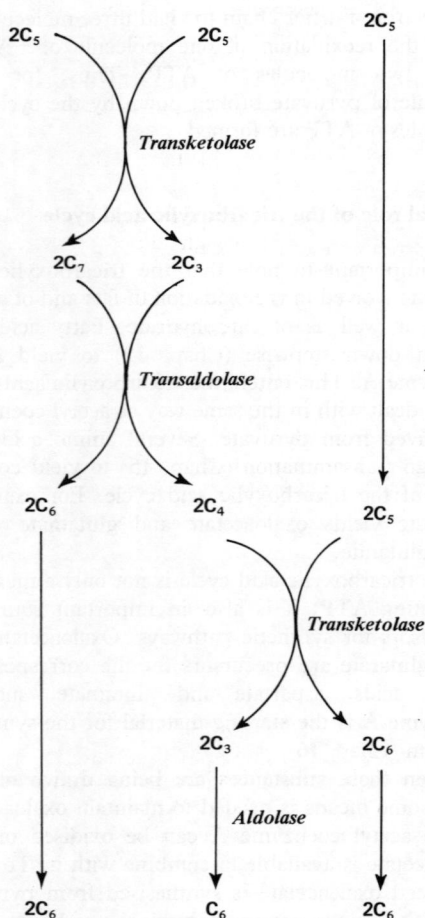

2C$_5$ 2C$_5$ 2C$_5$

Transketolase

2C$_7$ 2C$_3$

Transaldolase

2C$_6$ 2C$_4$ 2C$_5$

Transketolase

2C$_3$ 2C$_6$

Aldolase

2C$_6$ C$_6$ 2C$_6$

Fig. 9.16 The reactions in which six molecules of pentose phosphate (C$_5$) are converted to five of hexose phosphate (C$_6$)

Glucose

↓

Glucose 6-phosphate

5 hexose phosphates

6 hexose phosphates

Pi

12 $NADP^+$

12 $NADPH$ + 12 H^+

6 pentose phosphates

6 CO_2

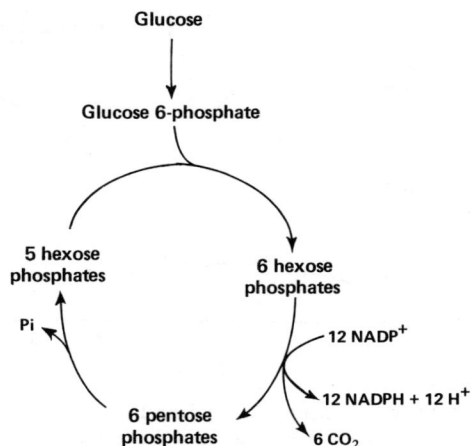

Fig. 9.17 A summary of the reactions of the pentose phosphate pathway

pentose monophosphates are then reconverted to five hexose monophosphates. The latter can then, in association with a fresh molecule of hexose monophosphate, go through the cycle again (fig. 9.17).

While the overall effect of the reaction is the complete oxidation of one glucose molecule, the pathway probably does not function primarily as a source of energy. In mammalian cells it is much more important as a means for converting NADP to NADPH since several important synthetic reactions require large amounts of NADPH. The pentose phosphate pathway is particularly active in the liver and the mammary glands where NADPH is needed for fatty acid synthesis. In the red cell the principal function of the pentose phosphate pathway is the generation of NADPH which reduces glutathione. In turn this prevents the inappropriate oxidation of –SH groups in haemoglobin and in membrane proteins to –S–S– bonds.

Glucose 6-phosphate dehydrogenase deficiency

This is by far the most common disorder of the pentose phosphate pathway. It is widespread in negro races; approximately 13 per cent of black men and 1 per cent of black women in the United States are affected. The disorder is X-linked and women are affected only if homozygous for the condition. Glucose 6-phosphate dehydrogenase (G6PD) de-

ficiency also occurs in the Mediterranean area and in some parts of South-east Asia.

G6PD deficiency does not cause any symptoms unless the patient takes one of a number of drugs, such as the anti-malarial primaquine, the anti-leprosy drug dapsone or the antibiotic nitrofurantoin. The administration of one of these causes rapid and often severe haemolysis and jaundice. In the Mediterranean form of the disease a wider range of drugs can precipitate haemolysis.

In G6PD deficiency the red cells are not completely lacking in the enzyme; younger cells contain some 6 per cent of the normal amount. This is sufficient to provide reducing power for everyday needs. When however a drug such as primaquine is administered the limited reducing power is overwhelmed, oxidation of intracellular proteins takes place and haemolysis, particularly of the older red cells, follows.

It is likely that G6PD deficiency is advantageous to individuals in malarious areas. The malarial parasite may depend for its optimal growth on the host's pentose phosphate pathway and reduced glutathione. If this is so G6PD deficiency represents in such areas a 'balanced polymorphism' in the same way as does sickle-cell trait (p. 208).

Metabolism of other carbohydrates

The polysaccharides in the diet are broken down within the lumen of the gut by digestive enzymes, notably pancreatic amylase, to disaccharides. These, together with the disaccharides ingested in the diet, are taken up by the microvilli (brush border) of the mucosal cells where they are broken down by disaccharidases, such as sucrase, lactase and maltase. In this way the carbohydrates of the diet yield glucose, galactose and fructose. Fructose and galactose may be oxidised or converted into glucose or glycogen.

Fructose metabolism

Several different metabolic pathways can be used for the metabolism of fructose but in the liver the principal pathway appears to be that shown in figure 9.18. Since glyceraldehyde 3-phosphate and dihydroxyacetone phosphate are intermediates in the glycolytic pathway they can be oxidised or reconstituted to yield glucose or glycogen.

101

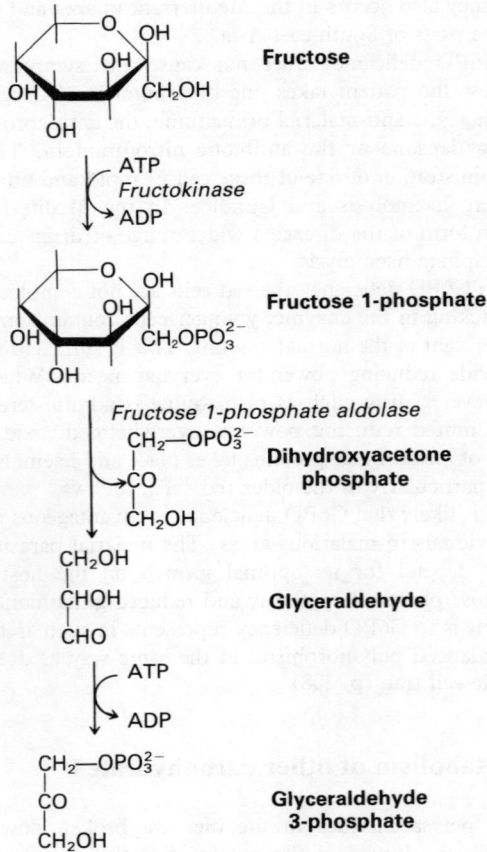

Fructose

$\downarrow$ ATP
Fructokinase
$\downarrow$ ADP

Fructose 1-phosphate

Fructose 1-phosphate aldolase

$CH_2-OPO_3^{2-}$
|
CO **Dihydroxyacetone**
| **phosphate**
CH_2OH

CH_2OH
|
$CHOH$ **Glyceraldehyde**
|
CHO

$\downarrow$ ATP
$\downarrow$ ADP

$CH_2-OPO_3^{2-}$
|
CO **Glyceraldehyde**
| **3-phosphate**
CH_2OH

Fig. 9.18 Pathway for the metabolism of fructose in the liver. The dihydroxyacetone phosphate and glyceraldehyde 3-phosphate can either enter the glycolytic pathway (fig. 9.1) or be combined to give fructose 1,6-bisphosphate and hence glucose or glycogen as described in figure 9.7

Fructose metabolism is abnormal in *fructose intolerance*. This inherited disorder is caused by lack of fructose 1-phosphate aldolase and leads to accumulation of fructose 1-phosphate in the liver. In turn this inhibits the formation of glucose from glycogen, from lactate and from amino acids. This results in hypoglycaemia. Most patients have only mild symptoms which coincide with the introduction of fructose into the diet. A few children have convulsions, an enlarged liver and jaundice. These symptoms diminish rapidly after the withdrawal of sucrose from the diet (fig. 9.19) and may be precipitated again by the administration of sucrose or fructose (fig. 9.20).

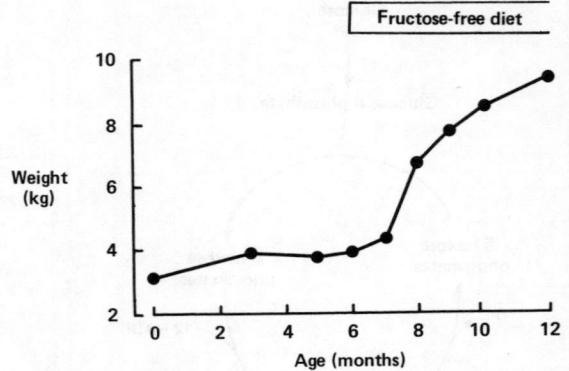

Fig. 9.19 Weight chart of a boy with fructose intolerance. He was breast-fed for two days and then given bottle-feeds with added sucrose. After seven weeks he began to have repeated vomiting and failed to gain weight. When a fructose-free diet was instituted his weight and general condition improved rapidly. (After Black, J. A. and Simpson, K. (1967) *British Medical Journal* **4**, 138)

Galactose metabolism

In the liver galactose is phosphorylated at position 1 by a specific enzyme galactokinase and the subsequent reactions are shown in figure 9.21.

Galactosaemia is an uncommon disorder caused by the hereditary lack of the enzyme uridyl transferase. Galactose and galactose 1-phosphate accumulate in

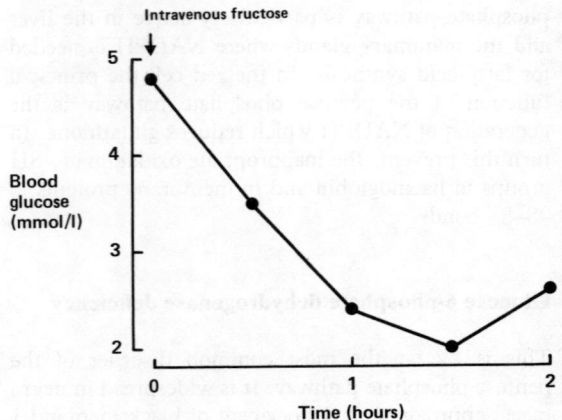

Fig. 9.20 Effect of a single intravenous injection of fructose on blood glucose in a child with fructose intolerance. (After Schwarz, V. (1978) *Clinical Companion to Biochemical Studies*. Reading: Freeman)

Galactose

$\downarrow$ ATP
Galactokinase
$\rightarrow$ ADP

Galactose 1-phosphate

$\downarrow$ UDP glucose
(*Uridyl transferase*) *Epimerase*
UDP galactose

Glucose 1-phosphate $\rightleftharpoons$ Glycogen

$\uparrow\downarrow$ *Phosphoglucomutase*

Glucose 6-phosphate $\rightleftharpoons$ Glucose

$\downarrow$

Glycolysis
and the tricarboxylic acid cycle
to CO_2 and H_2O

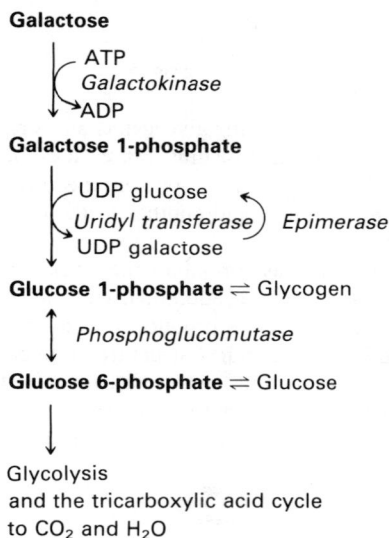

Fig. 9.21 The conversion of galactose to glycogen or glucose. UDP = uridine diphosphate

the blood and are excreted in the urine. Some infants with this condition first show symptoms within a day or two of birth after feeding has started. They begin to vomit, develop diarrhoea and rapidly become jaundiced. The liver enlarges and cataracts form in the lenses (fig. 9.22). It is thought that the damage to the lens is caused by the accumulation of galactose itself but the other abnormalities are caused by the accumulation of galactitol, a metabolite of galactose 1-phosphate. The condition can be controlled by the

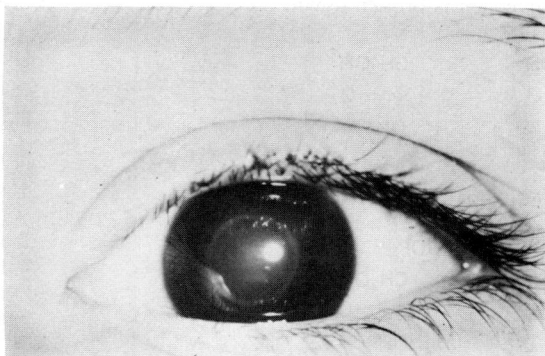

Fig. 9.22 Cataract in the lens of a child with galactosaemia. This patient had had a partial removal of the iris in an attempt to improve her vision. (Courtesy of G. N. Donnell)

Table 9.1 Uridyl transferase activity in lysed red cells in galactosaemia, in the parents of galactosaemic children (heterozygotes for the condition) and in normal subjects

	Uridyl transferase activity (mean ±1 standard deviation)
Child with galactosaemia	0
Parents of galactosaemic children	2.8 ± 0.7
Normal adults or children	5.9 ± 1.0

Results are expressed as μmol UDP glucose consumed per hour per gram of haemoglobin.
(Data of Donnell, G. N., Bretthauer, R. K. and Hansen, R. G. (1960) *Pediatrics* **25**, 572).

use of a lactose-free diet. Untreated, a child dies within a few months. Galactosaemia can be suspected by the demonstration of a reducing sugar in the urine; galactose can be identified by chromatography. The uridyl transferase activity can be measured in lysed red cells (Table 9.1). The early diagnosis of this condition is important in order to prevent permanent liver damage.

Glycogen

The metabolic pathways described earlier are all concerned with the metabolism of monosaccharides to yield energy. In addition a mechanism is needed for the storage of carbohydrate; if none existed, the tissues would be flooded with excess glucose after a meal and starved of it at other times. The accumulation of large quantities of such a small molecule would lead to large changes in the osmotic pressure within cells. It is, therefore, a great advantage to the organism to store its glucose as glycogen (p. 31) which has a high molecular weight and correspondingly low osmotic pressure.

Glycogen synthesis

Glycogen molecules are built up from units of uridine diphosphoglucose (UDP glucose) which are formed from glucose as shown in figure 9.23. The UDP-glucose can then transfer the glucose molecule to the end of an existing glycogen chain (fig. 9.24). Finally UDP can be reconverted to UTP by transfer of

103

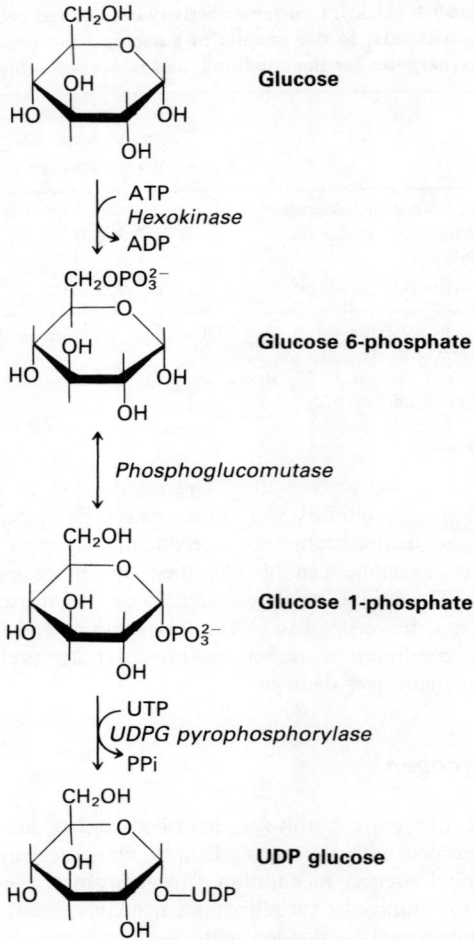

Glucose

$$\downarrow \begin{array}{l} \text{ATP} \\ \textit{Hexokinase} \\ \text{ADP} \end{array}$$

Glucose 6-phosphate

$$\updownarrow \textit{Phosphoglucomutase}$$

Glucose 1-phosphate

$$\downarrow \begin{array}{l} \text{UTP} \\ \textit{UDPG pyrophosphorylase} \\ \text{PPi} \end{array}$$

UDP glucose

Fig. 9.23 Formation of uridine diphosphoglucose

phosphate from ATP

$$\text{UDP} + \text{ATP} \rightleftharpoons \text{UTP} + \text{ADP}$$

In this way an existing glycogen chain is repeatedly extended by one glucose unit. For each extension two ATP molecules are used: one in the formation of glucose 6-phosphate and another in the regeneration of UTP.

Glycogen synthase promotes the formation only of the α-1,4 linkages; an additional enzyme, *branching enzyme*, is needed to produce the α-1,6 linkages required for the branched structure of glycogen (fig. 3.3, p. 32). Branching enzyme breaks off a terminal segment containing about seven glucose residues and reattaches them to another chain by an α-1,6 linkage (fig. 9.25).

Glycogen breakdown

The principal enzyme concerned in glycogen breakdown is *phosphorylase*, the activity of which is regulated as described in Chapter 12. Phosphorylase attacks the α-1,4 linkages at the ends of chains (fig. 9.26). It cannot attack the α-1,6 linkages nor the α-1,4 linkages near a branch point. The further degradation of the chains involves two further enzymes a *transferase* and a *debranching enzyme* as shown in figure 9.27.

While the synthesis of glycogen requires the expenditure of ATP, no ATP is generated when glycogen is broken down.

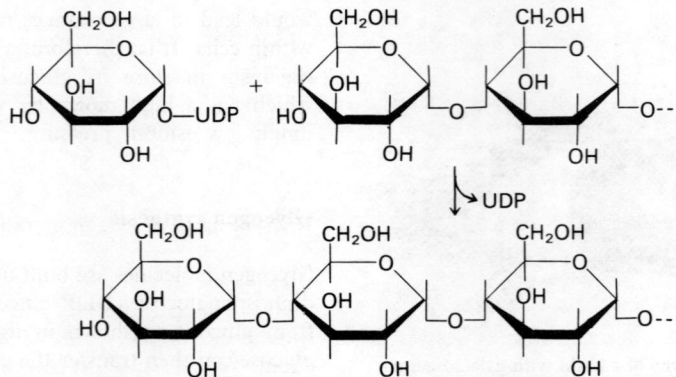

Fig. 9.24 Elongation of a glycogen chain in the presence of glycogen synthase. The control of glycogen synthase activity plays a major part in the control of glycogen metabolism (Chap. 12)

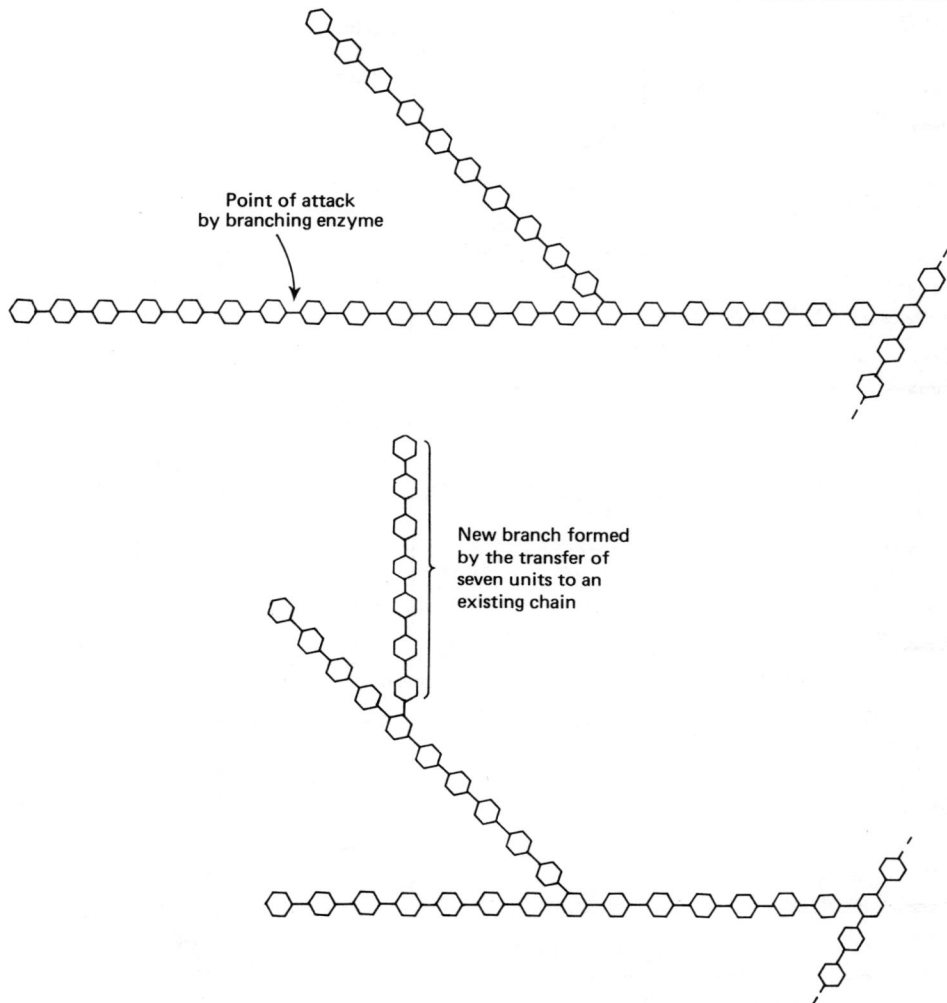

Point of attack by branching enzyme

New branch formed by the transfer of seven units to an existing chain

Fig. 9.25 Role of branching enzyme in transferring a segment of an existing glycogen chain to form a new branch

Lysosomal glycogen breakdown. A second pathway of glycogen breakdown is provided by the lysosomal enzyme, α-glucosidase. Although its glycogen-degrading activity is only a fraction of that of phosphorylase in muscle, its importance is indicated by the fact that its deficiency causes a severe glycogen storage disease and death within the first year of life (p. 108).

Regulation of glycogen metabolism

The synthesis and breakdown of glycogen, and the rate at which glycolysis and the tricarboxylic acid cycle proceed, are regulated to maintain the energy supply of cells and to prevent undue fluctuations in plasma glucose levels. The regulatory mechanisms involved are discussed in Chapter 12.

105

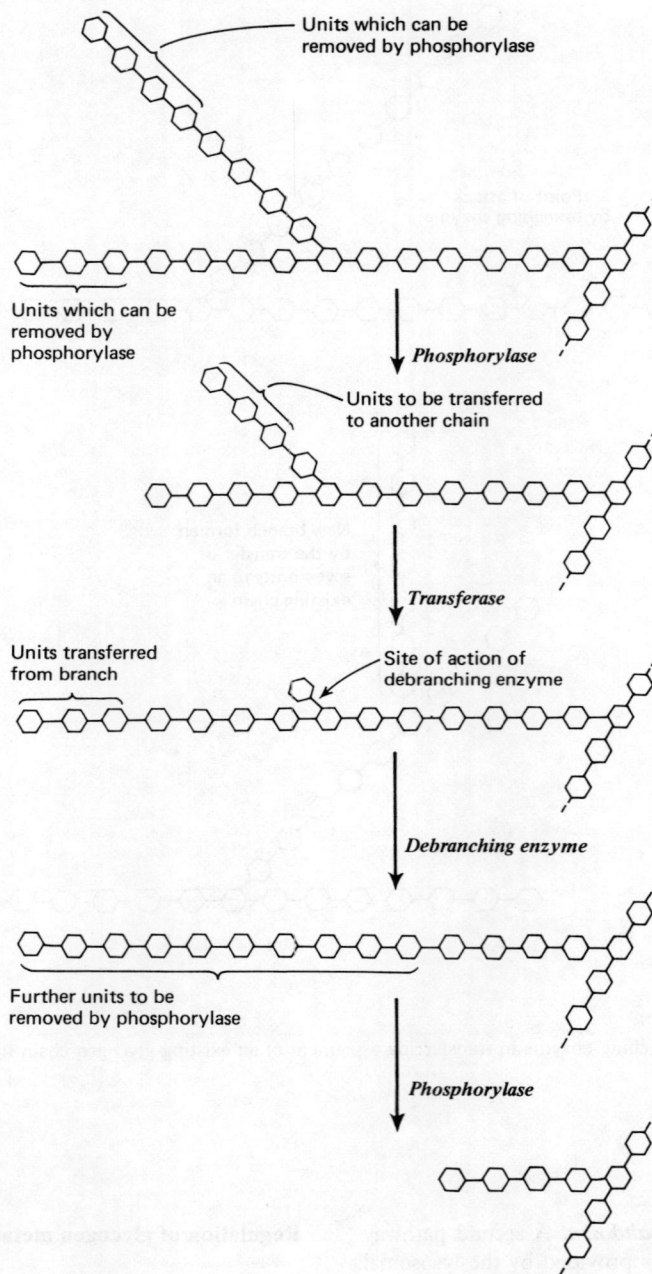

Fig. 9.26 Coordinated action of phosphorylase, transferase and debranching enzyme in removing glucose units from glycogen

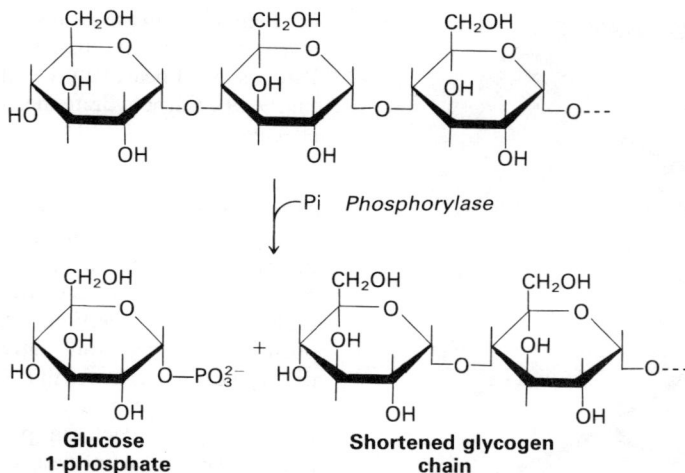

Fig. 9.27 Role of phosphorylase in releasing glucose units from glycogen chains

Disorders of glycogen metabolism

At least eight different enzyme defects which lead to the accumulation of glycogen in tissues have been described. While all are rare the combined incidence of the glycogen storage diseases is approximately 1 in 40 000 births.

The most common disorders are glucose 6-phosphatase deficiency, debranching enzyme deficiency and deficiency of lysosomal α-glucosidase. Deficiency of phosphorylase kinase (p. 132) also causes a glycogen storage disease.

Glucose 6-phosphatase deficiency. This is also known as Von Gierke's disease or type I glycogen storage disease. Since the production of glucose from glucose 6-phosphate is impaired, the patients develop severe hypoglycaemia if they fast. Glycogen is broken down to lactate and the liver is unable to reconstitute glucose from lactate. The liver becomes greatly enlarged (fig 9.28). About 50 per cent of patients die in childhood.

Debranching enzyme deficiency. This disorder (type III glycogen storage disease) causes the accumulation of a dextrin-like glycogen with very short outer

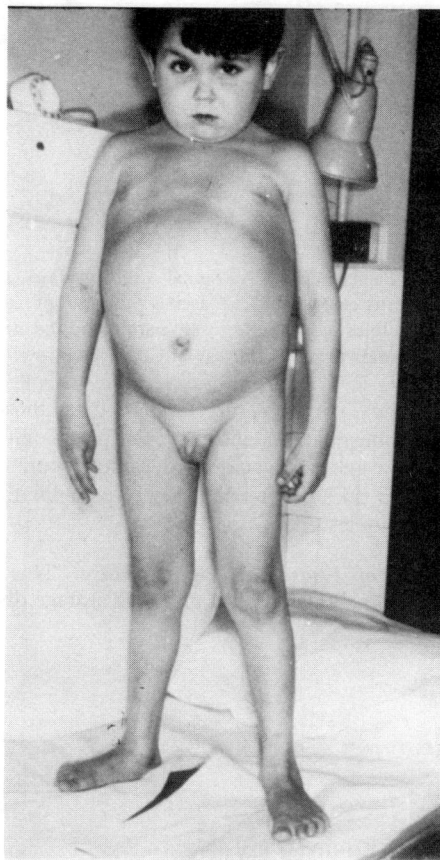

Fig. 9.28 Glucose 6-phosphatase deficiency to show the abdominal distension due to gross enlargement of the liver. (Courtesy of R. F. Mahler)

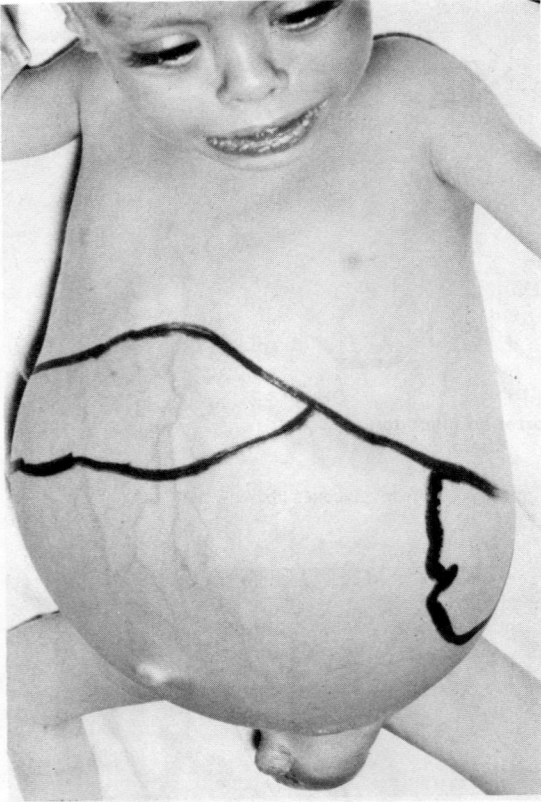

Fig. 9.29 Deficiency of lysosomal α-glucosidase. Gross abdominal distension due to ascites because of cardiac failure. The lines indicate the costal margin and the outlines of the liver and spleen. (Courtesy of G. M. Komrower)

chains. The symptoms are milder than those of glucose 6-phosphatase deficiency; the plasma glucose is not as low and lactic acidosis does not occur. Liver enlargement does occur but in many patients diminishes with adolescence.

Deficiency of lysosomal α-glucosidase. This disorder (also known as type II glycogen storage disease

or Pompe's disease) causes the accumulation of glycogen in many tissues including the heart. Patients die of heart failure within the first year of life (fig. 9.29). Carbohydrate metabolism generally is not affected.

Muscle phosphorylase deficiency. This is also known as type V glycogen storage disease or McArdle's disease. The glycogen accumulation is confined to muscle and patients are symptomless unless they undertake heavy exertion. The resting muscle can obtain its energy from fatty acid oxidation, but during heavy exertion, ATP levels fall severely and muscle cells are damaged. The patients develop pain in the muscles; myoglobin (p. 202) from the damaged muscle may appear in the urine.

Further reading

Beutler, E. (1978) *Haemolytic Anaemia in Disorders of Red Cell Metabolism.* New York: Plenum

Cohen, R. D. and Woods, H. F. (1976) *Clinical and Biochemical Aspects of Lactic Acidosis.* Oxford: Blackwell

Hers, H. G. and Hue, L. (1983) Gluconeogenesis and related aspects of glycolysis. *Annual Review of Biochemistry* **52**, 617–653

Howell, R. R. and Williams J. C. (1983) The glycogen storage diseases. In *Metabolic Basis of Inherited Disease*, 5th edn, eds Stanbury, J. B., Wyngaarden J. B., Fredrickson, D. S., Goldstein, J. L., Brown, M. S. pp. 141–166. New York: McGraw-Hill

Miwa, S. (1981) Pyruvate kinase deficiency and other enzymopathies of the Embden–Meyerhof pathway. *Clinics in Haematology* **10**, 57–80

Segal, S. (1983) Disorders of galactose metabolism. In *Metabolic Basis of Inherited Disease*, 5th edn, eds Stanbury, J. B., Wyngaarden, J. B., Fredrickson, D. S., Goldstein, J. L., Brown, M. S. pp. 167–191. New York: McGraw-Hill

RYT DGN DBW CRP

10 Biological oxidations

MAN'S need for oxygen is obvious. A man can survive for weeks without food, for days without water, but he dies within minutes if deprived of oxygen. The brain is particularly dependent on a continuous supply of oxygen. If the supply is interrupted as a result, for example, of cardiac arrest, consciousness is lost within seconds and permanent damage occurs within minutes.

The body needs oxygen principally to convert carbohydrate, fat and some protein to carbon dioxide and water and thus to obtain energy. Typically, for glucose

$$C_6H_{12}O_6 + 6O_2 \rightarrow 6CO_2 + 6H_2O$$

$$\Delta G^{0\prime} = -2900 \text{ kJ/mol}$$

and for a long-chain fatty acid

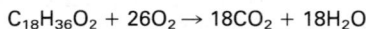

$$C_{18}H_{36}O_2 + 26O_2 \rightarrow 18CO_2 + 18H_2O$$

$$\Delta G^{0\prime} = -11\ 200 \text{ kJ/mol}$$

The enzymes of glycolysis are in the cytosol but the subsequent oxidation of pyruvate to CO_2 and H_2O takes place within the mitochondria, as does the complete oxidation of long-chain fatty acids (Chap. 11). The mitochondrion (fig. 10.1) has been termed the 'powerhouse of the cell', since it is able not only to oxidise the major energy-yielding substrates, but also to conserve a high proportion of the energy made available by the synthesis of ATP from ADP and phosphate.

In the cytosol the ratio of ATP to ADP is about 100 to 1, greater by a factor of 10^9 than that predicted from the equilibrium of the reaction

$$ADP + Pi \rightleftharpoons ATP + H_2O$$

The energy available from the hydrolysis of ATP is a function of the displacement of the reaction from equilibrium. For each ten-fold displacement 5.7 kJ/mol may be obtained; for the hydrolysis of ATP in the cytosol about 50 kJ/mol is therefore available to power muscle contraction, to pump ions across membranes or to drive reactions in otherwise unfavourable directions. The daily turnover of ATP in the body is about 150 mol, or about 75 kg.

In order to resynthesise the ATP used up in these metabolic processes, the mitochondrion requires an energy input of at least 50 kJ/mol. This energy comes from the oxidation of substrates, and the overall efficiency of *oxidative phosphorylation* (substrate oxidation and ATP synthesis) can be estimated. For example, the complete oxidation of 1 mol of 18-carbon fatty acid has a theoretical yield of 148 mol of ATP, equivalent to 7500 kJ/mol fatty acid, an overall thermodynamic efficiency of almost 70 per cent.

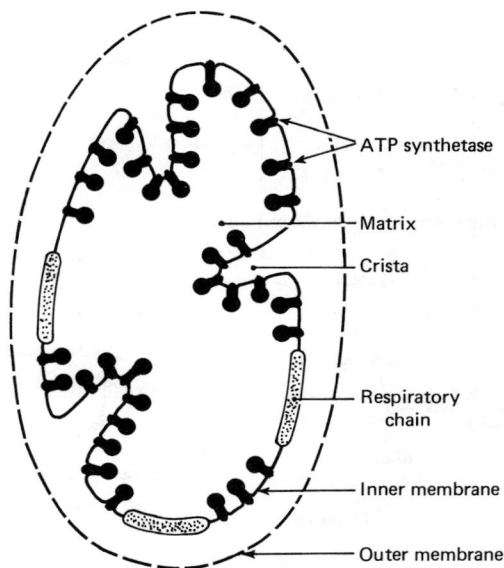

Fig. 10.1 Cross-section of a mitochondrion to show the position of the respiratory chains and the 'knobs' of ATP synthetase

Mitochondria and the synthesis of ATP

The pores of the outer membrane of the mitochondrion allow molecules with a mass less than 10 000 to penetrate. In contrast the inner membrane has a very limited permeability. The protein complexes responsible for the synthesis of ATP are buried in the inner membrane. The space enclosed by the inner membrane, the matrix, contains the enzymes of the tricarboxylic acid cycle and fatty acid β-oxidation (Chaps. 9 and 11).

The complete oxidation of pyruvate and long-chain fatty acids requires the enzymes of the tricarboxylic acid cycle and β-oxidation together with pyruvate dehydrogenase. These pathways contain a total of seven dehydrogenase enzymes, each of which extracts two electrons from its substrate leaving a more oxidised product (fig. 10.2). The electrons are initially transferred either to *nicotinamide adenine dinucleotide* (NAD^+) (p. 53), a low molecular weight soluble coenzyme which gains two electrons and one proton to become NADH (fig. 10.3), or to *flavin adenine dinucleotide* (FAD) (p. 54), a protein-bound coenzyme or prosthetic group which gains two electrons and two protons to become $FADH_2$ (fig.

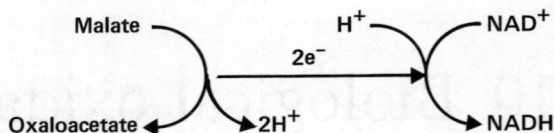

Fig. 10.3 Mechanism of reduction of NAD^+ coupled with the oxidation of malate

10.4). NADH shuttles electrons to the *respiratory chain* complex of proteins bound to the inner mitochondrial membrane. The FAD-linked dehydrogenases are themselves bound to the membrane and transfer electrons directly to the respiratory chain.

The potential of the electrons transported to the respiratory chain by NADH is about 300 mV more negative than that of a reference hydrogen electrode. The respiratory chain transfers these electrons to oxygen to form water:

$$2e^- + \tfrac{1}{2}O_2 + 2H^+ \rightarrow H_2O$$

The potential of the electrons in this reaction is about 800 mV positive to the hydrogen electrode. Since electrons are negatively charged this represents a *fall* in energy of 1.1 volts per electron, equivalent to about 200 kJ for the electrons transferred from 1 mol of NADH to form 1 mol of H_2O.

The most important function of the mitochondrion is to conserve as much as possible of this energy by the synthesis of ATP. Experiments with isolated mitochondria show that the transfer of two electrons from NADH to $\tfrac{1}{2}O_2$ can lead to the synthesis of up to three molecules of ATP. The efficiency of the overall process of 'oxidative phosphorylation' is therefore very high, since 3 mol ATP represent $3 \times 50 = 150$ kJ/mol of the 200 kJ/mol available.

The potential of electrons entering the respiratory chain through FAD is close to 0 mV, and the drop of two electrons through 800 mV to form H_2O can be conserved in the synthesis of two molecules of ATP.

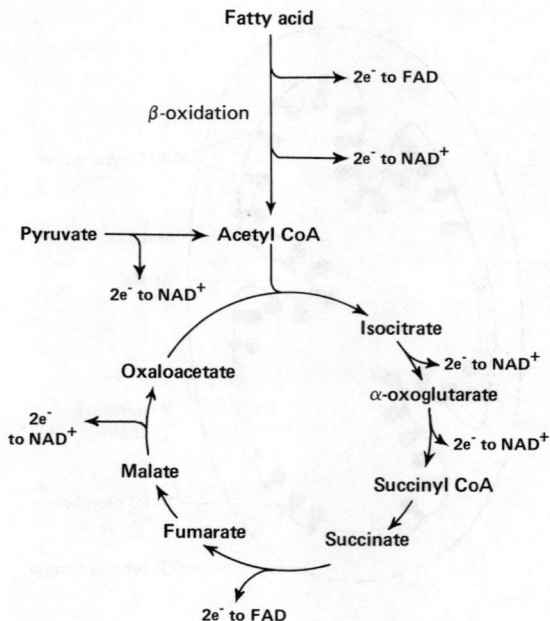

Fig. 10.2 Sources of electrons in the catabolism of pyruvate and of fatty acids by β-oxidation

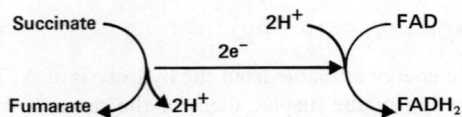

Fig. 10.4 Mechanism of reduction of FAD coupled with the oxidation of succinate by the enzyme succinate dehydrogenase

The *Chemiosmotic Theory* has explained successfully the mechanism of mitochondrial ATP synthesis. The respiratory chain functions as a complex proton pump in which the transfer of electrons is tightly coupled to the expulsion of protons from the matrix (fig. 10.5). This movement of positively charged protons generates both a pH gradient (the cytoplasm having a lower pH than the matrix) and an electrical potential, or *membrane potential*, across the membrane. The pH gradient and the membrane potential both try to drive the protons back into the matrix. The total driving force is a combination of these factors—the *proton electrochemical potential* or *protonmotive force* which in functioning mitochondria is about 200 mV. The protons cannot leak back into the matrix since the inner membrane is impermeable to protons, except at a specialised complex which is described next.

The mitochondrial inner membrane contains a second proton pump, the enzyme ATP synthetase, which is independent of the respiratory chain. On its own this enzyme would hydrolyse ATP in the matrix and pump protons out across the inner membrane in the same way as the respiratory chain. However the proton electrochemical potential generated by the respiratory chain forces this second pump to work in reverse, allowing protons to re-enter the matrix and cause the *synthesis* of ATP (fig. 10.5). In this way the

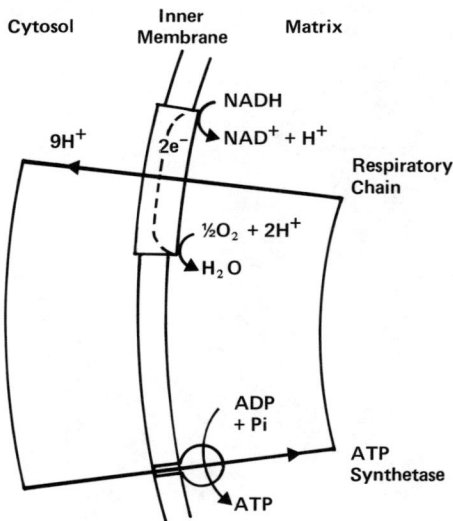

Fig. 10.6 Similarities between the proton circuit and a simple electrical circuit

energy in the proton electrochemical potential is largely conserved in the synthesis of ATP.

The combined action of the respiratory chain and ATP synthetase is to set up a circuit of protons across the inner membrane. This circuit is in many ways analogous to an electrical circuit in which the respiratory chain replaces the battery, the ATP synthetase replaces a machine doing useful work, the proton electrochemical potential replaces the electrical potential difference across the battery and a current of protons replaces the current of electrons (fig. 10.6).

The nature of respiratory control

The respiration of a tissue adjusts automatically to the energy requirement of its cells. For example in an exercising muscle respiration can increase ten-fold above resting levels. The control of mitochondrial respiration allows mitochondrial ATP synthesis and cellular ATP demand to be precisely matched. Respiratory control can be understood by reference to the proton circuit (fig. 10.5). Since proton re-entry into the matrix depends on a tightly coupled synthesis of ATP, it follows that when cellular demand for ATP ceases mitochondrial ATP synthesis stops, protons are no longer able to enter the matrix and the proton circuit becomes open circuited as though a switch had been opened in the analogous

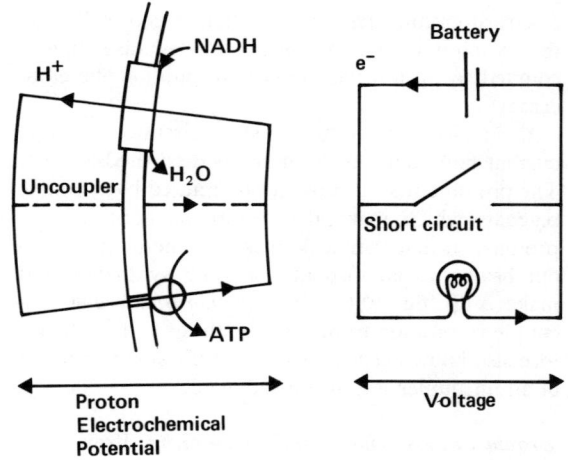

Fig. 10.5 The proton circuit which couples the respiratory chains with the synthesis of ATP

111

electrical circuit. This in turn stops electron flow in the respiratory chain, since this is also tightly coupled to proton movements; respiration therefore ceases.

If an electrical circuit is short-circuited, a large current flows and much energy is dissipated as heat. The proton circuit can be short-circuited by synthetic reagents which make the membrane permeable to protons, so that they leak back into the matrix without having to go through the ATP synthetase and make ATP (fig. 10.6). These *proton translocators* uncouple respiration from ATP synthesis and are therefore also known as *uncouplers*. A well-known example of an uncoupler is 2,4-dinitrophenol.

Brown adipose tissue and uncoupling. Brown adipose tissue (fig. 10.7) is specialised for the generation of heat and is particularly important in the newborn. The numerous mitochondria in the cells of this tissue are equipped with a unique protein in their inner membranes which provide a pathway for the leakage of protons back into the mitochondrial matrix. This uncoupling enables the rapid oxidation of the fatty acids derived from the triglyceride droplets without being limited by the cell's need for ATP. Since brown adipose tissue can oxidise triglyceride at a high rate, it has been suggested that it is important in the regulation of body weight. Some cases of obesity may be associated with deficiency of brown adipose tissue.

The respiratory chain

The respiratory chain is a complex of at least thirty proteins which are buried to various extents in the mitochondrial inner membrane. By gentle detergent action the chain can be fractionated into four separate complexes (fig. 10.8). With the exception of complex II which is, in part, the tricarboxylic acid cycle enzyme succinate dehydrogenase, each complex acts as an autonomous proton pump.

Complex I transfers electrons from NADH (at a potential of about -300 mV) to ubiquinone which has a potential of about 0 mV. The transfer of two electrons through complex I is coupled to the

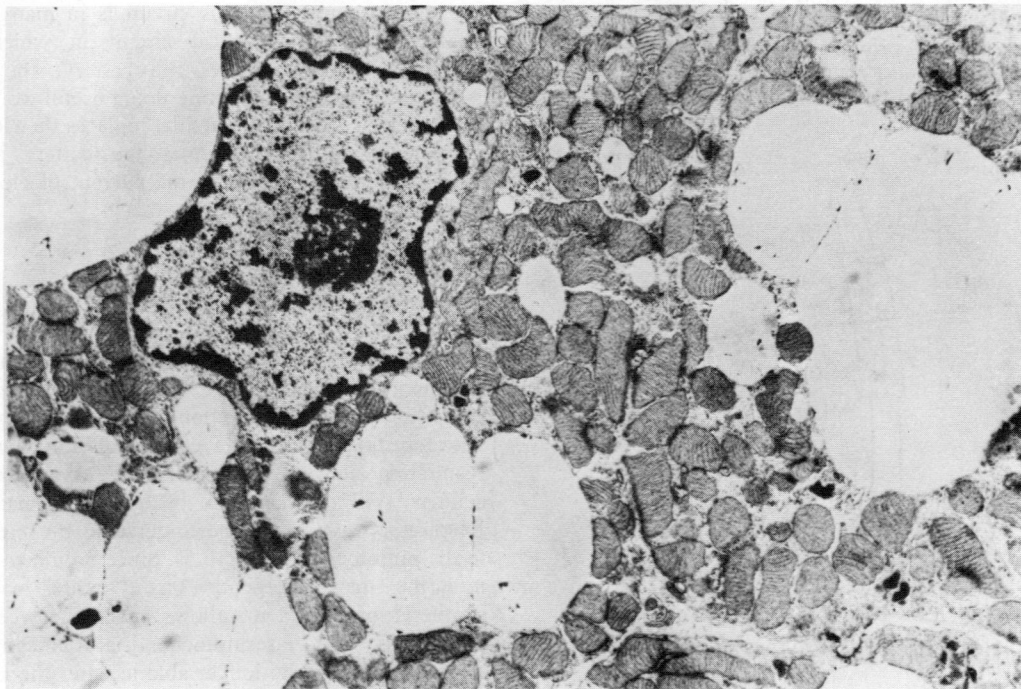

Fig. 10.7 Electron micrograph of brown adipose tissue to show the lipid droplets and the numerous mitochondria

112

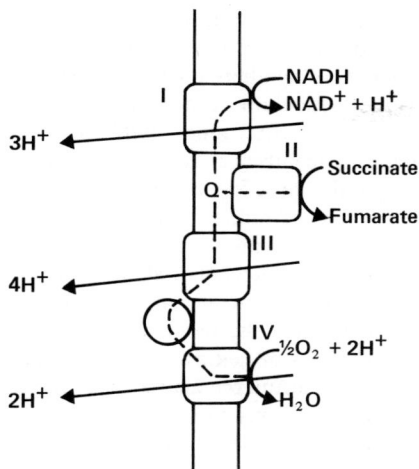

Fig. 10.8 The four complexes into which the respiratory chain can be fractionated

pumping of two or three protons across the membrane. The energy transfer is thus highly efficient; two electrons falling through 300 mV are able to pump as many as three protons up a potential of 200 mV.

Ubiquinone accepts two electrons and two protons to become ubiquinol (fig. 10.9). Ubiquinone, which can also accept electrons from complex II, transfers electrons to complex III. The passage of two electrons through complex III is coupled to the pumping of four protons across the membrane. The electrons then reduce cytochrome c.

Cytochrome c is a protein of 12 000 molecular weight which has a haem prosthetic group similar to that of haemoglobin. The haem group of cytochrome c does not bind oxygen; instead the iron atom at the centre of the haem can alternate between the ferrous and ferric states. Thus each cytochrome c can be

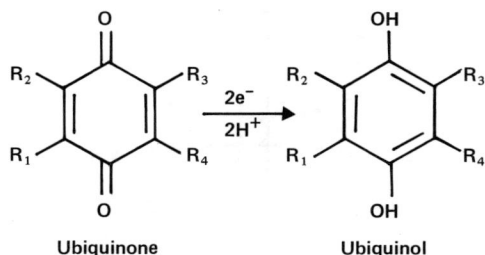

Fig. 10.9 Ubiquinone (UQ) and ubiquinol (UQH_2)

reduced from the ferric state to ferrous by the addition of one electron from complex III.

In the final stage of the respiratory chain electrons are transferred from ferrous cytochrome c to O_2, mediated by complex IV, also called cytochrome c oxidase. Four electrons are needed to reduce one O_2 to $2H_2O$

$$4e^- + O_2 + 4H^+ \rightarrow 2H_2O$$

Complex IV also pumps protons across the membrane. Two protons appear in the medium for every two electrons transferred to oxygen.

The molecular mechanism by which the complexes couple electron flow to proton pumping is not yet known. Each complex has a number of prosthetic groups which can undergo reversible oxidation and reduction and are therefore involved in the pathway of electron transfer. The major components of complexes III and IV are cytochromes. Complex III contains one or more cytochromes b plus a cytochrome c_1, while complex IV contains cytochromes a and a_3. Cytochrome a_3 can bind O_2 and is the terminal oxidase of the respiratory chain. It can also bind CO, CN^- and N_3^-; this accounts for the extreme toxicity of such compounds.

The nature of the ATP synthetase

If mitochondria are sectioned for electron microscopy and negatively stained with phosphotungstate, a succession of knobs, each about 9 nm in diameter, can be seen on the matrix face of the inner membrane (fig. 10.1). These are the catalytic units of ATP synthetase (the F_1-ATPase). If mitochondria are sonicated (subjected to high amplitude ultra-sonic sound) they fragment to form small inverted vesicles with the knobs on the outside. The F_1-ATPase can now be dissociated from the membrane by mild treatments: it is found to catalyse the hydrolysis of ATP to ADP and Pi, but not the reverse reaction, since in a soluble system there is no way to apply the energy from a trans-membrane proton electrochemical potential.

The vesicles which have lost their F_1-ATPase knobs behave as if they are leaky to protons. This is because a second component of the complete ATP synthetase has been left behind in the membrane. This fraction, called F_0, acts as a proton channel and normally functions to deliver protons across the inner

113

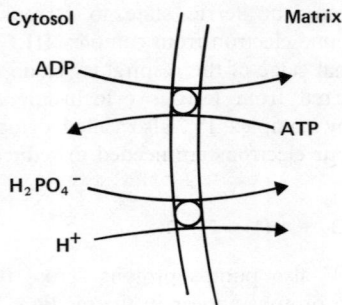

Fig. 10.10 The function of adenine nucleotide translocator in the transfer of ADP into and ATP out of the mitochondrion. The function of the phosphate carrier in transporting Pi into the mitochondrion

membrane to the F_1-ATPase, which fits tightly onto the matrix side of the channel. The F_0-channel can be blocked by the antibiotic *oligomycin*.

ATP synthetase produces ATP in the mitochondrial matrix, even though the cytosol has by far the greatest requirement for ATP. The highly charged nature of the adenine nucleotides prevents their free passage across the inner membrane: instead a specific transport protein is required. The adenine nucleotide translocator catalyses the 1:1 exchange of extra-mitochondrial ADP for mitochondrial ATP (fig. 10.10) so that the total adenine nucleotide content of the matrix does not change. The continuous need for Pi in the matrix is met by a second transport protein, the phosphate carrier.

Two naturally occurring inhibitors of the adenine nucleotide translocator were characterised as a result of the biochemical investigation of mysterious cases of poisoning. The Mediterranean thistle, *Atractylis gummifera* has poisonous leaves from which the alkaloid *atractylate* was purified, and found to be a potent inhibitor of the translocator. The second inhibitor *bongkrekate* was discovered as a toxin in contaminated coconut products.

Metabolite transport across the mitochondrial inner membrane

The major metabolite traffic across the inner membrane is the transport of Pi and adenine nucleotides to allow the continuous synthesis of ATP. However it is also necessary for metabolites to cross from the cytosol to the matrix for oxidation, and to cross in the

reverse direction for a number of synthetic processes.

Fatty acids are activated to acyl CoA by the enzyme acyl CoA synthetase which is located on the endoplasmic reticulum and outer mitochondrial membrane. Acyl CoA cannot itself cross the inner mitochondrial membrane to be oxidised by β-oxidation in the matrix. Instead acyl CoA is first converted to acyl carnitine by the enzyme acyl CoA: carnitine acyl transferase. This enzyme is present both on the outer mitochondrial membrane and on the inside of the inner membrane. The acyl carnitine is transported across the inner membrane and then the inner acyl CoA: carnitine acyl transferase regenerates acyl CoA in the matrix (fig. 11.5, p. 121).

The enzymes of glycolysis are in the cytosol and, in order that glycolysis should be linked to the tricarboxylic acid cycle for the complete oxidation of glucose, pyruvate must cross the inner membrane. Since there is a high membrane potential across the inner membrane, negative in the matrix, a negatively charged particle such as the pyruvate anion would be repelled from the matrix. This problem is overcome by a specific transport protein, the pyruvate carrier which allows pyruvate to cross as the electrically neutral pyruvic acid.

Dicarboxylic acids such as malate and tricarboxylic acids such as citrate cross the mitochondrial inner membrane in a cascade system involving the phosphate carrier (fig. 10.11).

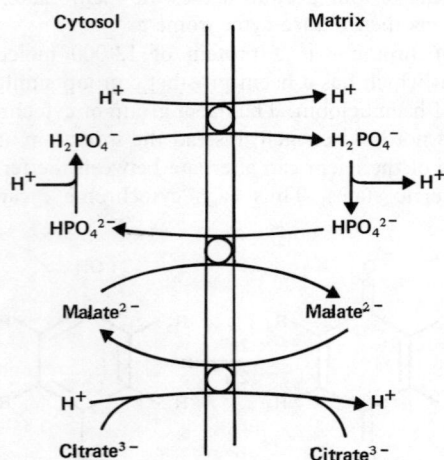

Fig. 10.11 Mechanism for the transport of malate and citrate into and out of the mitochondrion

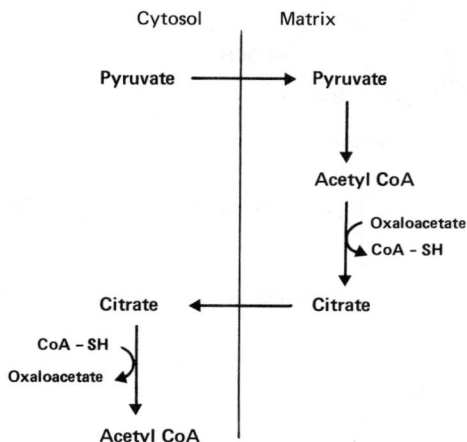

Fig. 10.12 Mechanism of transport of acetyl groups out of the mitochondrion

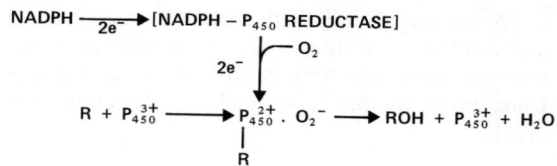

Fig. 10.13 Role of cytochrome P_{450} on the hydroxylation of substances such as drugs or steroids

These transport proteins are not only involved in transporting substrates into the matrix for subsequent oxidation, they may also be used to transport substrates out of the mitochondria to take part in synthetic pathways in the cytosol. For example the acetyl groups required for fatty acid synthesis come originally from the matrix and are transported out of the mitochondrion as citrate (fig. 10.12).

Electron-transfer pathways associated with the endoplasmic reticulum

Endoplasmic reticulum possesses short electron-transfer pathways involved in the hydroxylation of drugs and steroids, in the production of ions and radicals toxic to microorganisms and in the desaturation of long-chain fatty acids. These pathways do not transport protons and are not associated with ATP synthesis.

Cytochrome P_{450}. The pathway for drug and steroid hydroxylation (fig. 10.13) extracts electrons from NADPH in the cytosol and transfers them to the enzyme NADPH-cytochrome P_{450} reductase. Cytochrome P_{450} is unique to the endoplasmic reticulum. Its name is derived from the absorption band at 450 nm which appears when CO binds to the cytochrome. Cytochrome P_{450} forms a complex with the drug or steroid to be hydroxylated (R in fig. 10.13), molecular oxygen binds to the cytochrome,

and two electrons are received from the reductase. In the resulting hydroxylation one oxygen atom goes to hydroxylate the substrate and one to form H_2O. This is an example of a mixed-function oxidase.

Cytochrome P_{450} plays an essential role in the hydroxylation steps involved in synthesis of steroid hormones (p. 263), calcitriol (p. 245) and bile salts (p. 129). In addition cytochrome P_{450} is important for detoxication of foreign substances (xenobiotics) such as drugs. Many drugs and many other substances which contaminate the environment and food, such as pesticides and food additives, are lipid-soluble. To render them water-soluble, and therefore easier to excrete, the first stage may be a hydroxylation involving cytochrome P_{450}. This is often followed by the coupling of the hydroxy group with a polar molecule such as glutathione, glucuronate or sulphate (fig. 10.14).

The existence of this pathway has to be borne in mind in the design of drugs, since some products might be carcinogenic. The synthesis of cytochrome P_{450} may be increased greatly by chronic administration of a drug. As a result a steadily increasing amount of a drug may be needed to maintain a constant plasma concentration.

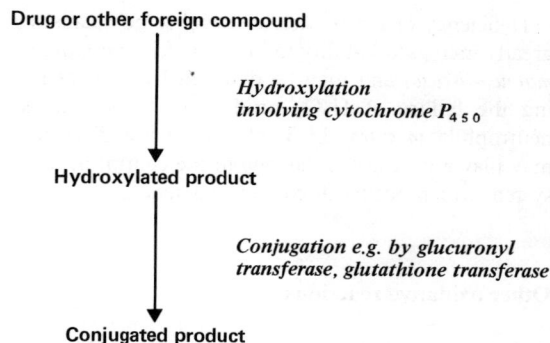

Fig. 10.14 The two phases in the metabolism of many lipid-soluble drugs and other foreign chemicals

The fatty acid desaturation pathway uses NADH and a further cytochrome, b_5. This pathway is not however completely separate from the drug hydroxylation pathway since a cross-flow of electrons can occur.

Cytochrome b and neutrophil function. The oxygen consumption of neutrophil leucocytes increases 10 to 20 fold during phagocytosis of microorganisms. This increase, which is not mitochondrial, is essential for the killing of the organisms. It is now clear that the pentose phosphate pathway, NADP and an electron transport chain including a cytochrome of the *b* group are involved and lead to the production of toxic radicals and ions such as superoxide (O_2^-)

$$O_2 + e^- = O_2^-$$

Superoxide in turn may react with hydrogen ions to give oxygen and hydrogen peroxide under the influence of the copper-containing enzyme superoxide dismutase.

$$2O_2^- + 2H^+ = H_2O_2 + O_2$$

The reduction products of oxygen found in stimulated neutrophils include hydroxyl (OH^-), the hydroxyl radical (OH) and the hydroperoxy radical (HO_2^-). Hydrogen peroxide and all these unstable radicals probably operate by oxidising or reducing certain molecules of the microorganism, notably the lipids of its cell wall. For example

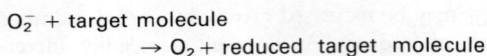

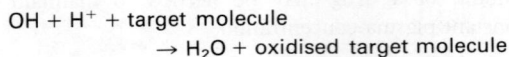

$$O_2^- + \text{target molecule}$$
$$\rightarrow O_2 + \text{reduced target molecule}$$

$$OH + H^+ + \text{target molecule}$$
$$\rightarrow H_2O + \text{oxidised target molecule}$$

Deficiency of the particular cytochrome *b* causes a greatly increased liability to infection (*chronic granulomatous disease*) and may be detected by demonstrating the failure of H_2O_2 production by stimulated neutrophils *in vitro*. Lack of superoxide dismutase may play a part in the damage to the central nervous system which occurs in copper deficiency.

Other oxidative reactions

Although the electron transport system is the main mechanism by which oxidations are catalysed in the living organism, it is not the only one. Several

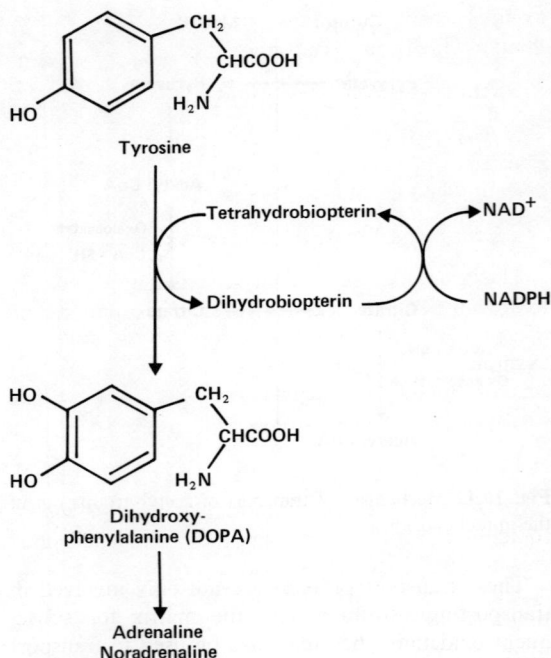

Fig. 10.15 The mode of action of tyrosine hydroxylase in the oxidation of tyrosine and its role in the synthesis of the catecholamines

enzymes which catalyse simple oxidations have been isolated. One of these is tyrosine hydroxylase, responsible for the oxidation of tyrosine which is the first step in the production of catecholamines (fig. 10.15).

The term *oxidase* is also applied to certain flavoproteins capable of catalysing oxidations in which molecular oxygen acts as a hydrogen acceptor. Such flavoproteins are sometimes called aerobic dehydrogenases. A well-known example is D-*amino acid oxidase* which catalyses the oxidation of D-amino acids:

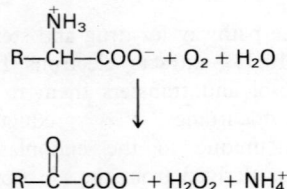

$$R{-}\overset{\overset{+}{N}H_3}{\underset{|}{C}H}{-}COO^- + O_2 + H_2O$$

$$\downarrow$$

$$R{-}\overset{O}{\overset{\|}{C}}{-}COO^- + H_2O_2 + NH_4^+$$

The product of the reaction is not water but hydrogen peroxide. The significance of such reactions is uncertain, except in the case of *xanthine oxidase* (Chap. 13), a flavoprotein containing molybdenum which catalyses the reaction

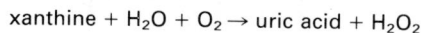

$$xanthine + H_2O + O_2 \rightarrow uric\ acid + H_2O_2$$

The flavoprotein *glucose oxidase* (from the mould *Penicillium notatum*) is used as a specific means of detecting and estimating glucose:

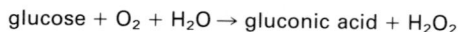

$$glucose + O_2 + H_2O \rightarrow gluconic\ acid + H_2O_2$$

The *hydroperoxidases* catalyse oxidations in which hydrogen peroxide acts as hydrogen acceptor and in the process is reduced to water. The best-known examples are the *peroxidase* of horse-radish and the *catalase* found in mammalian tissues. The latter catalyses the reaction:

$$2H_2O_2 \rightarrow 2H_2O + O_2$$

Catalase is a protein of molecular weight 248 000 containing four iron atoms per molecule. The iron is incorporated in the protein as a ferriprotoporphyrin (Chap. 16) prosthetic group. Catalase from liver tissue and erythrocytes has been purified and crystallised. It is one of the most powerful enzymes known; one molecule can decompose 2 640 000 molecules of hydrogen peroxide per minute at 0 °C.

These enzymes probably serve to dispose of hydrogen peroxide produced in the cell by the flavoprotein oxidases, or arising as a result of the metabolism of drugs and other foreign chemicals. In the red cell the most important enzyme with this function is *glutathione peroxidase* but catalase is also present.

Further reading

Amzel, L. M. and Pedersen, P. L. (1983) Proton ATPases: structure and function. *Annual Review of Biochemistry* **52**, 801–824

Nicholls, D. G. (1982) *Bioenergetics*. London: Academic Press

Ortiz de Montellano, P. R. and Correia, M. A. (1983) Suicidal destruction of cytochrome P_{450} during oxidative drug metabolism. *Annual Review of Pharmacology and Toxicology* **23**, 481–503

DGN

11 Lipid metabolism

THE triglyceride of adipose tissue is the major energy store in the body (Chap. 12); the other lipids, cholesterol and phospholipids, have an essential role in all biological membranes. This chapter describes the pathways involved in the synthesis and catabolism of these substances. Since lipids are insoluble in water they are transported between tissues as *lipoproteins*, complexes of proteins, cholesterol, triglycerides and phospholipids. Free fatty acids are transported in plasma bound to albumin.

The lipoproteins

Lipoproteins have a structure like micelles (p. 3). The outer layer consists of proteins and polar lipids such as cholesterol and the interior contains non-polar lipids such as triglyceride and esterified cholesterol. Details of the principal lipoproteins are given in Table 11.1. They can be classified in a number of ways but the most generally accepted nomenclature is based on their density as estimated by the preparative ultracentrifuge. The proteins associated with lipoproteins are known as *apoproteins*.

Chylomicrons

Chylomicrons are formed in the intestine from dietary lipids. They have a half-life of about 10 minutes and are the principal form in which absorbed lipid is carried in plasma. These are the largest lipoprotein particles with a 'molecular mass' of up to 10^{10} and a diameter of up to 100 nm. Although chylomicrons only contain small amounts of protein, largely apoprotein B (β-apoprotein), it is essential for their formation. Chylomicron formation does not occur in *abetalipoproteinaemia*, a rare inherited disorder in which apoprotein B is deficient (p. 120).

Chylomicrons being large particles render the plasma turbid or milky if present in appreciable amounts; when plasma containing chylomicrons is allowed to stand overnight they rise to the top to form a cream-like layer. Chylomicrons are formed in the mucosal cells of the intestine by the esterification of dietary fatty acids; their fatty acid composition reflects that of the diet. Chylomicrons are not normally present in blood collected from subjects who have fasted for more than 14 hours.

Table 11.1   The lipoproteins of human plasma

	Chylomicrons	VLDL	LDL	HDL
Physical features				
Density (g/ml)	<0.95	0.95–1.01	1.01–1.06	1.06–1.21
Electrophoretic mobility	Remain at origin	Pre-β	β	α
Particle size (nm)	70–100	30–80	20–30	7–10
Average chemical composition (per cent)				
Triglyceride	90	65	10	2
Cholesterol	5	13	43	18
Phospholipid	4	12	22	30
Protein	1	10	25	50
Principal apoproteins*	B	B, C, E	B	A

* Apoprotein A has two components AI and AII. Apoprotein C has three components CI, CII and CIII.

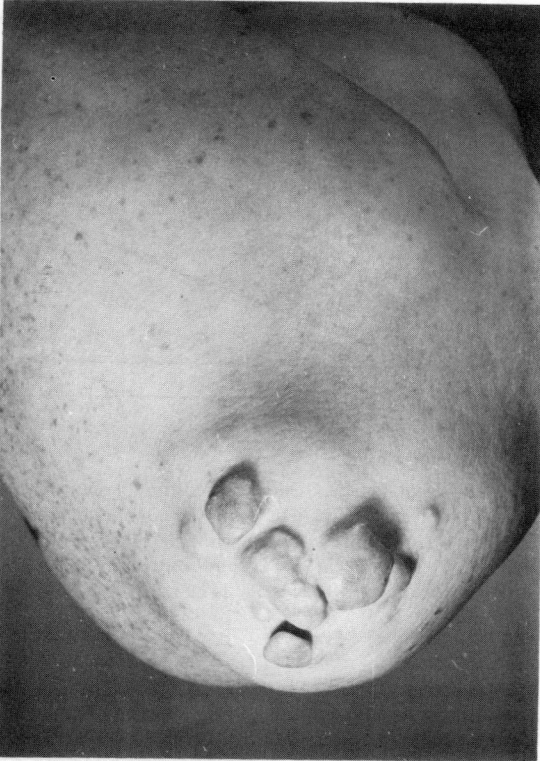

Fig. 11.1 Lipid deposits (xanthomata) on the elbow of a patient with hyperlipoproteinaemia. (Courtesy of E. J. Raffle)

Chylomicrons are removed from the circulation by lipoprotein lipases (p. 137) mainly in the liver but also in muscle and adipose tissue. Persistence of chylomicrons in the circulation is characteristic of a group of rare inherited disorders known as *exogenous hypertriglyceridaemias*. These result from deficiency of lipoprotein lipase or, in one recently described variant, lack of the particular apoprotein (C-II) required for the activation of lipoprotein lipase. These disorders are characterised by the formation of fatty deposits in the skin (fig. 11.1), abdominal pain and enlargement of the liver.

Very low density lipoproteins (VLDL)

These lipoproteins are formed between meals in the liver and, to a small extent, in the intestinal mucosa.

They are the main form in which triglyceride is transported from the liver to other tissues especially adipose tissue and muscle (fig. 12.10, p. 137). There, like chylomicrons, they release their lipid as a result of the action of lipoprotein lipases. The fatty acids enter the cells and the remaining cholesterol, phospholipid and protein reform into intermediate density lipoproteins (IDL).

VLDL have a half-life in the plasma of about 12 hours. When present in excess the plasma is opalescent. High plasma levels of VLDL are found in an uncommon inherited disorder *familial endogenous hypertriglyceridaemia* but also, more commonly, in a number of diseases such as diabetes mellitus (p. 144), thyroid insufficiency and also after an excessive intake of alcohol. In all these conditions patients have an increased liability to vascular disease, particularly of the coronary arteries, and to xanthomata (fig. 11.1) and abdominal pain.

Low density lipoproteins (LDL)

The LDL fraction of the lipoproteins is principally concerned with the transport of cholesterol; some 70 per cent of the plasma cholesterol is in this form. Two main fractions of LDL have been identified: the intermediate density lipoproteins (IDL) and a lower density LDL_2 fraction. IDL is formed from the remnants of VLDL and chylomicrons after the triglyceride has been removed (fig. 12.10, p. 137). In the liver IDL is converted to LDL_2 which is responsible for the transport of cholesterol from the liver to tissues which require it for the production of cell components.

A receptor is involved in the uptake of LDL by peripheral tissues. After entering the cell by endocytosis LDL is broken down by lysosomal enzymes to give free cholesterol and amino acids.

Disorders. High plasma levels of LDL (particularly LDL_2) are associated with a greatly increased risk of coronary artery disease. An approximate guide to LDL levels is given by the plasma cholesterol and diets low in cholesterol are often prescribed for patients with high values (fig. 11.2). It remains controversial whether such diets starting late in life improve the prognosis.

One cause of high LDL levels is *familial hypercholesterolaemia* which is caused by a defect in the receptor for LDL. The plasma cholesterol is greatly

119

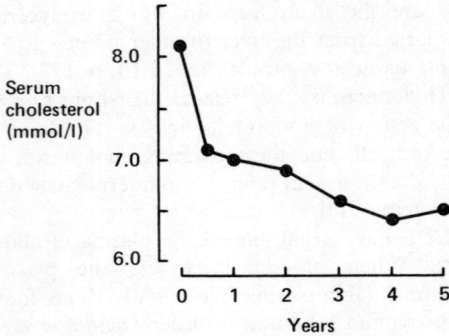

Fig. 11.2 Mean plasma cholesterol values in a group of 52 patients, with high initial levels, while taking a diet with a restricted fat and cholesterol content. (After Reid, V. *et al.* (1981) In *Recent Advances in Clinical Nutrition I*, ed. Howard, A. and Baird, I. M., p. 259. London: Libbey)

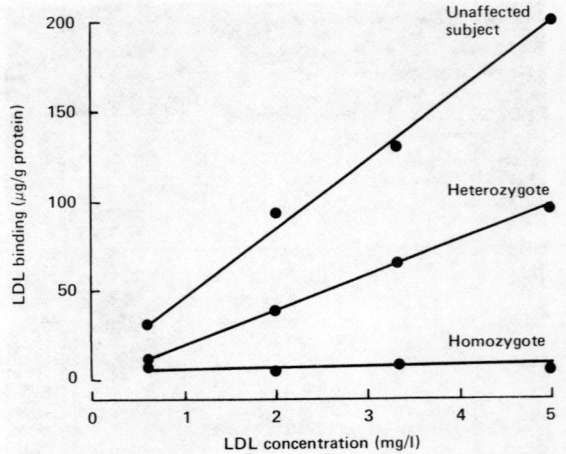

Fig. 11.4 Binding of LDL to cultured fibroblasts from one normal subject, one patient with (heterozygous) familial hypercholesterolaemia and one with the homozygous disorder. (After Goldstein, J. L. and Burns, M. S. (1975) *American Journal of Medicine* **58**, 147)

raised, xanthomas occur and the incidence of coronary artery disease is high. In this disorder, the most common familial hyperlipoproteinaemia, has an incidence of about 1 in 500; among Afrikaners in South Africa the incidence is 1 in 100. It is inherited as an autosomal dominant. Patients homozygous for the disorder have even higher plasma LDL and cholesterol levels; rapidly progressive vascular disease, involving particularly the coronary vessels, causes symptoms in childhood (fig. 11.3). Death often occurs before the age of 30.

The defect in the binding of LDL to cells is illustrated in fig. 11.4. One consequence of the defective uptake of LDL by cells is that cholesterol synthesis by the cells, normally inhibited by its cholesterol content (p. 127), continues at a high rate particularly in the homozygous disease.

Lack of apoprotein B (*abetalipoproteinaemia*) is a rare inherited disorder in which LDL is absent from the plasma. Fat is poorly absorbed because chylomicrons cannot be formed; red cells show bizarre shapes.

High density lipoproteins (HDL)

HDL contains 50 per cent protein, largely apoproteins AI and AII, together with phospholipids and cholesterol. HDL are formed in the intestinal mucosa and the liver by the combination of their lipids and proteins in the Golgi complex. The cholesterol is esterified by the enzyme *lecithin-cholesterol acyl transferase* (LCAT). The importance of this enzyme is illustrated by the rare disorder familial LCAT deficiency in which the plasma level of free cholesterol is greatly raised and cholesterol is deposited in tissues causing corneal opacities, anaemia and renal damage.

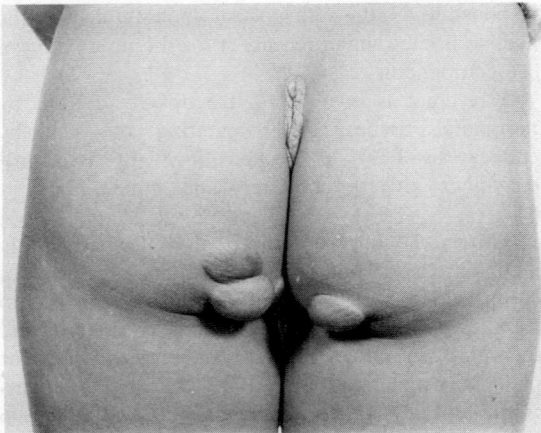

Fig. 11.3 Extensive xanthomata on the buttocks of a twelve-year-old girl with homozygous familial hypercholesterolaemia. (Courtesy of H. Seftel)

120

The function of HDL is thought to be the transport of cholesterol and phospholipids from tissues, to the liver for excretion in the bile (p. 129). High levels of HDL (particularly the HDL_2 fraction) are associated with a *diminished* liability to coronary artery disease. A rare familial anomaly with this result is known as the 'longevity syndrome'. Deficiency of HDL occurs in *Tangier disease*, a rare inborn error of metabolism. Children with this disorder have few symptoms but cholesterol esters are deposited in the tonsils and spleen which become enlarged.

Metabolism of triglycerides and fatty acids

When triglyceride from adipose tissue is to be oxidised for the provision of energy, it is converted to free fatty acids (FFA) under hormonal control (Chap. 12). These are conveyed in the blood to tissues such as liver and muscle in which fatty acid oxidation takes place. In both tissues the essential part of the process consists of the oxidation in the mitochondria of the long-chain fatty acids. The glycerol of the triglyceride molecule reacts with ATP to form glycerol phosphate which is oxidised to glyceraldehyde 3-phosphate. This in turn may either be converted to glycogen by reversal of glycolysis or be converted to pyruvate (p. 94).

Fatty acid catabolism

The major pathway for fatty acid catabolism is *β-oxidation* or the stepwise removal of two-carbon units. The oxidation of fatty acids requires their preliminary activation by the formation of coenzyme A derivatives. The activation occurs either in the cytosol of the cell or in the sucrose-permeable space of the mitochondrion. The remainder of the stages of β-oxidation occur on the matrix side of the inner mitochondrial membrane (p. 109). The inner membrane is impermeable to acyl CoA derivatives; the acyl radical is transferred to a carrier molecule, carnitine, and crosses the membrane as acyl carnitine (fig. 11.5). The acyl radical is then transferred from carnitine to coenzyme A at the matrix side of the inner membrane by a reversal of this reaction. Short-chain fatty acids, such as those from breast

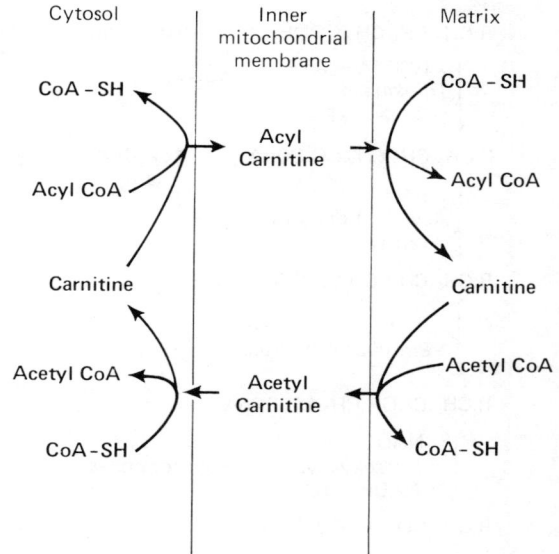

Fig. 11.5 The transport of acyl coenzyme A across the inner mitochondrial membrane. Carnitine deficiency, a rare inborn error of metabolism, is characterised by disorders of skeletal and cardiac muscle

milk, readily pass through the inner mitochondrial membrane.

The process of β-oxidation is shown in figure 11.6. One sequence of reactions produces one molecule of acetyl CoA and an acyl CoA containing two carbon atoms fewer than the original. The acyl CoA re-enters the sequence as a substrate for the first oxidation stage and loses a further two carbon atoms as acetyl CoA. In this way, a fatty acid containing an even number of carbon atoms is degraded two carbon atoms at a time to yield acetyl CoA.

The β-oxidation of a fatty acid with an odd number of carbon atoms gives acetyl CoA and eventually one molecule of propionyl CoA. This is carboxylated to yield methyl malonyl CoA which is converted to succinyl CoA and enters the tricarboxylic acid cycle (fig. 11.7).

There are two less important pathways of fatty acid oxidation, namely α- and ω- (omega) oxidation. In α-oxidation each sequence of reactions results in the loss of one carbon unit from the carboxyl end of the fatty acid as CO_2. No activation of the free fatty acid is required. This process produces fatty acids with an odd number of carbon atoms and α-hydroxy fatty

$$\overset{\gamma}{R.CH_2}.\overset{\beta}{CH_2}.\overset{\alpha}{CH_2}.COOH \qquad \textbf{Fatty acid}$$

HSCoA + ATP
Thiokinase
AMP + PPi

$$R.CH_2.CH_2.CH_2.CO.SCoA \qquad \textbf{Acyl CoA}$$

FAD
Acyl CoA dehydrogenase
FADH$_2$

$$R.CH_2.CH{=}CH.CO.SCoA$$

H$_2$O
Enoyl CoA hydratase

$$R.CH_2.CHOH.CH_2.CO.SCoA$$

NAD$^+$
β-hydroxyacyl CoA dehydrogenase
NADH + H$^+$

$$R.CH_2.CO.CH_2.CO.SCoA$$

HSCoA
Thiolase
CH$_3$.CO.SCoA Acetyl CoA

$$R.CH_2.CO.SCoA \qquad \textbf{Acyl CoA}$$

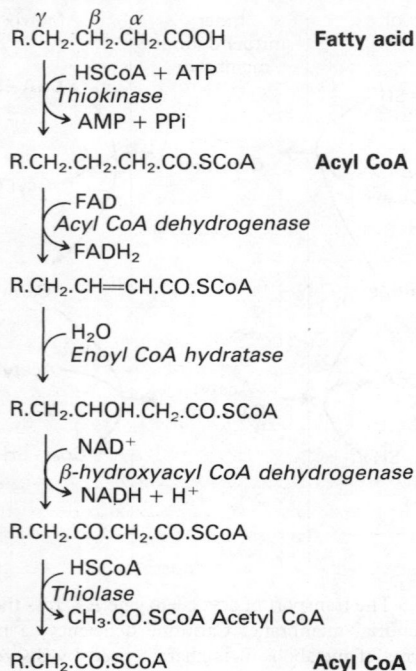

Fig. 11.6 The steps of β-oxidation. There are different dehydrogenases for acyl CoA molecules of different chain length. The FADH$_2$ and NADH are used to generate ATP as described in Chapter 10

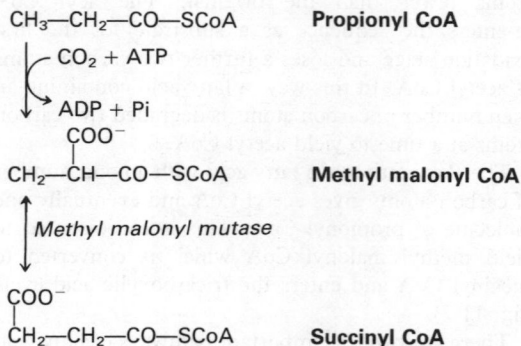

$$CH_3{-}CH_2{-}CO{-}SCoA \qquad \textbf{Propionyl CoA}$$

CO$_2$ + ATP

ADP + Pi

$$\begin{array}{c} COO^- \\ | \\ CH_2{-}CH{-}CO{-}SCoA \end{array} \qquad \textbf{Methyl malonyl CoA}$$

Methyl malonyl mutase

$$\begin{array}{c} COO^- \\ | \\ CH_2{-}CH_2{-}CO{-}SCoA \end{array} \qquad \textbf{Succinyl CoA}$$

Fig. 11.7 Metabolism of propionyl CoA produced from the β-oxidation of fatty acids with odd numbers of carbon atoms. Biotin is a coenzyme in the first step and 5′-adenosyl cobalamin, a derivative of vitamin B$_{12}$ (p. 153), is needed for the second

acids both of which are components of the cerebroside fraction of the brain lipids.

In ω-oxidation, which takes place in the endoplasmic reticulum, the terminal methyl group (ω-carbon atom) is oxidised to a carboxyl group to yield a dicarboxylic acid. This is oxidised by β-oxidation to yield acetyl CoA and eventually succinyl CoA which can enter the tricarboxylic acid cycle. This pathway appears to be important as a source of succinyl coenzyme A. In starvation and in uncontrolled diabetes this succinyl coenzyme A sustains the tricarboxylic acid cycle and limits the extent to which ketosis develops.

The fate of acetyl CoA. Acetyl coenzyme A is produced by β-oxidation and is also formed from carbohydrate by way of pyruvate as described in Chapter 9. Providing that the supply of oxaloacetate is adequate most of the acetyl CoA combines with oxaloacetate to form citrate which is oxidised by the tricarboxylic acid cycle. The coenzyme A released on citrate formation can be used to activate another molecule of fatty acid.

The ketone bodies. Acetoacetyl coenzyme A is formed in the liver from the last four carbon atoms of long-chain fatty acids which have been oxidised by β-oxidation with the successive removal of acetyl coenzyme A. It is also produced by the condensation of two molecules of acetyl coenzyme A under the influence of the enzyme thiolase:

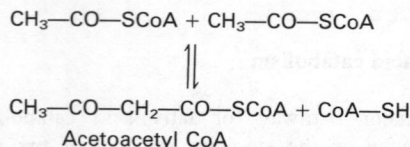

$$CH_3{-}CO{-}SCoA + CH_3{-}CO{-}SCoA$$

$$\Updownarrow$$

$$CH_3{-}CO{-}CH_2{-}CO{-}SCoA + CoA{-}SH$$
Acetoacetyl CoA

The acetoacetyl coenzyme A can then react with a further molecule of acetyl coenzyme A to form β-hydroxy-α-methyl-glutaryl coenzyme A (HMG.CoA) which is also an intermediate in steroid synthesis (p. 128). The HMG.CoA breaks down to yield acetoacetate and acetyl coenzyme A (fig. 11.8). The free acetoacetate so formed may be reduced to β-hydroxybutyrate:

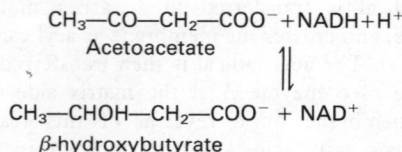

$$CH_3{-}CO{-}CH_2{-}COO^- + NADH + H^+$$
Acetoacetate

$$\Updownarrow$$

$$CH_3{-}CHOH{-}CH_2{-}COO^- + NAD^+$$
β-hydroxybutyrate

CH$_3$.CO.CH$_2$.CO.SCoA   **Acetoacetyl CoA**

    CH$_3$.CO.SCoA   **Acetyl CoA**

    HSCoA

    CH$_2$—COO$^-$ **β-hydroxy-**

CH$_3$.COH.CH$_2$.CO.SCoA   **β-methylglutaryl CoA**

    CH$_3$.CO.SCoA + H$^+$
    Acetyl CoA

CH$_3$.CO.CH$_2$.COO$^-$ **Acetoacetate**

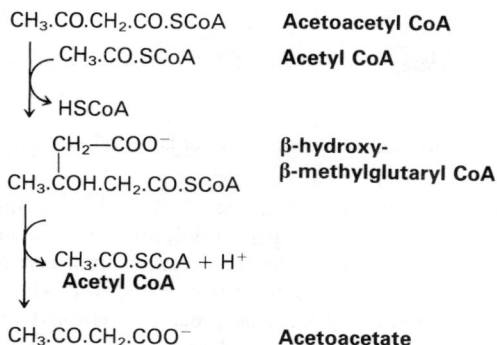

Fig. 11.8 Formation of acetoacetate

The acetoacetate and β-hydroxybutyrate leave the mitochondria and pass into the plasma. A small amount may be decarboxylated to yield acetone:

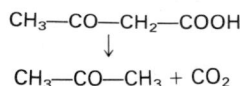

$$CH_3—CO—CH_2—COOH$$
$$\downarrow$$
$$CH_3—CO—CH_3 + CO_2$$

Acetoacetate and β-hydroxybutyrate are formed only in the liver. Acetone is formed spontaneously mainly in the lungs. All three are known as *ketone bodies*. Their concentration in the blood is normally less than 280 μmol/l in adults. The blood carries them to peripheral tissues in which β-hydroxybutyrate is oxidised to acetoacetate which in turn is converted into acetoacetyl coenzyme A either by reaction with succinyl coenzyme A,

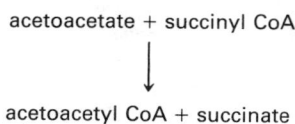

acetoacetate + succinyl CoA

$$\downarrow$$

acetoacetyl CoA + succinate

or by activation with ATP and coenzyme A,

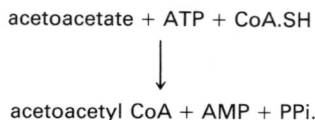

acetoacetate + ATP + CoA.SH

$$\downarrow$$

acetoacetyl CoA + AMP + PPi.

In normal circumstances, the acetoacetyl coenzyme A is then converted into two molecules of acetyl coenzyme A under the influence of the enzyme thiolase.

acetoacetyl CoA + CoA.SH ⇌ 2 acetyl CoA

The acetyl coenzyme A so formed may then enter the tricarboxylic acid cycle. The ability of the extrahepatic tissues to utilise ketone bodies is considerable. In starvation for example (p. 142) the fuel needs of many tissues are met by fatty acids and by ketone bodies.

Muscle, kidney, heart and brain (unlike the liver) possess the necessary enzymes to convert acetoacetate to acetoacetyl coenzyme A.

Although the concentration of ketone bodies in the blood is normally quite low, it may be greatly increased under certain circumstances. During starvation (p. 142) or during heavy exertion (p. 143), glycogen stores are rapidly depleted and there is a great increase in the use of fatty acids as a source of energy. Similarly in uncontrolled diabetes mellitus (p. 144) the utilisation of glucose is impaired and fatty acid production is increased, with the formation of ketone bodies in the liver in excess of the oxidative capacity of the peripheral tissues. Their concentration in the blood increases (*ketosis*) and they appear in the urine (*ketonuria*). The mechanisms controlling the production of ketone bodies are outlined in Chapter 12.

Fatty acid synthesis

Apart from dietary fat, the major source of fatty acids in the animal body is acetyl CoA derived from carbohydrate (p. 97). Acetyl CoA is the starting material for the synthesis of fatty acids which occurs not by reversal of β-oxidation but by an entirely separate process in the cytosol.

The first, and rate limiting, step in fatty acid synthesis is the formation of malonyl coenzyme A by carboxylation of acetyl coenzyme A by the biotin-containing enzyme *acetyl CoA carboxylase:*

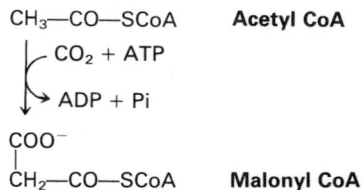

CH$_3$—CO—SCoA   **Acetyl CoA**

    CO$_2$ + ATP

    ADP + Pi

COO$^-$
CH$_2$—CO—SCoA   **Malonyl CoA**

All the subsequent steps are carried out by the enzyme complex *fatty acid synthase*. In man this is thought to consist of a single very large molecule (molecular weight 410 000) with at least seven different enzymic

$$\text{Serine residue—}PO_3^- \text{—} CH_2 \text{—} \underset{\underset{CH_3}{|}}{\overset{\overset{CH_3}{|}}{C}} \text{—} CHOH \text{—} CO \text{—} NH \text{—} CH_2 \text{—} CH_2 \text{—} CO \text{—} NH \text{—} CH_2 \text{—} CH_2 \text{—} SH$$

Serine residue of ACP

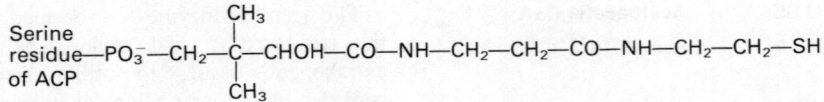

Fig. 11.9 The structure of the phosphopantetheine group attached to the acyl carrier protein (ACP) of fatty acid synthetase

activities. The active areas are ranged round a non-enzymic protein (*the acyl carrier protein*) to a serine residue of which is attached a phosphopantetheine group similar to that of coenzyme A (fig. 11.9).

The –SH group at the tip of the pantetheine is known as the *central thiol*, to distinguish it from the *peripheral thiol*, a part of the first enzyme in the sequence. The reaction sequence leads to the production of a butyryl radical (fig. 11.10). This is then transferred to the peripheral thiol, a further malonyl radical is attached to the freed central thiol and the above sequence of reactions is repeated to produce a six-carbon acyl radical. The process is repeated with further molecules of malonyl CoA until a fatty acid with 16 carbon atoms, palmitic acid, is produced; it is then released from the complex.

The synthesis of a single molecule of palmitate

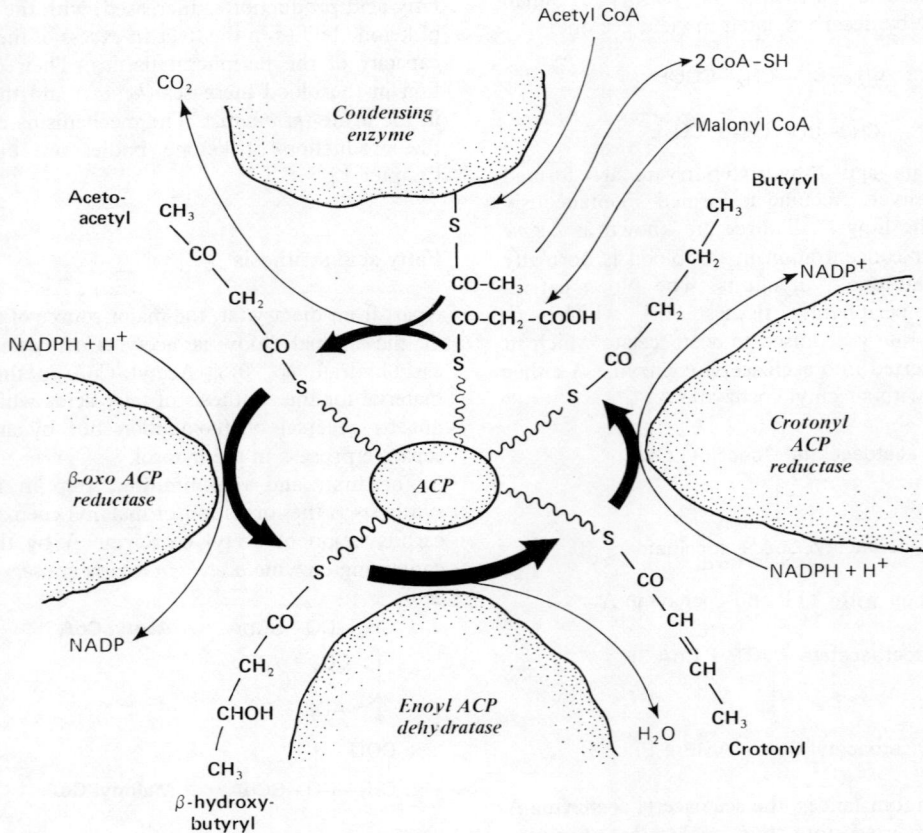

Fig. 11.10 First steps in fatty acid synthesis. Acetyl CoA becomes attached to the peripheral thiol of the condensing enzyme and malonyl CoA to the central thiol attached to ACP. Free coenzyme A is eliminated. Under the influence of the condensing enzyme the two radicals combine to yield acetoacetyl-ACP. The subsequent reactions lead in turn to β-hydroxybutyryl, crotonyl and butyryl radicals attached in each case to the ACP

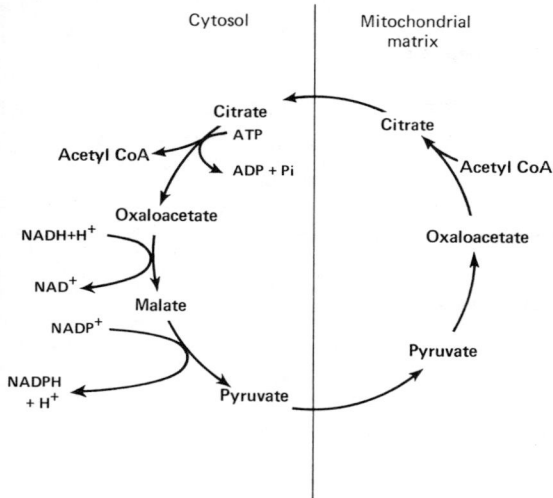

Fig. 11.11 The process whereby acetyl CoA, produced in the mitochondria from pyruvate or from fatty acid breakdown, is transported out into the cytoplasm. This cycle also yields some of the NADPH needed for fatty acid synthesis

requires 8 molecules of acetyl CoA, 14 of NADPH and 7 of ATP. The acetyl CoA is derived from the mitochondria but, since the inner mitochondrial membrane is only slightly permeable to acetyl CoA, it is transported out of the mitochondria as citrate (fig. 11.11)

The cyclical process shown in figure 11.11 also provides some of the NADPH required for fatty acid synthesis; the remainder is produced by the pentose phosphate pathway.

Diglyceride and triglyceride formation

The glycerol backbone of the glycerides is derived from glycerol 3-phosphate. This is formed either by the phosphorylation of glycerol itself by glycerokinase and ATP, or by the reduction of the glycolytic intermediate, dihydroxyacetone phosphate, by NADH and glycerophosphate dehydrogenase. The second reaction is the more important in most tissues.

Glycerol 3-phosphate is able to react with two acyl CoA molecules to give phosphatidic acid. This in turn is the precursor for the synthesis of triglycerides and glycerophospholipids (fig. 11.12).

Phospholipid metabolism

Phospholipid metabolism takes place within the cell's membranes, particularly the endoplasmic reticulum. Most of the enzymes involved are bound to membranes. All tissues, with the exception of the blood and the skin, are able to synthesise their own phospholipids by two basic pathways, one starting with diglyceride and the other starting with cytidine diphosphate diglyceride (CDP-diglyceride).

Synthesis of phospholipid from diglyceride. Phosphatidyl choline (lecithin) and phosphatidyl ethanolamine are synthesised by this pathway which is sometimes known as the *base-activated pathway*. The phosphate and base groups are first combined in an activated form by phosphorylation of the base and production of a CDP derivative (fig. 11.13). Phosphatidyl choline may also be formed by the stepwise methylation of phosphatidyl ethanolamine, by the methyl donor, *S*-adenosyl methionine (p. 188).

Phosphatidyl serine can only be synthesised by a base-exchange reaction, usually involving phosphatidyl ethanolamine. Any of the other glycerophosphatides may also be formed in this way.

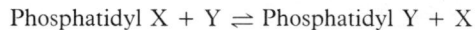

$$\text{Phosphatidyl X} + \text{Y} \rightleftharpoons \text{Phosphatidyl Y} + \text{X}$$

Synthesis of phospholipid from CDP-diglyceride. This pathway produces phosphatidyl inositol, phosphatidyl glycerol and cardiolipin. CDP-diglyceride is first formed by the condensation of CTP with phosphatidic acid. CDP-diglyceride then reacts with free inositol to produce phosphatidyl inositol or with another molecule of glycerol 3-phosphate to yield the intermediate phosphatidyl glycerol phosphate. A specific phosphatase then converts this intermediate to phosphatidyl glycerol. Cardiolipin is formed in mitochondria by the reaction of phosphatidyl glycerol and a further molecule of CDP-diglyceride (fig. 11.14). The phosphorylation of phosphatidyl inositol to yield di- and tri-phosphoinositides is brought about by specific kinase enzymes employing ATP as phosphate donor.

Phospholipid breakdown

The complete breakdown or modification of phospholipids is brought about by enzymes known as *phospholipases* which are hydrolases that attack ester bonds in phospholipid molecules.

125

$$CH_2-OH$$
$$CH-OH$$
$$CH_2-PO_3^{2-}$$ **Glycerol 3-phosphate**

$$R_1-CO-SCoA$$
$$R_2-CO-SCoA$$ } **Acyl CoA**
Acyl transferases
HSCoA

HSCoA

$$CH_2-O-CO-R_1$$
$$CH-O-CO-R_2$$ **Phosphatidic acid**
$$CH_2-PO_3^{2-}$$

CTP
PPi

Phosphatidate Phosphohydrolase
Pi

Diglyceride

$$CH_2-O-CO-R_1$$
$$CH-O-CO-R_2$$
$$CH_2-O-PO_3^--PO_3^--Cytidine$$
CDP-diglyceride

$$R_3CO.SCoA$$
Acyl transferase
HSCoA

Triglyceride

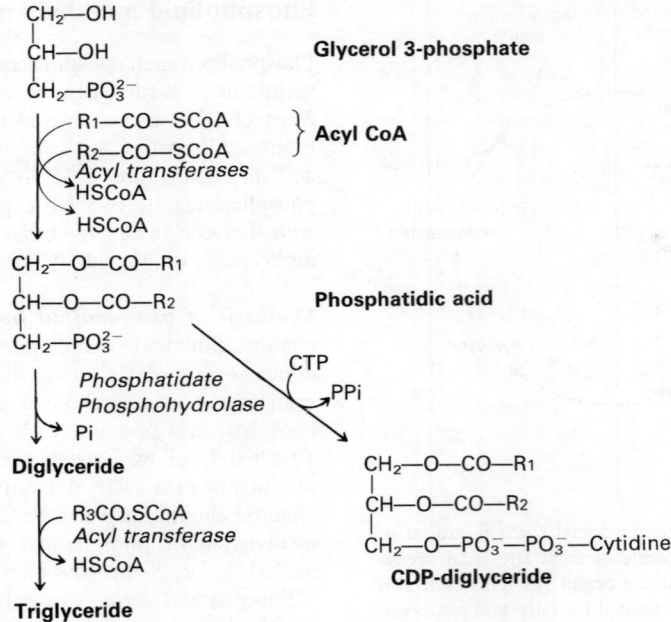

Fig. 11.12 Formation of diglycerides, triglycerides and CDP-diglycerides. Phospholipids are formed from diglycerides and CDP-diglycerides

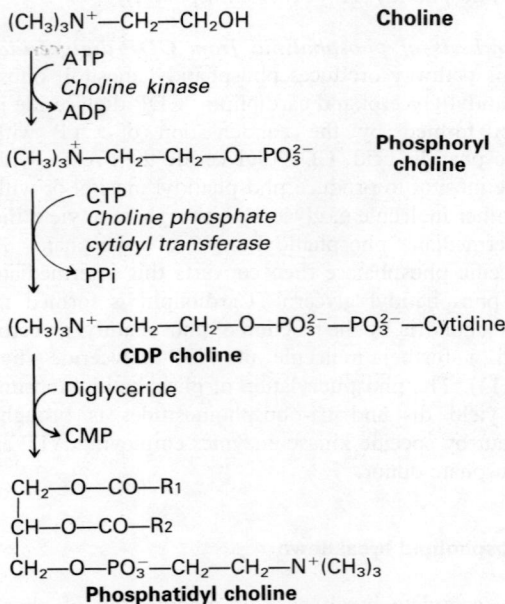

$$(CH_3)_3N^+-CH_2-CH_2OH$$ **Choline**

ATP
Choline kinase
ADP

$$(CH_3)_3\overset{+}{N}-CH_2-CH_2-O-PO_3^{2-}$$ **Phosphoryl choline**

CTP
Choline phosphate cytidyl transferase
PPi

$$(CH_3)_3N^+-CH_2-CH_2-O-PO_3^{2-}-PO_3^{2-}-Cytidine$$
CDP choline

Diglyceride
CMP

$$CH_2-O-CO-R_1$$
$$CH-O-CO-R_2$$
$$CH_2-O-PO_3^--CH_2-CH_2-N^+(CH_3)_3$$
Phosphatidyl choline

Fig. 11.13 Formation of phosphatidyl choline from choline. A similar reaction is used for the synthesis of phosphatidyl ethanolamine

The phospholipases are widely distributed in tissues. A full complement of phospholipases is found only in small intestinal secretions, where phospholipid molecules in the food are broken down, and in the spleen where the membranes of aged erythrocytes are degraded. In other tissues only one or two types of phospholipases are present.

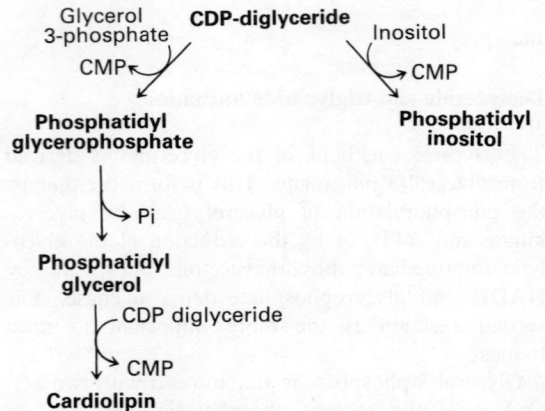

Glycerol 3-phosphate   **CDP-diglyceride**   Inositol
CMP   CMP

Phosphatidyl glycerophosphate   **Phosphatidyl inositol**

Pi

Phosphatidyl glycerol

CDP diglyceride
CMP

Cardiolipin

Fig. 11.14 Formation of cardiolipin and phosphatidyl inositol

All glycerophospholipids are attacked to some extent by these enzymes, although a particular phospholipase, depending on its source, usually exhibits a preference for a particular class or type of phospholipid substrate. For example phospholipase A_2 from snake venom shows a preference for phosphatidyl choline while that from the pancreas attacks the acidic phospholipids phosphatidic acid and phosphatidyl serine.

Phospholipase A_1 and *phospholipase A_2* hydrolyse the ester bonds linking the glycerol and fatty acyl groups at position 1 and 2 respectively, leaving either a 1-lysophosphatide or a 2-lysophosphatide and a free fatty acid in each case.

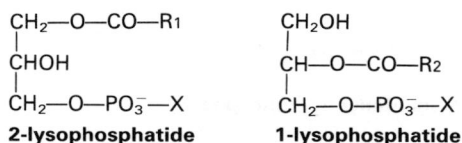

```
CH2—O—CO—R1              CH2OH
|                        |
CHOH                     CH—O—CO—R2
|                        |
CH2—O—PO3⁻—X             CH2—O—PO3⁻—X
2-lysophosphatide        1-lysophosphatide
```

While lysophosphatides are produced normally during the degradation of phospholipid molecules, their concentration in body fluids is very low because further breakdown or reacylation take place rapidly. Lysophosphatides are very toxic to cells especially erythrocytes because they have powerful detergent properties and cause disruption of the membrane. The lethal effect of snake venoms containing phospholipase A_2 can be attributed, at least in part, to the haemolytic properties of the lysophosphatides. Lysophosphatides also accumulate in the nervous system in demyelinating diseases such as multiple sclerosis. Lysolecithin levels are high in the gastric juice of many patients with gastric ulcers; lysolecithin may play a part in the development of the ulcer. It is thought that in these cases the lysolecithin is formed in the duodenum by the action of phospholipase A from pancreatic juice on lecithin from bile.

Phospholipases A are particularly important for the turnover of the fatty acid side-chains of phospholipids in cell membranes in processes such as exocytosis and endocytosis. Phospholipase A_2 is important for the mobilisation of arachidonic acid from phospholipid stores for the synthesis of prostaglandins, thromboxanes and prostacyclin (p. 46).

Lysophosphatide acyl hydrolase (LPAH) is found in all tissues containing either type of phospholipase A. It acts on either type of lysophosphatide, releasing the remaining fatty acid and leaving a non-toxic glycerophosphoryl base.

Phospholipase C cleaves the ester linkage between glycerol and the phosphoric acid residue, releasing a diglyceride and phosphoryl base.

Cholesterol metabolism

The normal adult has about 150 g cholesterol in his body of which 4 to 6 g is in the plasma. Much of the tissue cholesterol is exchangeable with the plasma cholesterol; equilibration is completed within a few hours in the case of the liver and intestine but several weeks may be needed for equilibration with skin and arterial wall.

Turnover and absorption. Cholesterol is excreted in the bile into the duodenum where it mixes with the dietary cholesterol. Part of it is then reabsorbed in the jejunum in the presence of bile salts, while the remainder is excreted in the faeces. The net absorption of dietary cholesterol varies between 50 per cent when dietary cholesterol is less than 300 mg daily and 10 per cent when the dietary cholesterol is more than 3 g daily. After absorption cholesterol is incorporated into chylomicrons.

Some cholesterol is used in the synthesis of steroid hormones whose metabolites are excreted in the urine, but the main route for cholesterol excretion is in the bile, partly as cholesterol and partly as bile salts (p. 44). Much of the bile salt is reabsorbed (fig. 11.15) but the faecal excretion of sterols gives an approximate measure of the turnover of cholesterol. Since sterol excretion amounts to 1 or 2 g daily the same amount of cholesterol must be derived from intestinal absorption and synthesis. Synthesis is the more important; intestinal absorption contributes only 200 to 300 mg daily.

Synthesis. Almost all tissues are able to synthesise cholesterol but more than 90 per cent of the endogenous cholesterol is made by the liver and intestinal mucosa. The synthetic pathway, which starts with acetyl CoA, is outlined in figures 11.16 and 11.17. In most tissues the rate of synthesis is regulated by the capacity of the enzyme *hydroxymethylglutaryl CoA reductase* (HMGCoA reductase) which catalyses an early, rate limiting step.

HMGCoA reductase is subject to induction and repression in several ways. For example when laboratory animals are fed on a high cholesterol diet,

127

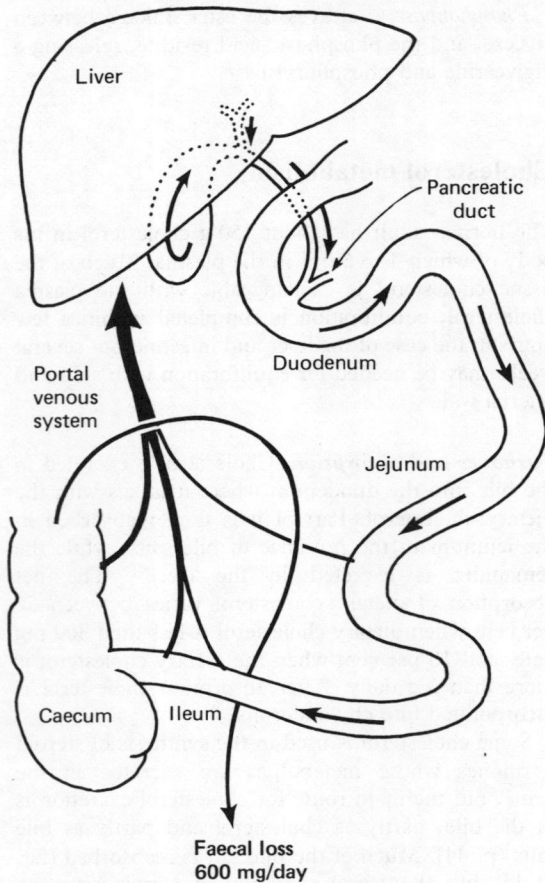

Fig. 11.15 Enterohepatic circulation of bile salts

$$CH_2{-}COO^-$$
$$CH_3{-}COH{-}CH_2{-}CO{-}SCoA$$

β-hydroxy-β-methyl glutaryl CoA (HMGCoA)

$$\Big\downarrow \begin{array}{l} NADPH + H^+ \\ \textit{HMGCoA reductase} \\ NADP + HSCoA \end{array}$$

$$CH_2{-}COO^-$$
$$CH_3{-}COH{-}CH_2{-}CH_2OH$$

Mevalonic acid

$$\Big\downarrow \begin{array}{l} 2ATP \\ 2ADP \end{array}$$

5-pyrophosphomevalonic acid

$$\Big\downarrow \; CO_2 + H_2O$$

Isopentenyl pyrophosphate

$$\Big\updownarrow$$

$$\begin{array}{l} CH_3 \\ C{=}CH{-}CH_2{-}O{-}PO_3^-{-}PO_3^{2-} \\ CH_3 \end{array}$$

3,3-dimethylallyl pyrophosphate

Fig. 11.16 The early steps in the synthesis of cholesterol. The β-hydroxy-β-methyl-glutaryl CoA (HMG.CoA) is derived from three molecules of acetyl CoA as shown in figure 11.8

endogenous cholesterol synthesis is inhibited because of repression of the synthesis of HMGCoA reductase. In this way the animal is able to compensate for changes in cholesterol intake. In man there are differences between races and even individuals in this respect. When cholesterol intake is increased some subjects show a marked suppression of synthesis while others do not. The Masai people of East Africa have a very high cholesterol intake in milk, meat and blood but they have low plasma cholesterol values.

The plasma cholesterol. The total cholesterol level in the plasma in western countries is normally between 3.6 and 7.8 mmol/l (140 to 300 mg/100 ml). About two-thirds of this is esterified to long-chain fatty acids

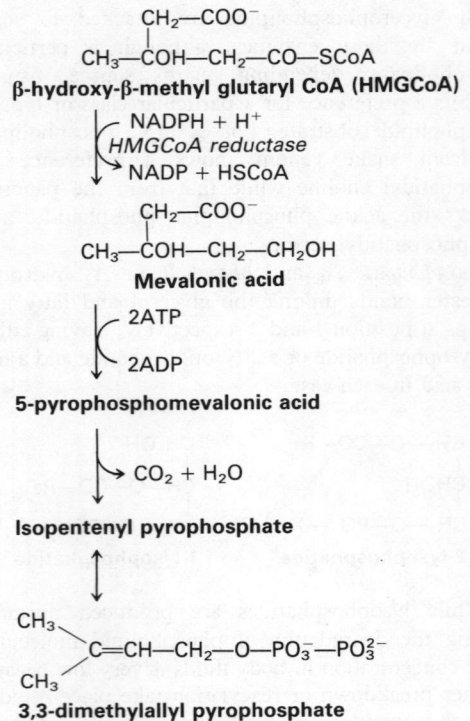

notably linoleic acid. These cholesterol esters are constantly being hydrolysed and resynthesised. The hydrolysis takes place in the liver but the resynthesis takes place within the plasma mainly by the transfer of a fatty acid residue from lecithin under the influence of *lecithin-cholesterol acyltransferase* (LCAT) (p. 120).

Cholesterol is present in all the plasma lipoproteins (Table 11.1); in a fasting subject about 60 per cent is carried in LDL. A high plasma level of cholesterol is associated with an increased incidence of coronary heart disease.

Apart from familial hypercholesterolaemia (p. 119), plasma cholesterol is raised in pregnancy and in diabetes mellitus, thyroid insufficiency and in obstructive jaundice (p. 201). Low values are found in malnutrition, hyperthyroidism and in severe liver cell damage caused by certain drugs.

**1 molecule of
3,3-dimethylallyl pyrophosphate
and 1 molecule of
isopentenyl pyrophosphate**

$\searrow$ PPi

Geranyl pyrophosphate

one further molecule of
3,3-dimethylallyl pyrophosphate

$\searrow$ PPi

CH_3
$C=CH-CH_2-CH_2-\overset{CH_3}{C}=CH-CH_2-CH_2-\overset{CH_3}{C}=CH-CH_2-O-PO_3^--PO_3^{2-}$
CH_3

Farnesyl pyrophosphate

$NADPH + H^+$ + a further molecule of
farnesyl pyrophosphate

$NADP^+ + 2PPi$

Squalene

Cholesterol

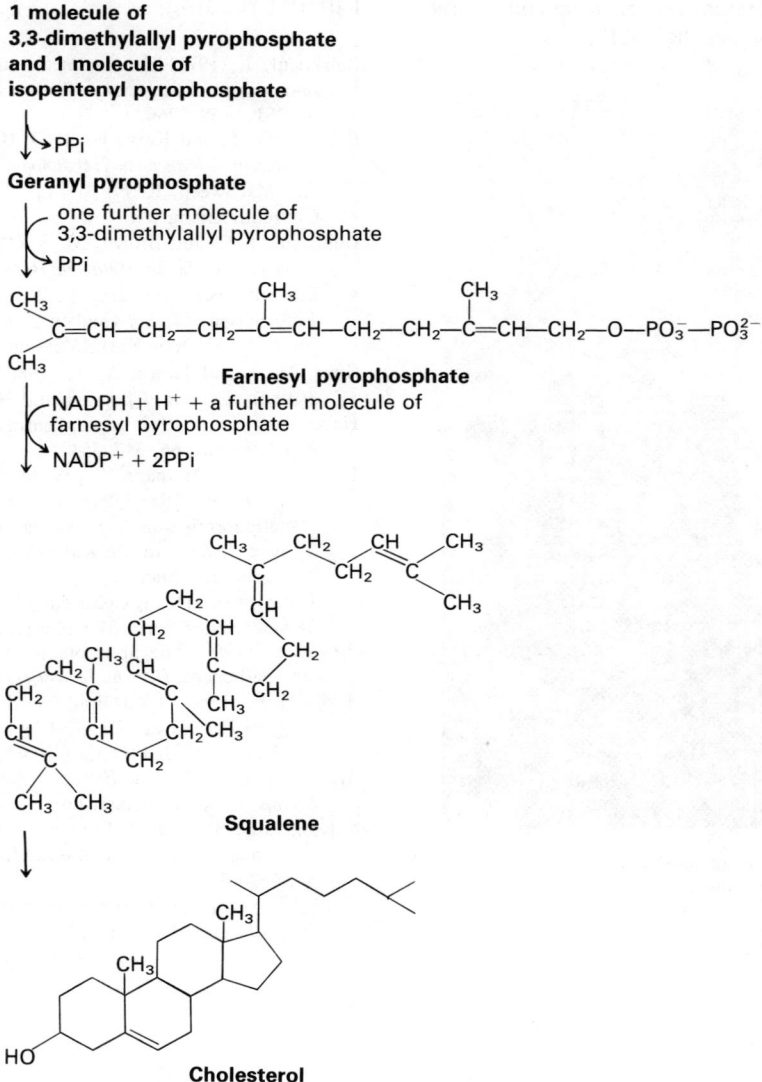

Fig. 11.17 Principal steps in the later stages of the synthesis of cholesterol. In the final step a series of enzymes converts squalene into cholesterol by closure of the rings and the loss of three methyl groups as CO_2

Cholesterol in bile. The principal constituent of most gall-stones is cholesterol derived from the free cholesterol in bile. In normal subjects about one-third of this has been synthesised in the liver and two-thirds is derived from the free cholesterol in HDL. The two primary bile acids, cholic acid and chenodeoxycholic acid, are also synthesised from cholesterol. Patients with a tendency to form gall-stones have bile in which the concentration of cholesterol is excessive in relation to the concentrations of bile salts and phospholipids, so that it is supersaturated with respect to cholesterol. This is most likely to occur in bile which has been concentrated in the gall-bladder. If some nucleating

129

or seeding agent is present crystals form and enlarge rapidly to give gall-stones (fig. 11.18).

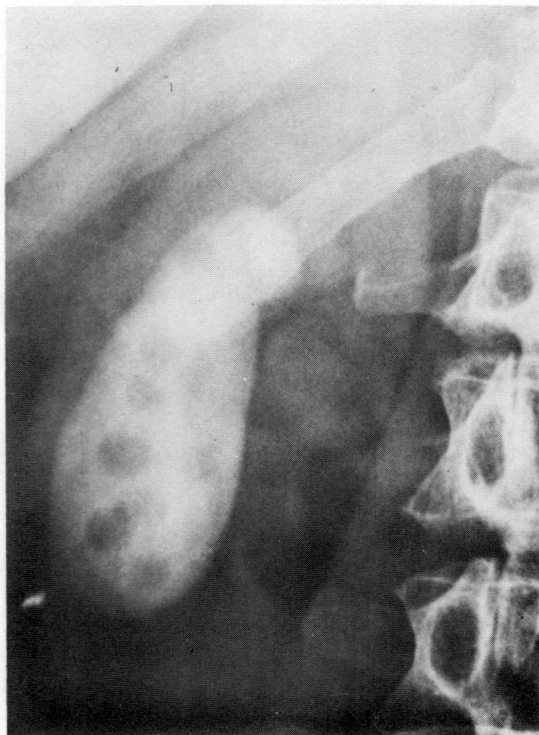

Fig. 11.18 Cholecystogram to show radiolucent gall-stones. (Courtesy of I. A. D. Bouchier)

Further reading

Björkhem, I. (1976) On the mechanism of regulation of ω-oxidation of fatty acids. *Journal of Biological Chemistry* **251,** 5259–5266

Everson, G. T. and Kern, F. (1983) Bile acid metabolism. In *Recent Advances in Hepatology*, eds Thomas, H. C. and MacSween, R. N. M., pp. 171–188. Edinburgh: Churchill Livingstone

Goldstein, J. L. and Brown, M. S. (1983) Familial hypercholesterolemia. In *Metabolic Basis of Inherited Disease*, 5th edn, eds Stanbury, J. B., Wyngaarden, J. B., Fredrickson, D. S., Goldstein, J. L., Brown, M. S. pp. 672–712. New York: McGraw-Hill

Gurr, M. I. and James, A. T. (1980) *Lipid Biochemistry*, 3rd edn. London: Chapman and Hall

Havel, R. J. (ed.) (1982) Lipid disorders. *Medical Clinics of North America* **66,** 317–553

Herbert, P. N., Assmann, G., Gotto, A. M. and Fredrickson, D. S. (1983) Familial lipoprotein deficiency: abetalipoprotinemia, hypobetalipoproteinemia, and Tangier disease. In *Metabolic Basis of Inherited Disease*, 5th edn, eds Stanbury, J. B., Wyngaarden, J. B., Fredrickson, D. S., Goldstein J. L., Brown, M. S. pp. 589–621. New York: McGraw-Hill

Lewis, B. (1983) The lipoproteins: predictors, protectors, and pathogens. *British Medical Journal* **287,** 1161–1164

Lloyd, J. K. (1984) Plasma lipid disorders. In *Chemical Pathology and the Sick Child*. eds Clayton, B. E. and Round, J. M., pp. 245–264. Oxford: Blackwell

Myant, N. B. (1981) *The Biology of Cholesterol and Related Sterols*. London: Heinemann

Nilsson-Ehle, P. (1980) Lipolytic enzymes and plasma lipoprotein metabolism. *Annual Review of Biochemistry* **49,** 667–693

Schroepfer, G. J. (1981) Sterol biosynthesis. *Annual Review of Biochemistry* **50,** 585–621

DT MISH CRP

12 Integration & regulation of energy metabolism

THE previous three chapters showed how ATP is produced from ADP and inorganic phosphate as a result of the oxidation of glucose, fatty acids and ketones. ATP synthesis can also be powered by the oxidation of the carbon chains of amino acids (Chap. 15). The pathways involved lead in each case to precursors or intermediates of the tricarboxylic acid cycle (fig. 12.1); the mechanism of ATP generation is identical whether acetyl coenzyme A is derived from glucose or from fatty acids.

This chapter is concerned with the ways in which the energy supply to each tissue of the body is ensured and how the storage of energy is regulated. The pathways for the storage and realisation of

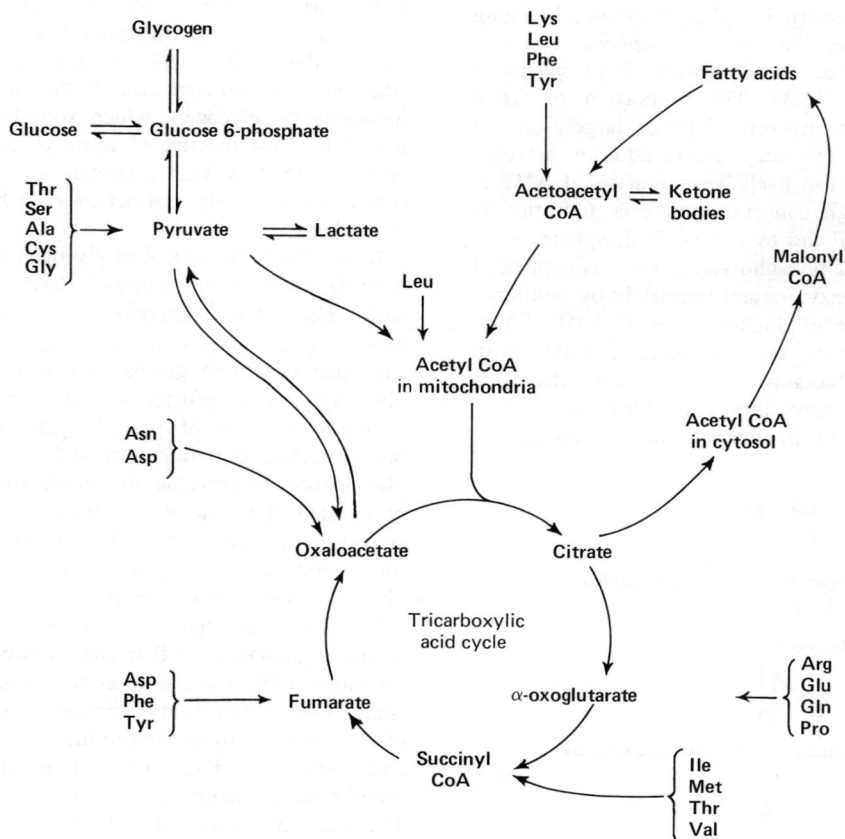

Fig. 12.1 Pathways for the production of energy from carbohydrates, fatty acids and amino acids. Note that the conversion of pyruvate to acetyl CoA is irreversible so that fatty acids and certain amino acids such as leucine cannot be used for glucose production

energy are controlled both by allosteric interactions (p. 87) and by various hormones, notably insulin, glucagon and adrenaline.

Control of individual pathways

Glycogen synthesis and breakdown

As described in Chapter 9, different enzymes are involved in the synthesis and the breakdown of glycogen (fig. 12.2). Each is controlled in a number of ways both by allosteric effects and by changes in activity associated with the addition or removal of a phosphate group.

Phosphorylase is not a single enzyme. It exists in two forms, phosphorylase a which is active and phosphorylase b which is ordinarily inactive. Phosphorylase b is converted to phosphorylase a by being phosphorylated by the enzyme *phosphorylase kinase*. The reverse process is catalysed by *phosphorylase phosphatase* (fig. 12.3). The proportion of active phosphorylase at any time depends largely on the balance of these two enzymes. In addition however phosphorylase b can itself become active if AMP is present at a high concentration; this activation is inhibited by ATP and by glucose 6-phosphate.

The activity of phosphorylase kinase is controlled by a *cascade* of enzymes and ultimately by hormones which act by the production of cyclic AMP (Chap. 22). In the liver the most important hormone with this effect is glucagon but in muscle adrenaline (epinephrine) is most important. This cascade (fig. 12.4) provides a means whereby minute amounts of

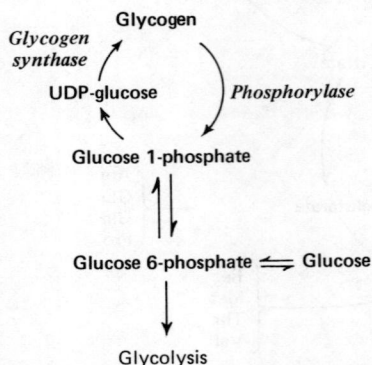

Fig. 12.2 Summary of pathways for the synthesis and breakdown of glycogen, described in detail in Chapter 9

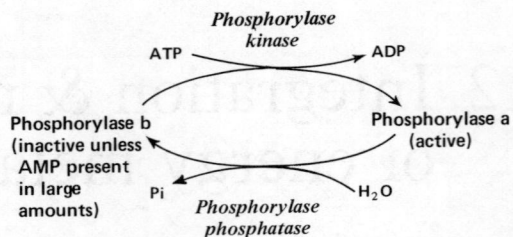

Fig. 12.3 Enzymes involved in the activation and inactivation of phosphorylase

cyclic AMP can cause the rapid transformation of large quantities of phosphorylase from the b to the a form.

The synthesis of glycogen is active when glycogen breakdown is inhibited and inactive when glycogen breakdown is stimulated. Glycogen synthase is rendered inactive by phosphorylation by the same cascade that activates phosphorylase (fig. 12.4). In addition glycogen synthase in muscle is inhibited indirectly by glycogen, which stimulates the accumulation of the inactive phosphorylated form of the enzyme. In this way glycogen storage does not continue indefinitely in a person on a high carbohydrate diet.

In the liver the control of glycogen metabolism is essential for the maintenance of plasma glucose levels within close limits, generally 4.0 to 8.0 mmol/l. It is now clear that in addition to the hormonal mechanisms just described glucose has a direct effect on phosphorylase a. Infusion of glucose into a rat leads to the conversion of phosphorylase a to inactive phosphorylase b within a minute (fig. 12.5). The phosphatase responsible for inactivating the phosphorylase is the same as that which *activates* glycogen synthase; an increase in synthase activity follows very soon after the fall in phosphorylase activity. It is likely that a similar mechanism applies in man.

In muscle, glycogen breakdown is stimulated by electrical excitation so that glucose 6-phosphate and therefore ATP is available at the time of a muscle contraction. Stimulation of muscle leads to the inflow of calcium ions; these bind to the proteins *calmodulin* and *troponin C* which in turn bind to and activate phophorylase kinase. This activates phosphorylase and inactivates glycogen synthase so that glycogen breakdown and ATP synthesis is increased. This coordination between muscle contraction and its energy supply is illustrated in figure 12.6.

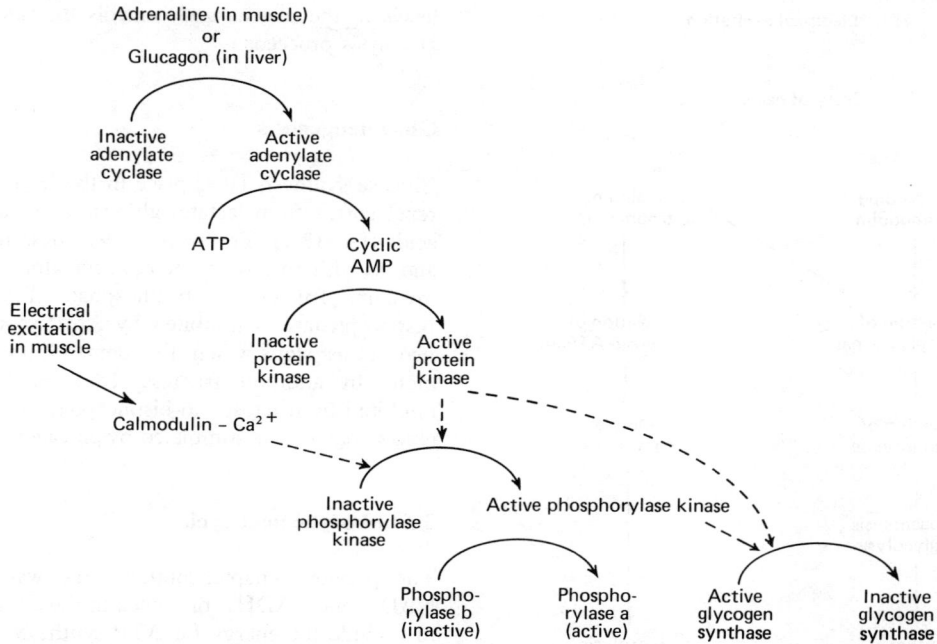

Fig. 12.4 The 'cascade' of enzymes involved in the hormonal control of glycogen synthesis and breakdown. One cause of glycogen storage disease (p. 107) is a deficiency of phosphorylase kinase in the liver

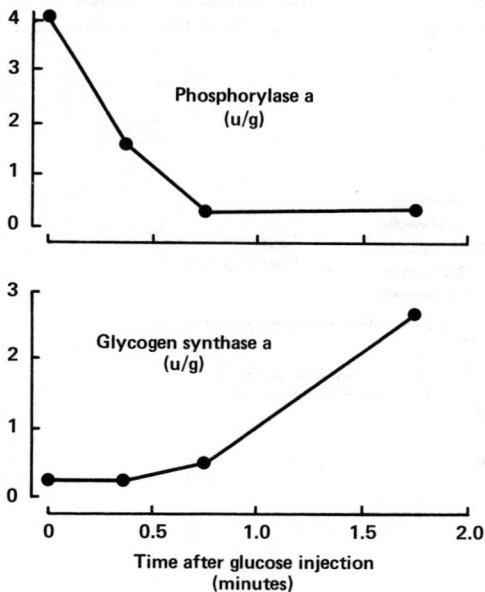

Fig. 12.5 Changes in levels of active phosphorylase and active glycogen synthase in the liver of a rat in response to an injection of glucose. (After Stalmans, W. *et al.* (1974) *European Journal of Biochemistry* **41**, 127)

Glycolysis

The key step controlling the rate at which glycolysis proceeds is that catalysed by phosphofructokinase. This is allosterically inhibited by ATP and activated by AMP, so that glycolysis is related to the requirements for energy. In addition phosphofructokinase is inhibited by citrate so that the rate of glycolysis is related to the concentrations of the intermediates of the tricarboxylic acid cycle; glycolysis is increased when the concentration of these intermediates is reduced: for example, when they are required as precursors for amino acids needed in protein synthesis.

It has recently been recognised that phosphofructokinase is also activated, at least in liver, by *fructose 2,6-bisphosphate*. This substance is produced from fructose 6-phosphate by a side-reaction inhibited by the cyclic AMP-dependent protein kinase described earlier and thus by glucagon. Its breakdown to fructose 6-phosphate is also stimulated by the protein kinase, and thus by glucagon (fig. 12.7). The concentration of fructose 2,6-bisphosphate within the hepatocyte therefore varies according to glucagon

133

Electrical excitation

Entry of calcium ions

Ca^{2+} binding to calmodulin

Ca^{2+} binding to troponin C

Activation of phosphorylase kinase

Activation of actomyosin ATPase

Activation of phosphorylase

Muscle contraction

Glycogenolysis and glycolysis

Synthesis of ATP

Hydrolysis of ATP

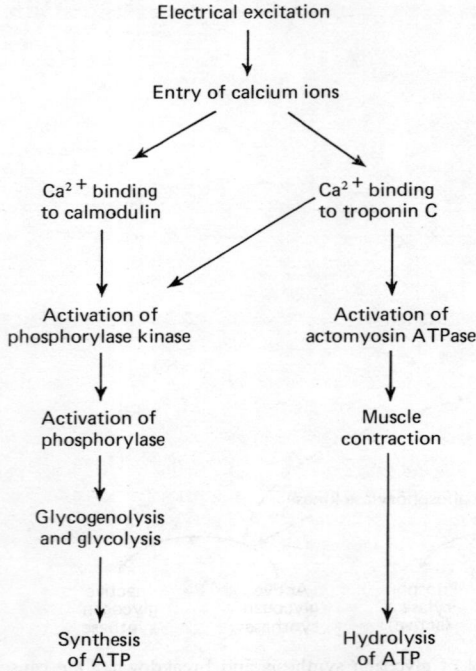

Fig. 12.6 Coordination of glycogen breakdown with contraction in skeletal muscle. (After Picton, C., Klee, C. B. and Cohen, P. (1981) *Cell Calcium* **2**, 281)

levels in the plasma and controls the rate at which glycolysis proceeds.

Gluconeogenesis

Glucose synthesis takes place in the liver and in the renal cortex from lactate, glycerol and many amino acids (fig. 12.1). The pathway was described earlier and, as with glycolysis, the key regulatory step is that involving fructose 1,6-bisphosphate. Fructose 1,6-bisphosphatase is inhibited by AMP so that gluconeogenesis stops when the demand for energy is high. In addition fructose 1,6-bisphosphatase is inhibited by fructose 2,6-bisphosphate and therefore gluconeogenesis is stimulated by glucagon (fig. 12.7).

Tricarboxylic acid cycle

The previous chapter outlined the way in which NADH and FADH$_2$, produced in the cycle, are used to provide the energy for ATP synthesis from ADP and Pi. When the concentration of ATP within a mitochondrion is high its synthesis stops; since proton entry into the matrix is coupled to ATP synthesis, this too ceases. In turn, the electron flow

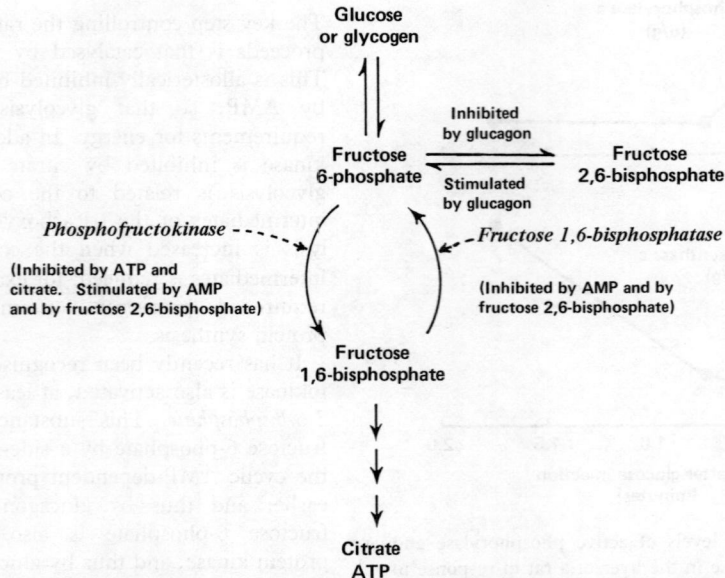

Glucose or glycogen

Fructose 6-phosphate

Inhibited by glucagon

Stimulated by glucagon

Fructose 2,6-bisphosphate

Phosphofructokinase

(Inhibited by ATP and citrate. Stimulated by AMP and by fructose 2,6-bisphosphate)

Fructose 1,6-bisphosphatase

(Inhibited by AMP and by fructose 2,6-bisphosphate)

Fructose 1,6-bisphosphate

Citrate ATP

Fig. 12.7 The factors controlling the interconversion of fructose 6-phosphate and fructose 1,6-bisphosphate, key reactions in the regulation of glycolysis and reverse glycolysis

in the respiratory chain stops. A rise in the ratio of NADH to NAD^+ also reduces the activity of isocitrate dehydrogenase and α-oxoglutarate dehydrogenase.

Pyruvate dehydrogenase

A further key step in metabolism is the irreversible oxidative decarboxylation of pyruvate to acetyl CoA within the mitochondria. This enzyme is controlled by a number of allosteric interactions; for example it is inhibited by acetyl CoA itself. In this way pyruvate is converted to acetyl CoA when ATP levels are low or when acetyl CoA is needed for the synthesis of fatty acids.

In addition pyruvate dehydrogenase is reversibly inactivated by phosphorylation in response to glucagon and activated by dephosphorylation in response to insulin. In this way carbohydrate can be converted to fatty acids for storage in adipose tissue after meals.

Pentose phosphate pathway

The key regulating step in the pentose phosphate pathway is that catalysed by glucose 6-phosphate dehydrogenase (p. 99). This is controlled by the $NADP^+/NADPH$ ratio (fig. 12.8). Since NADPH is needed for fatty acid synthesis, the rate at which the pentose phosphate pathway proceeds is related to the rate of fatty acid, and therefore of triglyceride, production. Glycolysis and the pentose phosphate pathway both take place in the cytosol but since one uses NAD^+ and the other $NADP^+$ they can be regulated independently.

Fatty acid synthesis

The process of fatty acid synthesis takes place in the cytosol and the details were given in Chapter 11. The first step, and the key step for the control of the pathway, is the synthesis of malonyl CoA from acetyl CoA (fig. 12.9). The enzyme concerned is allosterically activated by citrate and inhibited by the eventual products of the reaction particularly palmityl CoA. When ATP is present at high concentration, citrate is plentiful and fatty acid synthesis is stimulated.

In addition acetyl CoA carboxylase is controlled by phosphorylation. The cyclic AMP-dependent protein kinase, whose activation occurs in response to adrenaline or glucagon (p. 139), phosphorylates the carboxylase to a form which is inactive unless citrate levels are very high. In this way fatty acid synthesis is inhibited by glucagon in the liver and by adrenaline in adipose tissue.

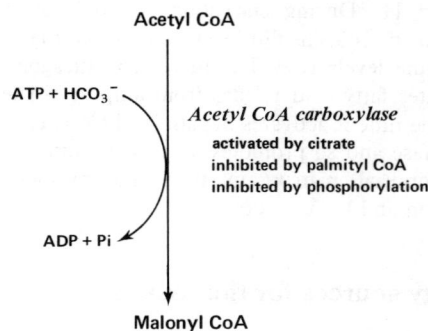

Fig. 12.9 The key enzyme step which regulates fatty acid synthesis

Fatty acid mobilisation and breakdown

The mobilisation of fatty acids from triglycerides in adipose tissue is carried out by lipase which exists in two forms, lipase a which is active and lipase b which is inactive. In response to adrenaline the cascade described earlier (fig. 12.4) operates and the active protein kinase catalyses the phosphorylation of lipase b to lipase a. In turn this leads to the conversion of

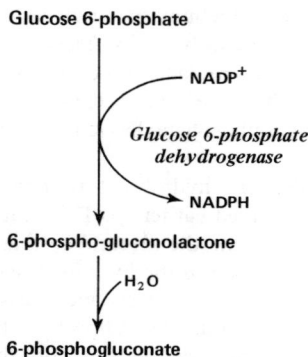

Fig. 12.8 The role of glucose 6-phosphate dehydrogenase in the controlling step in the pentose phosphate pathway

135

triglycerides to diglycerides, monoglycerides, free fatty acids and glycerol. At the same time re-esterification is inhibited by the phosphorylation (inactivation) of the enzyme glycerol-phosphate-acyl transferase.

Fatty acid breakdown takes place within the mitochondria (Chaps. 10 and 11) and, in most tissues other than the liver, leads to the production of acetyl CoA. In turn this is degraded by the tricarboxylic acid cycle with the production of $FADH_2$ and NADH. In the liver fatty acid breakdown leads to the production of ketone bodies (p. 122).

Both in the liver and in adipose tissue fatty acid synthesis and breakdown occur continually. Which pathway predominates depends on the levels of malonyl CoA which inhibits acyl carnitine transferase and so prevents the entry of fatty acids into the mitochondria. A rise in malonyl CoA levels decreases fatty acid breakdown.

Ketone body production

The role of the liver in producing the ketone bodies from the β-oxidation of fatty acids was described in Chapter 11. During starvation (p. 142) or heavy exertion (p. 143) insulin levels fall and glucagon and adrenaline levels rise. The increased glucagon level stimulates fatty acid release from adipose tissue. At the same time it activates hepatic acyl CoA: carnitine transferase and so promotes fatty acid entry into the mitochondrial matrix; in turn this promotes β-oxidation and ketone body production.

Energy sources for the tissues

The tissues of the body differ widely from each other in their patterns of metabolic activity and particularly in their principal fuel for energy-requiring processes. Certain tissues have large reserves of energy (Table 12.1); others have virtually none. This section outlines the main metabolic characteristics of each tissue.

Adipose tissue

Adipose tissue represents the major energy reserve in man (Table 12.1) and in many other animals. In migrating birds for example stores of triglycerides

Table 12.1 Energy reserves in a 70 kg man (in MJ)

	Tri-glyceride	Glycogen or glucose	Available protein	Total
Adipose tissue	564.3	0.3	0.2	565
Muscle	1.9	5.0	100.3	107
Liver	1.9	1.7	1.7	5

(After G. F. Cahill (1976) *Clinics in Endocrinology and Metabolism* **5**, 397–415.)

provide the energy needed for long distance non-stop flights.

Adipose tissue is found all over the body, but particularly in the subcutaneous tissues, and consists of large cells each containing a fat droplet within a thin rim of cytoplasm.

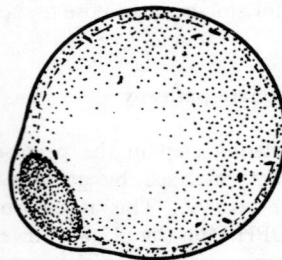

These are stacked together in an almost geometrical manner in a tissue richly supplied with capillaries.

Adipose tissue obtains its raw materials mainly as very low density lipoproteins (VLDL) synthesised in the liver, and partly as chylomicrons from the gut. At the capillary endothelium the triglycerides in these lipoproteins are hydrolysed by *lipoprotein lipase* and the free fatty acids are taken up by the adipose tissue cells. There they are re-esterified to triglycerides using glycerol 3-phosphate derived from glucose (fig. 12.10).

Triglycerides are hydrolysed to fatty acids and glycerol as described earlier (p. 121). The adipocyte cannot phosphorylate the glycerol so that this leaves the cell and is passed to the liver for resynthesis into glucose. Triglyceride synthesis and breakdown goes on continually within the adipocyte; the rate of synthesis or resynthesis depends on the level of glycerol 3-phosphate and hence on the glucose available within the cell. When this is low the free

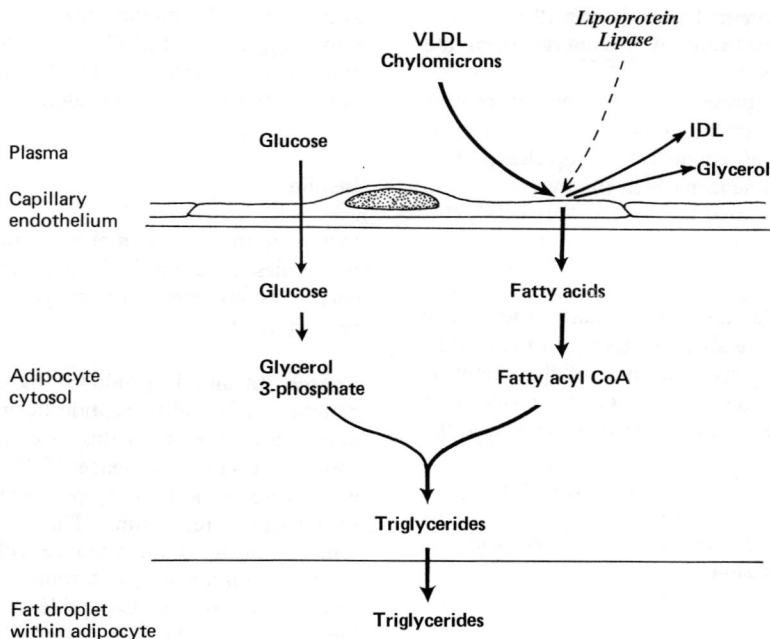

Fig. 12.10 Steps in the storage of energy in adipose tissue as triglycerides. While lipoprotein lipase provides the major route for the entry of lipids into the cell a small amount of triglyceride can be taken up by pinocytosis. VLDL = very low density lipoproteins, IDL = intermediate density lipoproteins. Defects in VLDL and/or chylomicron removal from plasma probably underlie several of the hyperlipidaemias (p. 119).

fatty acids are released into the plasma where, bound to albumin, they are transported to the liver and other tissues for utilisation.

Liver

The liver has a central role in the metabolism of carbohydrates, lipids and amino acids. The liver and the kidney are the only tissues able to carry out gluconeogenesis, the production of glucose from glycerol, lactate and certain amino acids, particularly alanine (p. 95). After a meal the liver can take up large quantities of glucose and store it as glycogen. During fasting the liver can produce glucose from the breakdown of glycogen and by gluconeogenesis.

The metabolism of lipids in the liver also varies with meals. After a meal, fatty acids are synthesised, esterified with glycerol and secreted into the plasma as VLDL. During fasting the liver breaks down fatty acids to ketones which pass into the plasma as an energy source for other tissues. Which pathway

predominates depends on the levels of malonyl CoA (p. 135).

For its own energy needs the liver mainly uses the carbon skeletons of amino acids from the diet or from the tissues. Although glycolysis does take place in the liver this is mainly for the production of intermediates for synthetic pathways.

Muscle

Muscle like liver contains a large store of glycogen which is built up after meals. Unlike liver glycogen, muscle glycogen cannot be exported as glucose to other tissues since muscle lacks glucose 6-phosphatase. During exertion the glycogen is converted directly to glucose 6-phosphate which enters the glycolytic pathway as does glucose 6-phosphate derived from glucose in the plasma. In vigorous exertion glycolysis yielding lactate provides most of the energy; relatively little is obtained from the tricarboxylic acid cycle. In muscle at rest much of the

137

energy needed is obtained from fatty acid oxidation. Heart muscle derives most of its energy from the oxidation of ketones.

Muscle protein represents a major energy reserve and in starvation or malnutrition the muscle mass is greatly diminished as amino acids are released into the plasma for gluconeogenesis in the liver.

Kidney

The renal medulla and the renal cortex have strikingly different metabolic patterns. In the medulla which is relatively hypoxic most of the energy is derived from glycolysis; the lactate produced is reconstituted to glucose in the renal cortex and in the liver.

The renal cortex is able to carry out gluconeogenesis like the liver but obtains most of its own energy from the oxidation of fatty acids or ketones from the plasma.

Brain

The brain normally obtains all its energy from plasma glucose which is broken down by glycolysis and the tricarboxylic acid cycle. In prolonged starvation an adaptation takes place whereby up to two-thirds of the brain's energy requirements can be obtained from ketones. The brain has a small store of glycogen which is sufficient to maintain glycolysis in anaerobic conditions for about four minutes. Anoxia for a longer period leads to irreversible damage.

Red cells

Since red cells contain no mitochondria most of their energy needs are obtained from glycolysis (p. 95). Each day some 35 g lactate are produced by the red cells and reconstituted to glucose in the liver and kidney. The pentose phosphate pathway operates in red cells to provide NADPH for the reduction of oxidised haemoglobin.

Hormones in the regulation of metabolism

The metabolism of fuel molecules is regulated to ensure constancy of the glucose supply to the brain

and several hormones play a part. The most important are insulin, glucagon, adrenaline, cortisol and growth hormone. Insulin lowers the blood glucose; the other four hormones raise it.

Insulin

Insulin is the key hormone organising the response by tissues to a meal. It promotes the storage of energy as glycogen and triglycerides and increases protein synthesis.

Source. Insulin is produced by the β-cells of the pancreas. Like other peptide hormones it is formed first as a precursor, in this case *pre-pro-insulin*. This contains a signal sequence of 16 residues (p. 171) which directs the growing polypeptide chain into the endoplasmic reticulum. The signal sequence is removed in the endoplasmic reticulum and *proinsulin* (with 84 amino acids) is transported to the Golgi complex where the C-peptide is removed to give insulin (fig. 12.11). This is stored in granules within the cytoplasm of the β-cell.

Secretion. Insulin is secreted in response to high levels of glucose in the plasma (fig. 12.12) or to high levels of amino acids, particularly arginine. The response to glucose is mediated by a rise in the concentration of calcium ions in the islet cells. Glucose by mouth has a more marked effect on insulin secretion than glucose given intravenously; oral glucose leads to the production of the gastrointestinal hormones, gastrin, secretin, pancreozymin and enteroglucagon which also stimulate insulin release.

Metabolism. Insulin is broken down rapidly by the liver and has a half-life in the plasma of about five minutes.

Actions. Insulin has many actions which are summarised in Table 12.2. In all cells glucose uptake is increased and also the uptake of the principal intracellular cations K^+ and Mg^{2+} is increased.

The mode of action of insulin at a cellular level has recently become clearer. It has been demonstrated that the receptor for insulin is a protein kinase with specificity for tyrosine residues. It is thought that the insulin receptor is a transmembrane protein with the binding site for insulin on the outer surface and the protein kinase activity on the inner surface of the

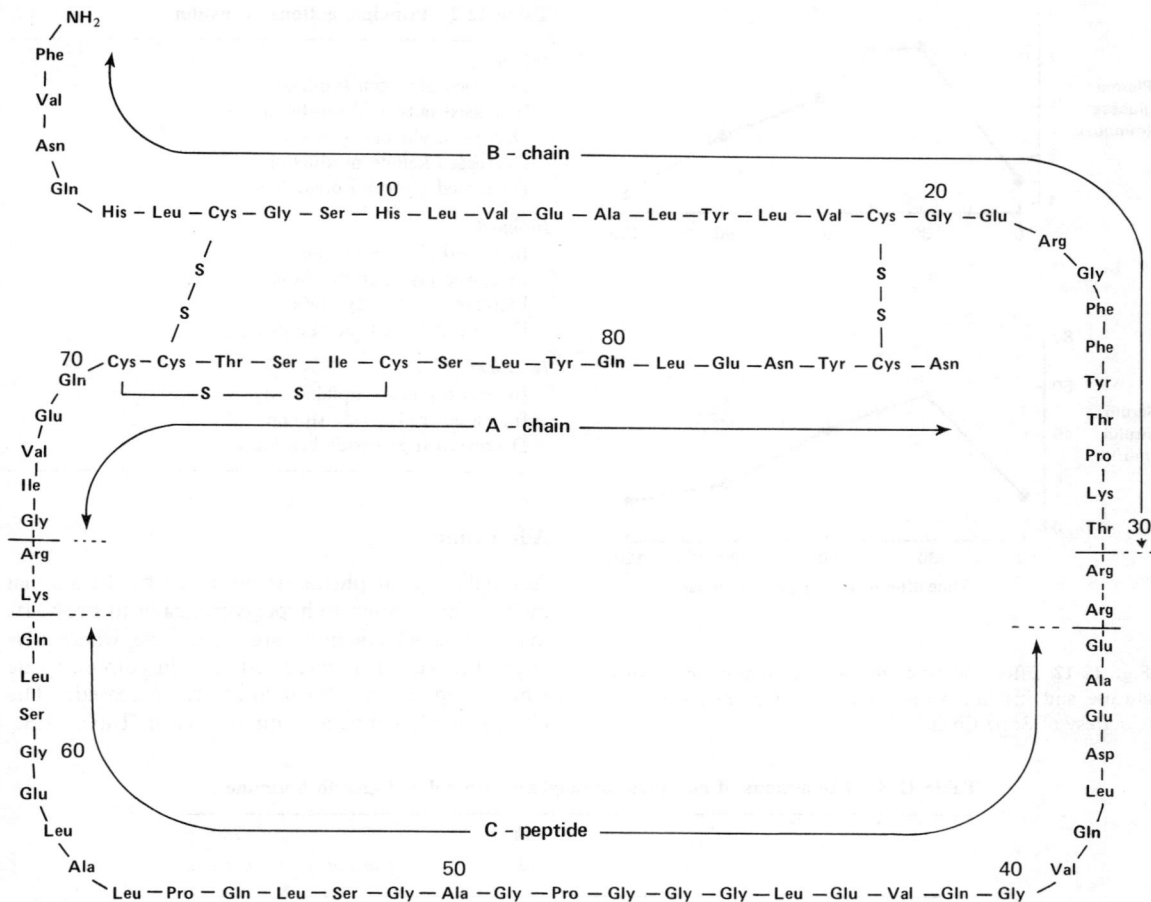

Fig. 12.11 Amino acid sequence of human proinsulin. The points of cleavage by proteolytic enzymes, to give insulin and the C-peptide, are indicated

plasma membrane. The insulin–receptor interaction activates the tyrosine kinase activity, which thus, in response to insulin, phosphorylates intracellular proteins. It seems likely that the action of insulin, and of other growth promoting hormones, does not involve a second messenger like cyclic AMP or calcium.

Glucagon

Glucagon is produced by the α-cells of the pancreatic islets and like insulin is initially produced as a precursor (*proglucagon*). It is secreted in response to hypoglycaemia and its principal action is the stimulus of glycogen breakdown by the liver as described earlier (p. 132). Additional actions in the liver include the stimulation of gluconeogenesis, an increase in amino acid catabolism and a decrease in fatty acid synthesis (Table 12.3). Glucagon also stimulates insulin release.

The principal role of glucagon is the prevention of hypoglycaemia during fasting. In addition glucagon and insulin are released together after a protein meal; in this way the insulin release stimulated by the rise in plasma amino acids does not cause hypoglycaemia.

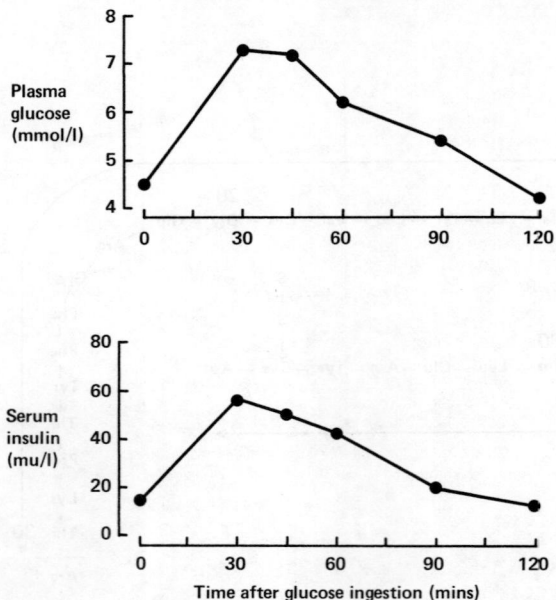

Fig. 12.12 Effect of 50 g glucose by mouth on plasma glucose and insulin. Mean values for 9 normal subjects. (Courtesy of Anna Challa)

Table 12.2 Principal actions of insulin

In liver
  Increased glycogen synthesis
  Increased fatty acid synthesis
  Decreased gluconeogenesis
  Decreased ketone production
  Decreased glycogen breakdown

In muscle
  Increased glucose uptake
  Increased glycogen synthesis
  Increased protein synthesis
  Decreased protein breakdown

In adipose tissue
  Increased glucose uptake
  Increased triglyceride storage
  Decreased triglyceride breakdown

Adrenaline

Adrenaline (epinephrine) is produced by the adrenal medulla in response to hypoglycaemia or to stress. Its major metabolic actions are in muscle where glycogen breakdown is increased, and in adipose tissue where triglyceride breakdown is increased. The actions of adrenaline are summarised in Table 12.3.

Table 12.3   The actions of glucagon, adrenaline, cortisol and growth hormone

	Glucagon	Adren-aline	Cortisol	Growth hormone
Liver				
Glycogen synthesis	↓	↓	↑	↑
Glycogen breakdown	↑	↑		
Gluconeogenesis	↑	↑		
Glucose release	↑	↑	↑	↑
Ketone body production	↑	↑		
Amino acid catabolism	↑			
Fatty acid synthesis	↓	↓		
Muscle				
Fatty acid utilisation		↑		↑
Glycogen breakdown		↑		
Glucose uptake		↓	↓	↓
Protein synthesis			↓	↑
Adipose tissue				
Fatty acid release		↑	↑	↑
Triglyceride synthesis				↓

Cortisol

Cortisol produced by the adrenal cortex (p. 263), provides a slow response to hypoglycaemia, stress or injury. It reduces protein synthesis and increases amino acid release from muscle. The amino acids can then be used for gluconeogenesis. The actions of cortisol on metabolism are summarised in Table 12.3.

Growth hormone

Growth hormone secretion is promoted by hypoglycaemia. While its principal action is the promotion of the growth of tissues, particularly bone and muscle, it also causes a rise in plasma glucose by increasing glucose release from the liver and decreasing glucose uptake by muscles, which use fatty acids to a greater extent as a source of energy (Table 12.3).

Integration of energy metabolism

Although the glycogen in liver is only a small fraction of the body's energy stores (Table 12.1), it is very important in the hour-to-hour regulation of blood glucose levels.

Effect of food and fasting

While a meal including carbohydrate is being digested and absorbed the plasma glucose rises. Insulin secretion increases and glucagon secretion diminishes (fig. 12.13). Glucose is taken up by the liver and glycogen is synthesised. Glucose entry into muscle and adipose tissue is also promoted; glucose is used to a greater extent as a source of energy for muscle and in adipose tissue promotes triglyceride synthesis (fig. 12.14).

At the same time some absorbed amino acids are taken up by the liver for the synthesis of glycogen; the remainder, including all the branch-chain amino acids, is taken up by muscle for conversion to protein. Triglycerides in the diet, largely reformed into chylomicrons, are partly stored in adipose tissue (p. 136), partly used for energy by muscle and other tissues, and partly taken up by the liver. There they are converted to VLDL which are passed on to adipose tissue. The high plasma insulin level acti-

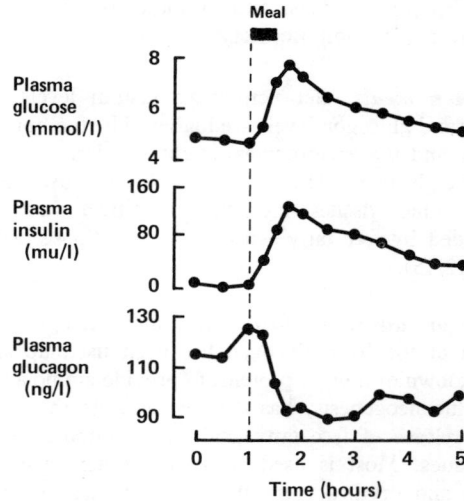

Fig. 12.13 Insulin and glucagon secretion after a meal containing carbohydrate. Mean values in 11 normal subjects. (After Ungar, R. (1970) *New England Journal of Medicine* **283**, 109)

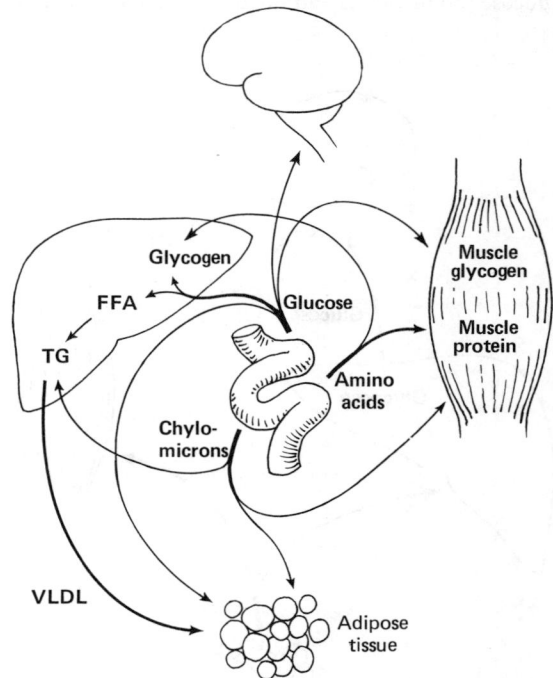

Fig. 12.14 The movements of fuel molecules during the digestion of a meal. TG = triglycerides, FFA = free fatty acids, VLDL = very low density lipoproteins

141

vates lipoprotein lipase and promotes the uptake of lipoproteins by adipose tissue.

Between meals. Between meals insulin levels are lower and glucagon levels are higher. The fuel for the brain (and the erythrocytes) is glucose derived from hepatic glycogen. The energy supply for muscle and most other tissues including the liver itself is provided by free fatty acids from the adipose tissue (fig. 12.15).

After an overnight fast. After an overnight fast much of the liver glycogen has been used up and breakdown of muscle protein, to provide amino acids for gluconeogenesis, has commenced. At the same time release of free fatty acids from adipose tissue continues. Most is used by tissues other than the brain and erythrocytes but ketone bodies are also produced in the liver; these are also used as the energy supply for muscle (fig. 12.16).

After prolonged fasting. After about two days of fasting glycogen stores in the liver are exhausted and glucose production is entirely due to gluconeogenesis

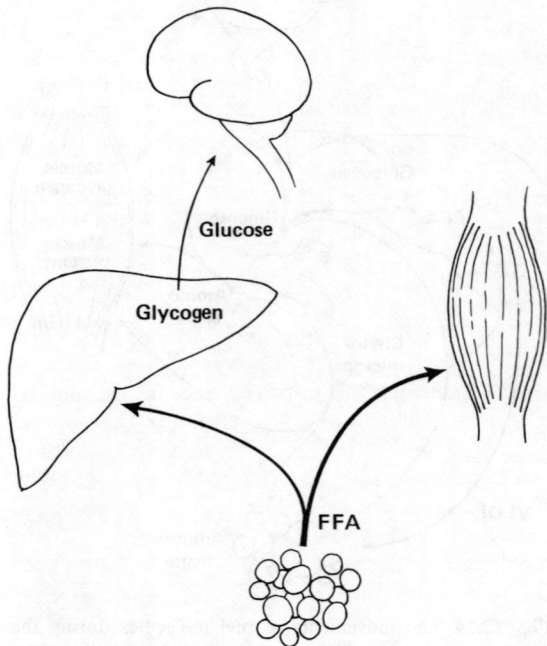

Fig. 12.16 Fuel metabolism after an overnight fast. The renal cortex also plays a major part in gluconeogenesis from amino acids. Some glycogen is still available in muscle. KB = ketone bodies

from amino acids, glycerol and lactate (fig. 12.17). While free fatty acids continue to be produced and utilised, an increasing proportion is converted to ketone bodies in the liver. As fasting continues the brain begins to use ketone bodies as a source of energy; after about ten days it derives a greater

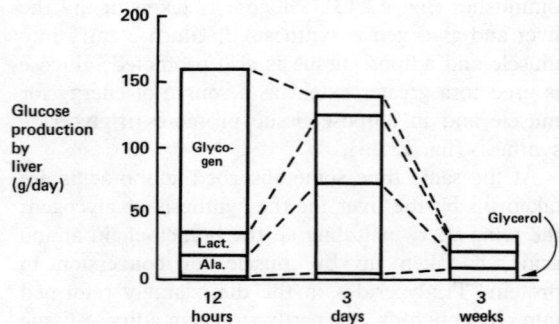

Fig. 12.15 Fuel metabolism between meals

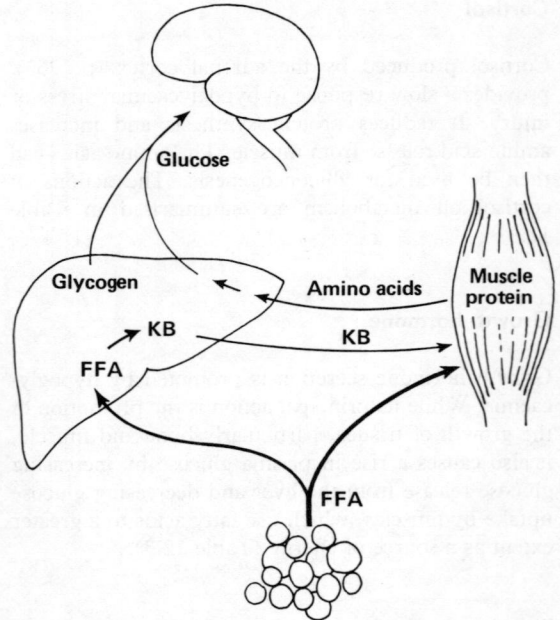

Fig. 12.17 Glucose production and its sources at different stages of fasting. (After Tygstrup, N. and Iversen, J. (1981) in *The Liver I*, eds Arias, I. M., *et al.*, p. 1. Amsterdam: Excerpta Medica)

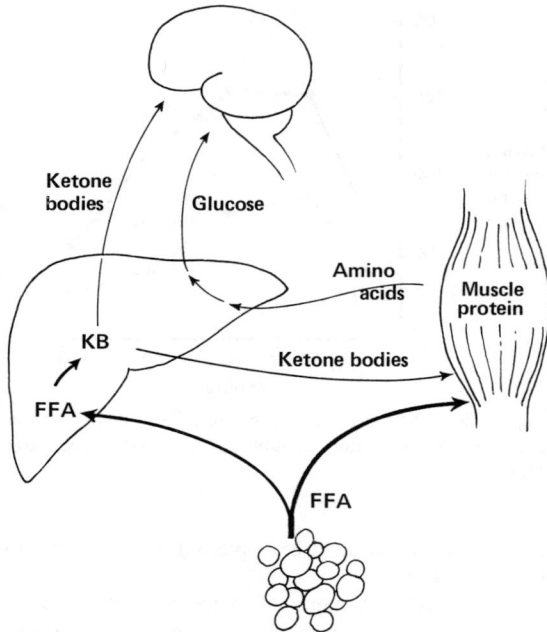

Fig. 12.18 Fuel metabolism after a prolonged fast. The brain derives an increasing proportion of its energy from ketone bodies

energy is obtained by the oxidation of free fatty acids derived from adipose tissue.

Metabolism after injury

In the first three or four days after an injury or surgical operation catecholamine release from sympathetic nerve endings and from the adrenal medulla is greatly increased. In turn this leads to an inhibition of insulin secretion and a stimulation of glucagon release. At the same time the secretion of ACTH, and therefore of cortisol, increases.

These hormonal changes result in the rapid breakdown of glycogen and muscle protein for the production of glucose. Free fatty acid mobilisation from adipose tissue increases and ketone bodies are formed in the liver.

Disorders of fuel metabolism

Obesity

An excess of energy intake over energy output leads to an increase in the amounts of triglyceride stored in adipose tissue. The amounts may be very large and exceed the weights of all the other tissues (fig. 12.19). Obesity is associated with an increased mortality and an increased liability to vascular and other diseases.

Although in many cases an excessive food intake is the principal causative factor in obesity, in some an unusually low energy output is important. The reasons for these variations between apparently normal people is not clear; variations in the amount of brown adipose tissue (p. 112) may play a part. In a small minority of obese subjects an endocrine disorder such as cortisol excess is responsible.

Many metabolic abnormalities are found in obesity. Plasma levels of insulin and cortisol are high. Glucose tolerance is impaired and the insulin response to a glucose load is greatly increased (fig. 12.20). Anabolic activity is increased both in liver and in adipose tissue. In the liver glycolysis and fatty acid synthesis are greatly increased; the increased triglyceride formed is mainly exported as VLDL but some may be deposited as lipid droplets within liver cells. In adipose tissue triglyceride storage is increased; the cells increase in size. It seems clear that most of the metabolic abnormalities seen in obesity are the result of rather than the cause of the disorder.

proportion of its energy from ketone bodies than from glucose (fig. 12.18). This has the effect of reducing the breakdown of muscle protein. While intellectual performance is unimpaired during a prolonged fast emotional changes are noted and may be the result of the change to ketone body utilisation by the brain.

Exercise

The first few minutes of any sudden exertion are characterised by the anaerobic breakdown of muscle glycogen to give lactate. This is reconverted to glucose by the liver (the Cori cycle, p. 96).

For the next 30 minutes or so the energy is obtained largely from blood glucose, some of it from the operation of the glucose–alanine cycle (p. 181). Glucose entry into the cell is facilitated by the enormously increased blood flow through muscle. If exertion continues an increasing proportion of the

143

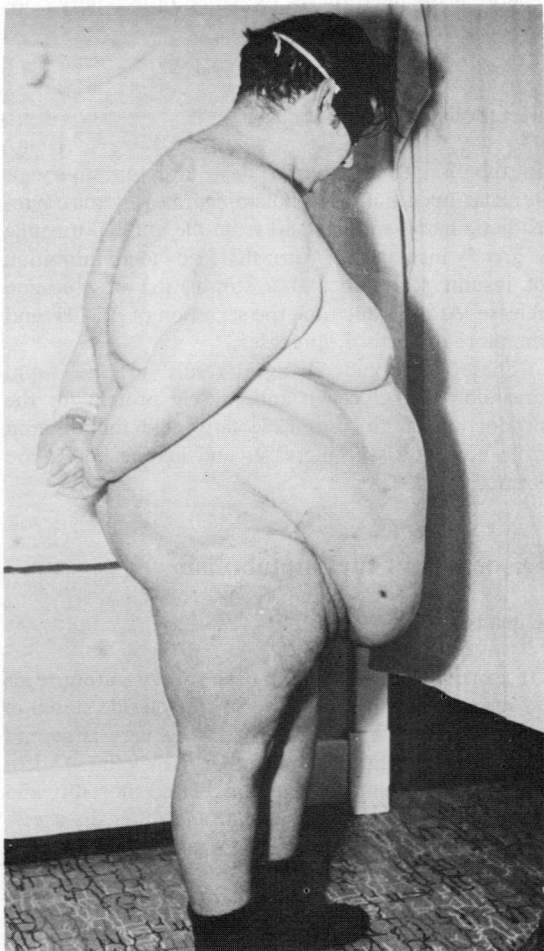

Fig. 12.19 Gross obesity. (Courtesy of W. K. Stewart)

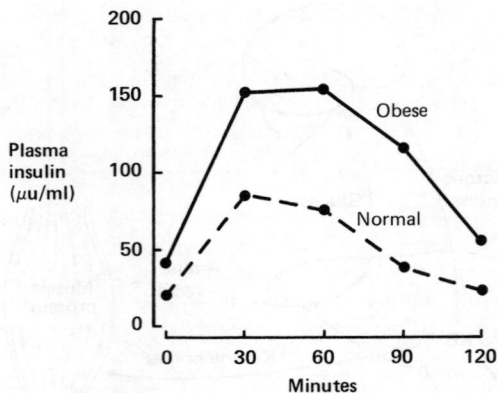

Fig. 12.20 Plasma insulin response to a 50 g glucose load in obese people and in normal subjects. (Courtesy of Margaret McKiddie)

from adipose tissue are converted in the liver to ketone bodies.

The initial symptom of diabetes mellitus is a greatly increased urine volume (*polyuria*) because large amounts of glucose are excreted in the urine and cause an osmotic diuresis. Despite drinking large amounts of water (*polydipsia*) the patient remains thirsty.

In milder cases these symptoms may continue for months but in severe cases, particularly in younger patients, the loss of muscle protein and the defective glucose uptake by muscle causes tiredness and weakness. The accumulation of ketone bodies (*ketosis*) leads to acidosis and overbreathing and eventually to loss of appetite, and vomiting.

If no treatment is given the continuing fluid losses lead to saline depletion (Chap. 19) and the increasing ketosis causes drowsiness, unconsciousness (*diabetic coma*) and death.

Treatment. Diabetes mellitus has two main variants. In one (type I or insulin-dependent type) plasma insulin levels are low and the patient requires insulin by injection to avoid ketosis and diabetic coma. In the other (type II) insulin levels are normal or high and many of the patients are obese. Most respond to weight reduction and dietary restriction.

Hypoglycaemia

Hypoglycaemia has a large number of causes; by far the most common is excessive insulin therapy. If

Diabetes mellitus

Diabetes mellitus is caused either by lack of insulin or by defective tissue responsiveness to circulating insulin. The biochemical features are hyperglycaemia, increased mobilisation of body fat and increased breakdown of protein. These abnormalities result from the lack of the normal effect of insulin in restraining gluconeogenesis, triglyceride hydrolysis and protein catabolism respectively. The amino acids from the protein breakdown lead to greatly increased glucose production in the liver and the fatty acids

severe, hypoglycaemia causes irreversible damage to the brain.

Insulin-producing-tumours of the pancreas (*insulinomas*) are rare but of considerable interest. A patient may be noted to behave oddly, as if drunk, particularly if a few hours have passed since a meal. Patients may learn to avoid symptoms by eating frequent carbohydrate-rich snacks; they tend to gain weight. Insulinomas can be suspected from the excessive fall in blood glucose in a prolonged fast and from the finding of plasma insulin levels inappropriately high in relation to the plasma glucose.

Glucagon excess

Glucagon secreting tumours of the pancreas are rare. They cause a variety of symptoms but particularly skin changes (fig. 12.21). This may be related to the characteristic biochemical abnormality, extremely low plasma levels of amino acids. The plasma glucose is often normal presumably because insulin secretion is increased.

Further reading

Atkins, G. L. (1981) *An Outline of Energy Metabolism in Man*. London: Heinemann

Bove, A. A. and Lowenthal, D. T. (1983) *Exercise Medicine*. New York: Academic Press

Cahill, G. F. (1976) Starvation in man. *Clinics in Endocrinology and Metabolism* 5, 397–415

Cohen, P. (1983) *Control of Enzyme Activity*, 2nd edn. London: Chapman and Hall

Cryer, P. (1984) Glucose counter-regulatory mechanisms in normal and diabetic man. In *Recent Advances in Diabetes*, eds Nattrass, M. and Santiago, J. V., pp. 73–90. Edinburgh: Churchill Livingstone

Dobbs, R. E. and Unger, R. H. (1982) Glucagon: secretion, function and clinical role. In *Contemporary Metabolism 2*, ed. Freinkel, N., pp. 61–118. New York: Plenum

Flier, J. S. (1983) Insulin receptors and insulin resistance. *Annual Review of Medicine* **34**, 145–160

Foster, D. W. (1983) Diabetes mellitus. In *Metabolic Basis of Inherited Disease*, 5th edn, eds Stanbury, J. B., Wyngaarden, J. B. Fredrickson, D. S., Goldstein, J. L., Brown, M. S. pp. 99–117. New York: McGraw-Hill

Garrow, J. S. (1981) Obesity and energy balance. In *Recent Advances in Medicine 18*, eds Dawson, A. M., Compston, N. and Besser, G. M., pp. 75–92. Edinburgh: Churchill Livingstone

Greenwood, M. R. C. (1983) *Obesity*. New York: Churchill Livingstone

Hers, H.-G., Hue, L. and Van Schaftigen, E. (1982) Fructose 2,6-bisphosphate. *Trends in Biochemical Sciences* 7, 329–331

Johnston, D. G., Pernet, A. and Nattrass, M. (1984) Hormonal regulation of fatty acid mobilisation in normal and diabetic man. In *Recent Advances in Diabetes*, eds Nattrass, M. and Santiago, J. V., pp. 91–106. Edinburgh: Churchill Livingstone

Marks, V. and Rose, F. C. (1981) *Hypoglycaemia*, 2nd edn. Oxford: Blackwell

McArdle, W. D., Katch, F. I. and Katch, V. L. (1981) *Exercise Physiology: Energy, Nutrition and Human Performance*. Philadelphia: Lea and Febiger

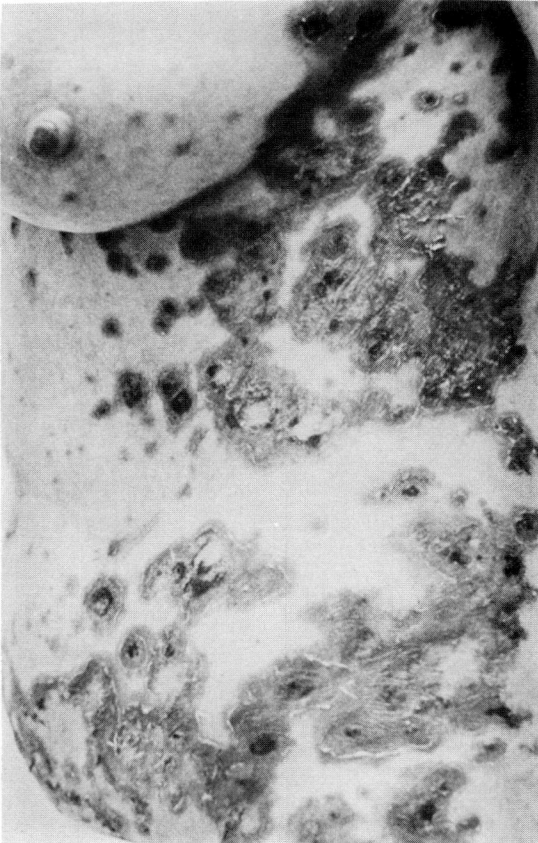

Fig. 12.21 Skin changes in a patient with a glucagonoma. (From Helland, S. *et al.* (1979) *Tidsskrift for den Norske Laegerforening* **99**, 638 by courtesy of authors and editor. Figure kindly provided by J. J. Holst)

Owen, O. E., Caprio, S., Reichard, G. A., Mozzoli, M. A., Boden, G. and Owen, R. S. (1983) Ketosis of starvation. *Clinics in Endocrinology and Metabolism* **12**, 359–379

Randle, P. J., Steiner, D. F. and Whelan, W. J. (1981) *Carbohydrate Metabolism and its Disorders 3*. London: Academic Press

Smith, R. (1980) The metabolic effects of injury. In *Scientific Foundations of Orthopaedics and Traumatology*, eds Owen, R., Goodfellow, J. and Bullough, P. London: Heinemann

White, D. A., Middleton, B. and Baxter, M. (1984) *Hormones and Metabolic Control*. London: Arnold

C R P   P C   D G H   T E I

13 Metabolism of nucleotides & nucleic acids

THE structures of the nucleic acids and of their constituent nucleotides were described in Chapter 5. This chapter outlines the pathways for the synthesis of nucleotides, the replication of DNA, the formation of RNA and the breakdown of nucleic acids. Chapter 14 describes the way in which the sequence of bases on DNA is applied to the synthesis of proteins.

Synthesis of purine nucleotides

The purine ring is built up on a molecule of ribose 5-phosphate derived from the pentose–phosphate pathway (p. 99). Ribose 5-phosphate is first converted to 5-phosphoribosyl-1-pyrophosphate (PRPP) (fig. 13.1). The next stage yielding 5-phosphoribosyl-1-amine is the key rate-controlling step. The enzyme involved is subject to allosteric inhibition (p. 87) by IMP, GMP and AMP. A defect in this inhibitory mechanism causes one form of gout (p. 159).

The remaining steps of purine synthesis are outlined in figures 13.2 and 13.3. It can be seen that the atoms which make up the purine ring are derived from several different precursors. Among them are two derivatives of the vitamin folic acid whose role is described on page 152).

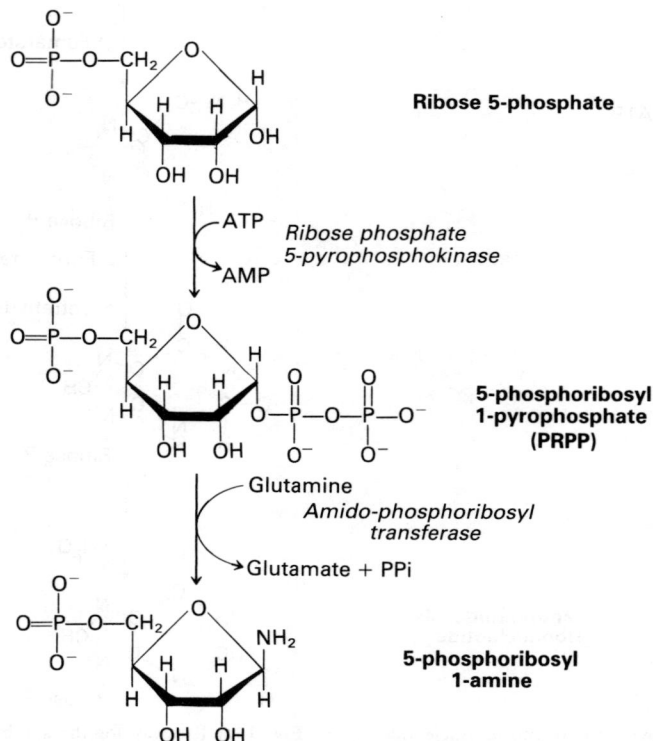

Fig. 13.1 The first two steps in purine synthesis. Amido-phosphoribosyl transferase is a key regulatory enzyme

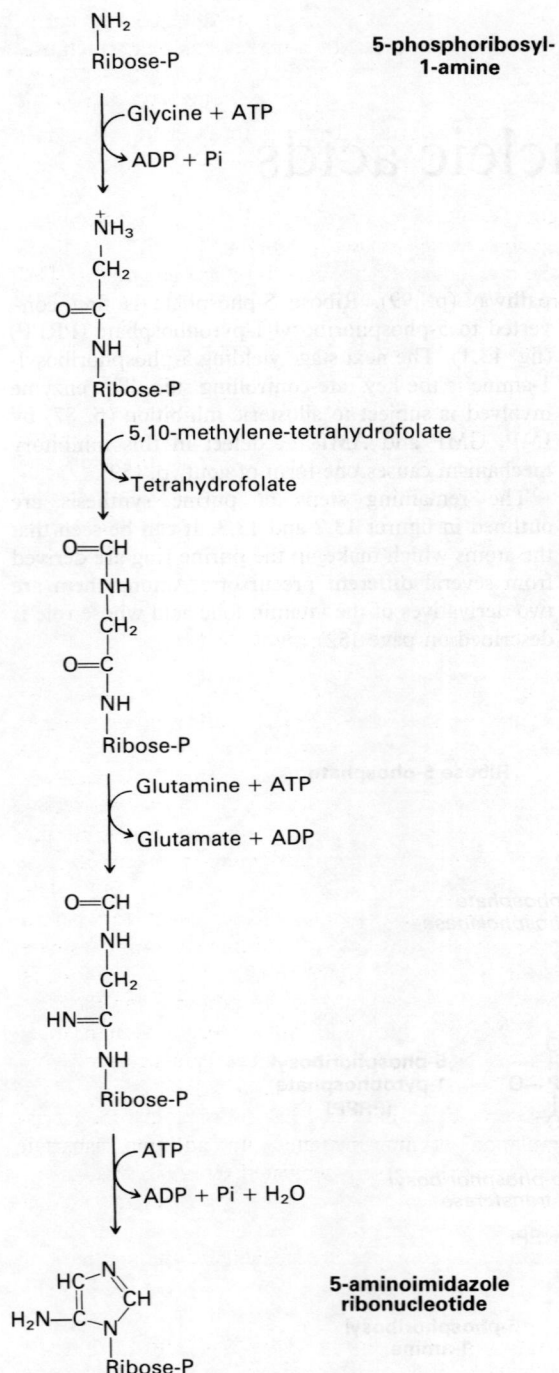

Fig. 13.2 Pathway for the synthesis of purine nucleotides, stage 2

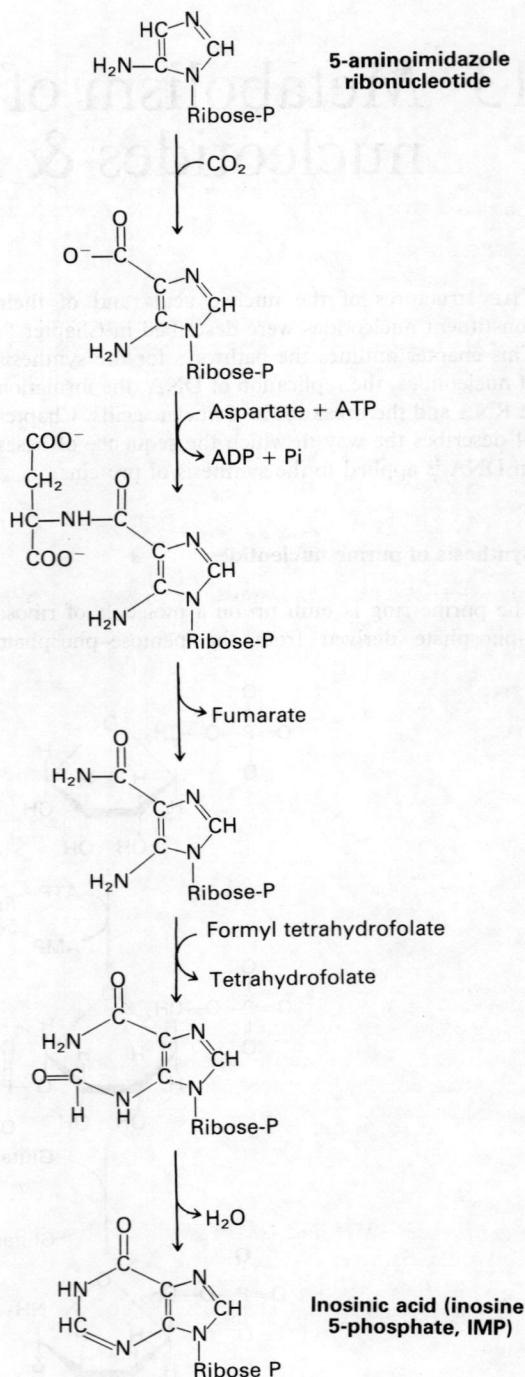

Fig. 13.3 Pathway for the synthesis of inosine monophosphate (IMP)

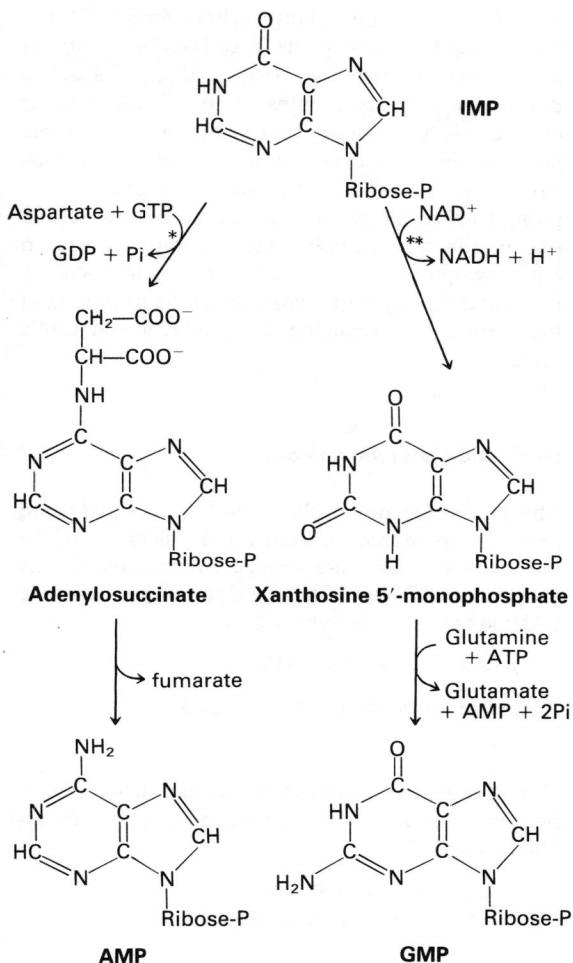

Fig. 13.4 The pathways for the production of adenosine monophosphate (AMP, adenylic acid) and guanosine monophosphate (GMP, guanylic acid) from IMP. The step marked * is inhibited by AMP and that marked ** is inhibited by GMP

Formation of AMP and GMP. With the formation of IMP, the pathway divides to give AMP and GMP (fig. 13.4). The first step in each case is rate limiting so that excessive amounts of the intermediates do not accumulate.

Pathways for the salvage of purine bases. Free purines are formed during the breakdown of nucleic

acids and nucleotides. Purine nucleotides can be reformed from these by a *salvage pathway* which uses PRPP (fig. 13.1). This pathway produces nucleotides at considerably less cost in metabolic energy than the synthesis described earlier. Two separate enzymes are involved in the salvage pathway. *Adenine phosphoribosyl transferase* (APRT) is responsible for the conversion of adenine to AMP while *hypoxanthine-guanine phosphoribosyl transferase* (HGPRT) catalyses the conversion of hypoxanthine and guanine to IMP and GMP respectively.

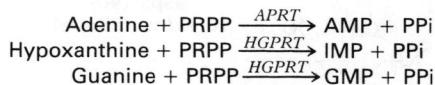

$$\text{Adenine} + \text{PRPP} \xrightarrow{APRT} \text{AMP} + \text{PPi}$$
$$\text{Hypoxanthine} + \text{PRPP} \xrightarrow{HGPRT} \text{IMP} + \text{PPi}$$
$$\text{Guanine} + \text{PRPP} \xrightarrow{HGPRT} \text{GMP} + \text{PPi}$$

Defects in these enzymes are well recognised causes of disease. Partial deficiency of HGPRT is one cause of gout while a complete deficiency causes a very unpleasant inborn error of metabolism, the Lesch–Nyhan syndrome (p. 160).

Synthesis of pyrimidine nucleotides

The starting compounds for the synthesis of the pyrimidines are aspartate and carbamyl phosphate. Although carbamyl phosphate is also used for the synthesis of urea (p. 178) the two processes are distinct; carbamyl phosphate used in pyrimidine synthesis is formed in the cytosol from glutamine and HCO_3^- while that for urea synthesis is produced in the mitochondria from NH_4^+ and HCO_3^-. The pathway leading to the production of the parent pyrimidine nucleotide, uridine monophosphate (UMP), is shown in figure 13.5. The pathway is controlled by the enzymes responsible for the first two steps; carbamyl phosphate production is inhibited by UMP while aspartate carbamyltransferase is inhibited by CTP. In these ways, as in other pathways, feed-back regulation prevents the accumulation of intermediates. In addition aspartate carbamyltransferase is activated by ATP; this activation serves to ensure that approximately equal amounts of pyrimidines and purines are produced since equal amounts are needed for the synthesis of DNA and RNA.

Defects of pyrimidine synthesis. The only well-recognised disorder of pyrimidine synthesis is the very rare *hereditary orotic aciduria*. This results from deficiency of the protein with the two enzyme

149

COO⁻
CH₂ **Aspartate**
CH—COO⁻
NH₂

HCO₃⁻ → **Glutamine + 2ATP**
Carbamylphosphate synthetase
→ **Glutamate + 2ADP + Pi**

H₂N—CO—O—PO₃ **Carbamyl phosphate**

Aspartate carbamyltransferase
→ Pi

COO⁻
CH₂
H₂N
CO   CH—COO⁻
NH

Carbamyl-aspartate (ureidosuccinate)

Dihydro-orotase

CO
NH   CH₂
CO   CH—COO⁻
NH

Dihydro-orotate

NAD⁺
Dihydro-orotic dehydrogenase
→ NADH + H⁺

CO
NH   CH
CO   C—COO⁻
N

Orotate

PRPP
Orotate phosphoribosyl transferase
→ PPi

CO
NH   CH
CO   C—COO⁻
N
Ribose-P

Orotidine 5′-monophosphate

Orotidine 5′-phosphate decarboxylase
→ CO₂

CO
NH   CH
CO   CH
N
Ribose-P

Uridine 5′-monophosphate (UMP)

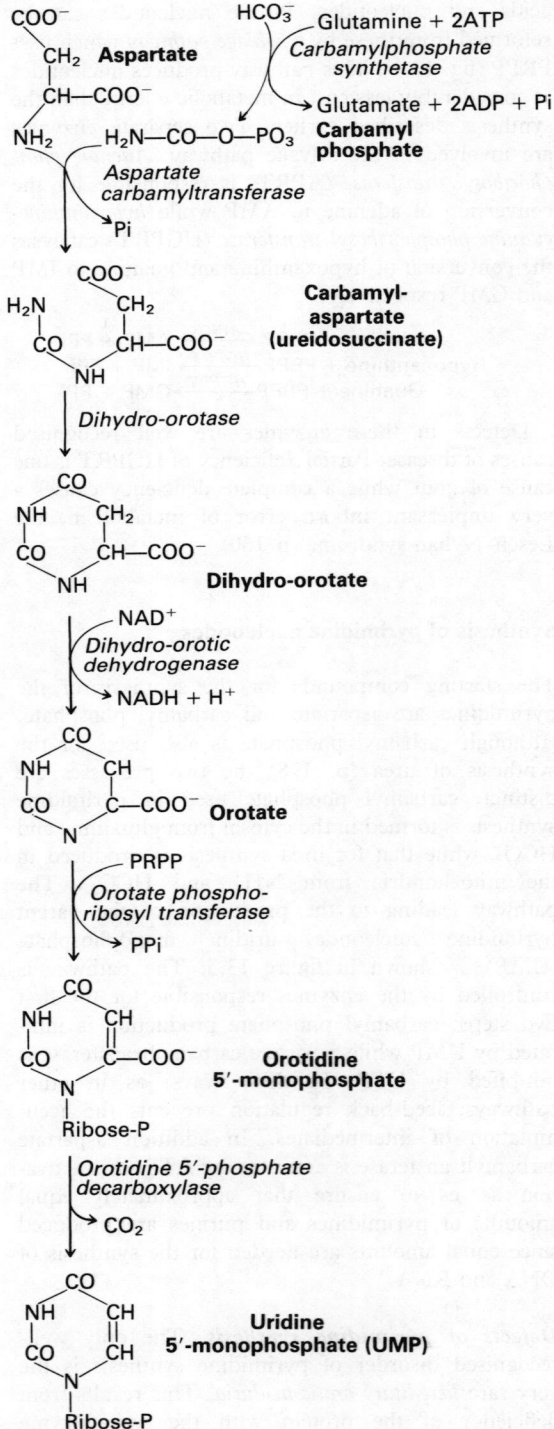

activities, orotate phosphoribosyltransferase (OPRT) and orotidine 5′-phosphate decarboxylase. Patients are normal at birth but become severely anaemic during the first six months of life. Some patients excrete crystals of orotic acid in the urine and some have a degree of mental retardation. Several patients have been treated with uridine or cytidine with a prompt fall in the urinary excretion of orotic acid, a rise in the haemoglobin, and an improvement in other symptoms. The uridine or cytidine acts by inhibiting the regulatory enzymes at the beginning of the pathway, so reducing the production of orotic acid.

Nucleotide interconversions

The nucleoside diphosphates and triphosphates are important as sources of energy and precursors of the nucleic acids. They are formed by interconversions involving ATP and kinase enzymes. For example UMP kinase phosphorylates UMP:

$$ATP + UMP \rightleftharpoons ADP + UDP$$

Adenylate kinase phosphorylates AMP:

$$ATP + AMP \rightleftharpoons 2ADP$$

A wide variety of nucleoside diphosphates can be phosphorylated by the enzyme nucleoside diphosphate kinase. Thus

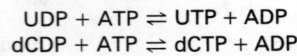

$$UDP + ATP \rightleftharpoons UTP + ADP$$
$$dCDP + ATP \rightleftharpoons dCTP + ADP$$

Cytidine nucleotide formation. The second major pyrimidine, cytidine, is formed as its triphosphate (CTP) by a transamination involving glutamine (fig. 13.6).

Formation of deoxyribose nucleotides. The deoxyribonucleotides needed for the synthesis of DNA are formed from the corresponding ribonucleoside diphosphates; NADPH is required. The enzyme

Fig. 13.5 Pathway for the synthesis of uridine monophosphate (UMP), the parent pyrimidine nucleotide. The first three enzymes are all parts of a single large protein with a molecular mass of 200 000. The last two enzymes are also parts of a single protein; both activities are defective in orotic aciduria

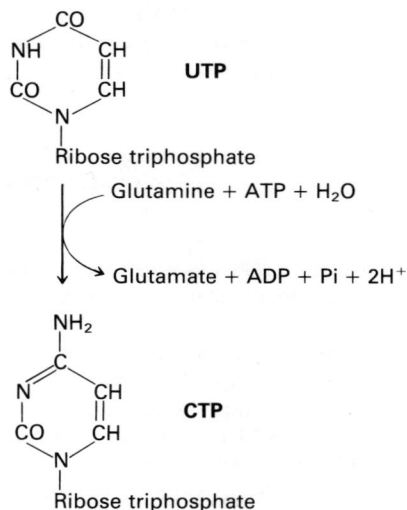

Fig. 13.6 Formation of cytidine triphosphate (CTP) from uridine triphosphate (UTP)

involved is *ribonucleotide reductase* a complex iron-containing enzyme. The overall reaction is

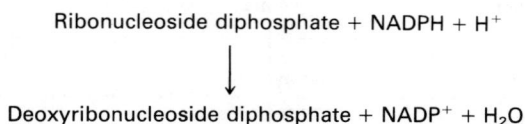

$$\text{Ribonucleoside diphosphate} + \text{NADPH} + \text{H}^+$$

$$\downarrow$$

$$\text{Deoxyribonucleoside diphosphate} + \text{NADP}^+ + \text{H}_2\text{O}$$

A protein known as *thioredoxin* is also involved in this reaction. It has two sulphydryl groups close together and transfers electrons from the NADPH to the sulphydryl groups in the catalytic site of the enzyme:

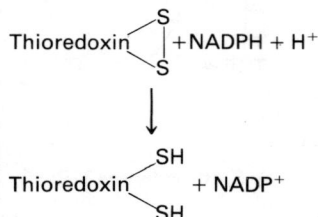

Ribonucleotide reductase requires the presence of specific nucleotides which act as allosteric effectors for the reduction of particular nucleotides. Thus the reduction of CDP and UDP requires ATP, that of ADP requires dGTP and ATP, and that of GDP requires dTTP and ATP. This allosteric regulation of ribonucleotide reductase is defective in patients with

a rare inborn metabolic error *adenosine deaminase deficiency*. Adenosine deaminase normally deaminates adenosine and deoxyadenosine to inosine and deoxy-inosine. In its absence deoxyadenosine accumulates in excessive amounts. The inhibition of ribonucleotide reductase in lymphoid tissue causes a severe disturbance of the immune response and recurrent infections.

Ribonucleotide reductase produces dCDP, dUDP, dGDP and dADP. All are phosphorylated further to give the triphosphates but dUTP is immediately hydrolysed to dUMP since dUTP is not used for DNA synthesis.

Thymine nucleotide formation. The essential step in the production of thymine nucleotides is the methylation of dUMP to dTMP by the enzyme thymidylate synthetase (fig. 13.7). This reaction involves the folic acid derivative 5,10-methylene tetrahydrofolate which is regenerated from the dihydrofolate by a process described later (p. 152).

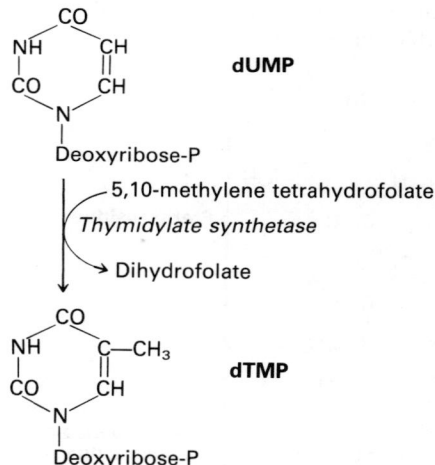

Fig. 13.7 The formation of deoxythymidine monophosphate (dTMP)

Antimetabolites

Many important drugs which are effective against neoplastic cells act by inhibiting the formation of dTMP, essential for DNA synthesis in rapidly dividing cells. An example is fluorouracil (fluoro-deoxyuridine) which is converted by the normal metabolic pathways to a fluorinated analogue of dUMP. In turn this is bound to thymidylate synthetase and acts as an irreversible inhibitor. Other effective anti-cancer drugs include aminopterin and amethopterin which block the regeneration of 5,10-methylene tetrahydrofolate (fig. 13.9) and therefore, indirectly, dTMP production.

5-Fluorocytosine (flucytosine) is an analogue of cytosine which inhibits the growth of many pathogenic fungi but is harmless to the patient.

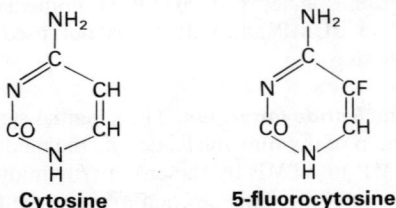

Cytosine   **5-fluorocytosine**

Folate coenzymes

Derivatives of folate are involved in the transfer of one-carbon units at two stages in purine synthesis

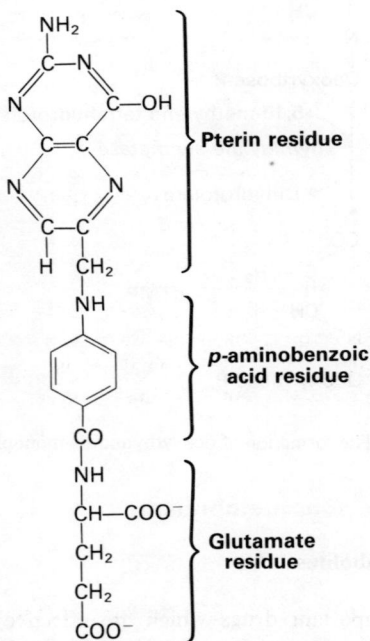

Fig. 13.8 Structure of the folate molecule. In plant and animal tissues folate occurs mainly as polyglutamates with as many as seven glutamate residues per molecule

and also in the formation of dTMP from dUMP. It is not surprising therefore that folate deficiency causes disturbances in nucleic acid formation.

Folate is a water-soluble vitamin found in fresh green vegetables and in liver. Its structure is shown in figure 13.8. In plant and animal tissues folate occurs mostly in the form of polyglutamates with as many as seven glutamate residues attached to each molecule.

Dietary folate is broken down in the lumen of the intestine to a monoglutamate which is taken up by the mucosal cells. There folate is converted to 5-methyl tetrahydrofolate (methyl THF) which is the principal form of folate in the plasma.

Methyl THF is taken up by cells and converted to THF by a reaction which involves methylcobalamin (p. 152). THF is an essential coenzyme; it can accept a one-carbon unit from serine and become 5,10-methylene THF. When this gives up its one-carbon unit THF can be regenerated as shown in figure 13.9. The structures of THF and its derivatives are given in figure 13.10. THF is also involved in the synthesis of serine and histidine.

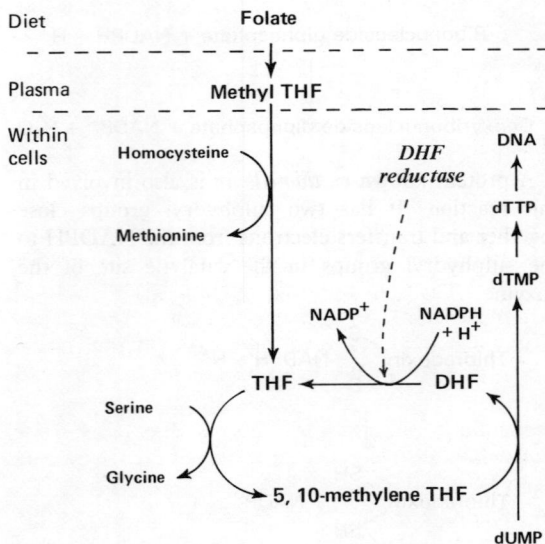

Fig. 13.9 Pathways of folate metabolism and its essential role in DNA synthesis. The enzyme DHF reductase is blocked by folate analogues such as amethopterin which therefore inhibit DNA synthesis. Such substances can be used in the treatment of certain forms of malignant disease such as leukaemia. THF = tetrahydrofolate; DHF = dihydrofolate

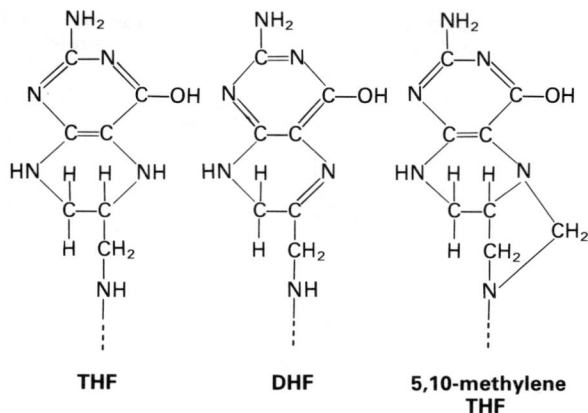

Fig. 13.10 Structures of tetrahydrofolate (THF), dihydrofolate (DHF) and 5,10-methylene THF. In each case only the pterin nucleus is shown

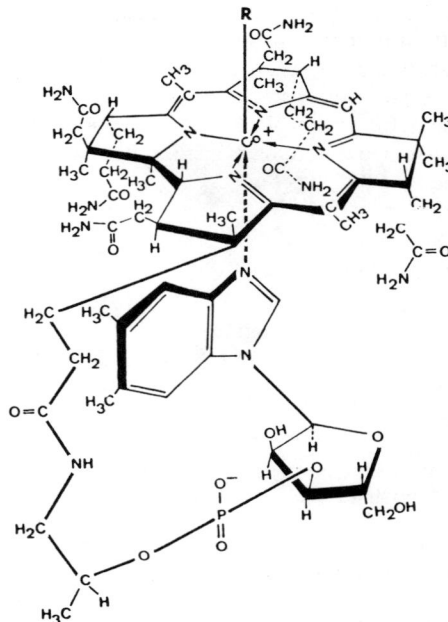

Fig. 13.11 General structure of the cobalamins. Each molecule consists of two main parts—a corrin ring containing cobalt, and a nucleotide. In hydroxocobalamin R = OH, in methylcobalamin R = CH_3 and in 5-deoxyadenosylcobalamin R is a 5'-deoxyadenosyl group

The cycle for the recovery of THF is not perfect and new THF must therefore be obtained from the plasma and, ultimately, from dietary folate. Non-pregnant adults need about 200 μg folate daily; more is required in pregnancy. Folate deficiency causes an anaemia characterised by large erythrocyte precursors in the marrow (megaloblastic anaemia).

The cobalamins

The cobalamins (vitamin B_{12}) are a group of complex substances with the general structure shown in figure 13.11. Three cobalamins are found in plasma (fig. 13.12). 5-Deoxyadenosylcobalamin is the major form within cells; it is a coenzyme in the conversion of methylmalonyl CoA to succinyl CoA. Methylcobalamin, the principal cobalamin in plasma, is a coenzyme in the methylation of homocysteine to methionine. Since this reaction is essential for the utilisation of folate (fig. 13.9), lack of cobalamin causes a megaloblastic anaemia similar to that caused by folate deficiency. In addition patients with cobalamin deficiency may have neurological changes including a peripheral neuropathy and damage to the posterior and lateral columns of the spinal cord. The anaemia responds rapidly to treatment with hydroxocobalamin but the neurological changes may be permanent.

The cobalamins are present in all foods of animal origin and dietary deficiency occurs only in very strict vegetarians (vegans). A much more common cause of deficiency is malabsorption. Cobalamins are absorbed in the ileum only when bound to a specialised glycoprotein secreted by the parietal cells of the stomach (*intrinsic factor*). *Pernicious anaemia* is the megaloblastic anaemia caused by lack of intrinsic factor due to atrophy of the parietal cells. Before the physiology of the cobalamins was understood pernicious anaemia was always fatal. It is now treated successfully with monthly injections of hydroxocobalamin.

Nucleic acid metabolism

DNA synthesis

A new strand of DNA is synthesised on a template of an existing strand by the assembly of nucleoside triphosphates. The enzymes involved are *DNA polymerases* (fig. 13.13). These enzymes only assemble the nucleotides which can form base pairs (p. 55)

153

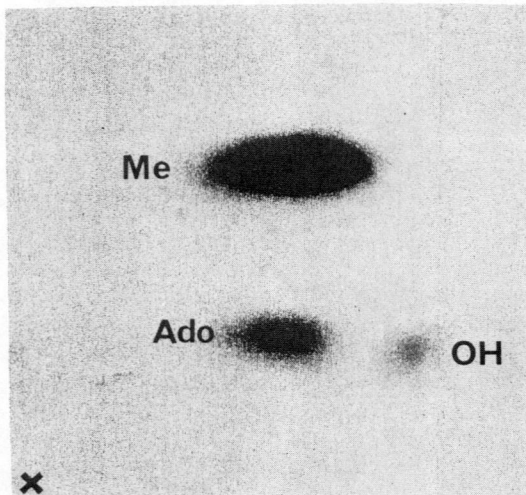

Fig. 13.12 The individual cobalamins of normal human plasma. Me = methyl cobalamin, Ado = 5′-deoxyadenosyl-cobalamin, OH = hydroxocobalamin. The cobalamins were separated by thin-layer chromatography and identified by a cobalamin-dependent strain of *E. coli*. X marks the origin. (From Matthews, D. M. and Linnell, J. C. (1978) *British Medical Journal* **2,** 533 by courtesy of authors and editor)

with the nucleotides on the existing strand; the old and the new strands are therefore complementary (fig. 5.8, p. 55). Before DNA polymerase can act the double helix has to be unwound. This process involves several enzymes including a 'helicase' which derives energy from the hydrolysis of ATP. The helix is prevented from reforming until replication is complete by specific *helix destabilising proteins* which bind to the single strands of DNA.

The two strands of the double helix of DNA are replicated simultaneously. The replication can proceed only in the 5′ to 3′ direction so that the formation of one strand is readily understandable. The other strand is formed initially as fragments of between 200 and 300 nucleotides (*Okazaki fragments*). These are then joined together by *DNA ligase* which uses ATP as its source of energy (fig. 13.14).

DNA polymerase requires a 'primer'. This has been shown to be a short segment of RNA, of about 10 bases, formed by an RNA polymerase. These sections of RNA allow DNA polymerase to begin to operate; later in the process they are removed and replaced with DNA.

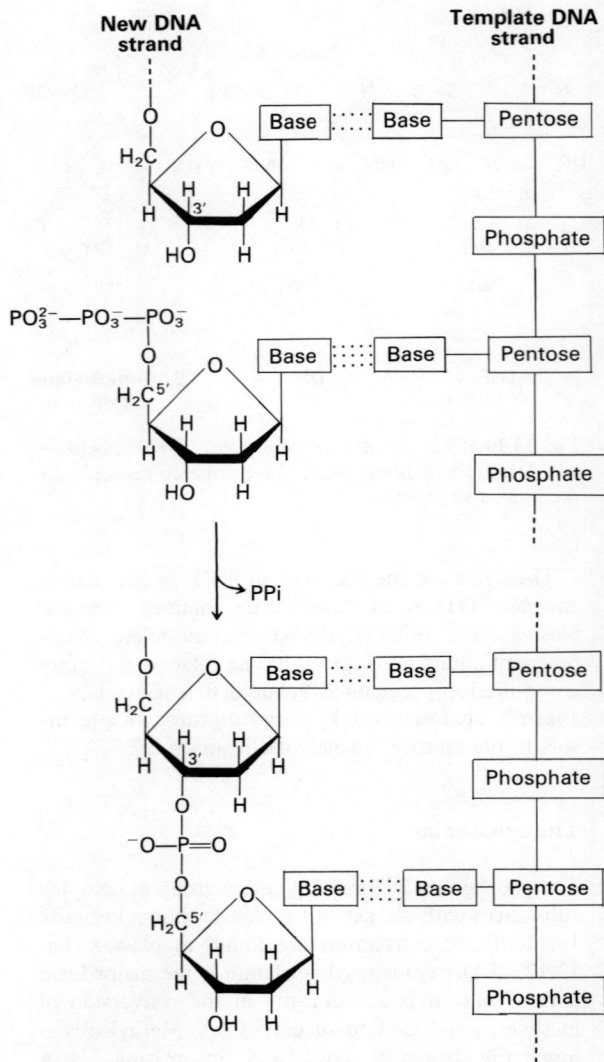

Fig. 13.13 The elongation of a new DNA strand catalysed by DNA polymerases. It can be seen that the molecule is built up in 5′ to 3′ direction

RNA transcription

RNA is synthesised by RNA polymerases which also use the DNA molecule as a template for the assembly of the ribonucleoside triphosphates. RNA polymerases also synthesise new strands in a 5′ to 3′ direction. Unlike DNA replication, however, RNA synthesis is selective in the points of initiation and

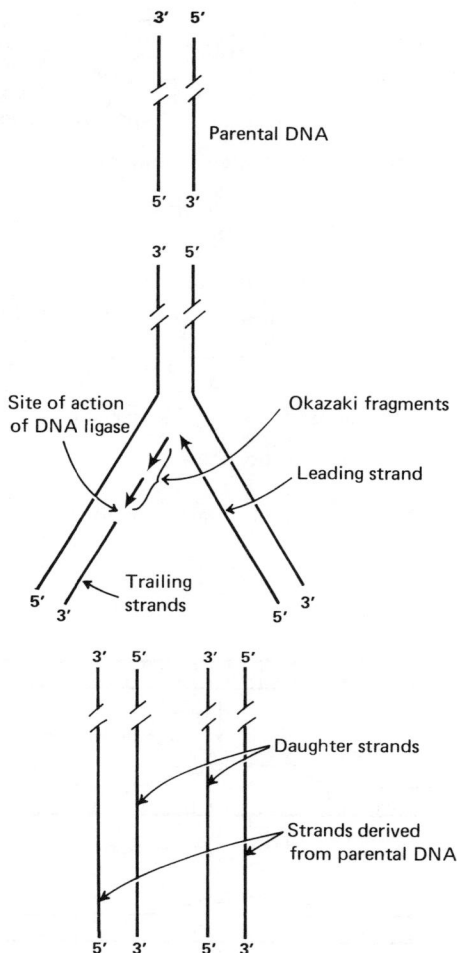

Fig. 13.14 Replication of double-stranded DNA. Each parental strand acts as a template for the assembly of the new polynucleotide chains. The daughter strands are assembled in a 5′ to 3′ direction; the 'trailing strand' is initially formed as fragments which are then joined by DNA ligase. It has recently become clear that both in the leading strand and in the trailing strand DNA assembly is preceded by the formation of an RNA 'primer' which is removed by a ribonuclease and replaced by DNA

termination on the DNA molecule. These positions are determined by particular sequences of bases which are recognised by the polymerase.

Post-transcriptional modifications. Transcription yields a *primary transcript* which in the case of ribosomal RNA is a large molecule with a sedimentation coefficient of 45 S. This is hydrolysed by various nucleases in the nucleolus to yield the ribosomal RNA molecules. While this process is continuing the proteins of the ribosomes enter the nucleolus and become associated with the appropriate RNA molecules to form ribosomal subunits.

Messenger RNA is also subject to considerable modification before it is used as a template for protein synthesis (fig. 14.4, p. 167). One frequent post-transcriptional modification of tRNA and rRNA is the methylation of certain bases and pentoses. The source of the methyl groups is *S*-adenosylmethionine (Chap. 15).

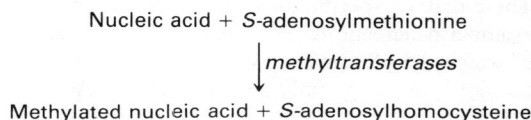

Nucleic acid + *S*-adenosylmethionine

$$\downarrow \textit{methyltransferases}$$

Methylated nucleic acid + *S*-adenosylhomocysteine

In the case of rRNA, methylation of the 2′ hydroxy group of the ribose plays an important part in the accurate conversion of the 45 S precursor to the 18 S and 28 S products. The parts of the precursor which are to be preserved are methylated to a substantial extent and so protected from attack by the ribonucleases in the nucleoli. Other post-transcriptional modifications include glycosylation, phosphorylation and acylation.

Inhibitors of RNA synthesis. The process of transcription is affected by a number of antibiotic drugs. Actinomycin D, a polypeptide-containing substance produced by *Streptomyces*, binds to the DNA template at guanine residues and prevents transcription. Since it affects both bacteria and higher animals it is not used clinically.

The synthetic drug *rifampicin* binds to RNA polymerase in bacteria but not in man. It prevents the binding of enzyme to DNA, and is used in the treatment of tuberculosis.

The toxin in the poisonous mushroom *Amanita phalloides* is the peptide *α-amanitin* which binds to the RNA polymerases responsible for the synthesis of mRNA and tRNA. Within 6–12 hours of eating the mushroom acute liver damage with abdominal pain, diarrhoea and jaundice develops. Many cases are fatal.

ESSENTIALS OF HUMAN BIOCHEMISTRY

Nucleic acid breakdown

The bonds between nucleotides are hydrolysed by *nucleases*. Some are specific for DNA, some are specific for RNA and some are capable of hydrolysing both types of nucleic acid.

Some nucleases, the *exonucleases*, hydrolyse only an internucleotide bond at the end of a strand of nucleic acid. Such enzymes release nucleotides from the ends one at a time. Some of these are specific for the 3′ ends and others for the 5′ ends. An exonuclease which attacks the 3′ ends of DNA is found in the venom of certain snakes.

Other nucleases, the *endonucleases*, hydrolyse internucleotide bonds throughout the nucleic acid chains. These may be specific for certain types of bond. For example pancreatic RNAase only hydrolyses a bond in which the phosphate residue is attached to the 3′ position on a pyrimidine nucleotide. Pancreatic nucleases are important digestive enzymes which break down nucleic acids in foods. Endonucleases are found in the venoms of certain jellyfish.

DNA repair

Various chemical and physical agents can cause alterations in the structure of DNA such as modifications and deletions of bases, and cross-linking between strands. These changes could lead to the misinterpretation or complete loss of the original genetic information. An example is provided by the effect of ultraviolet radiation on the skin in causing the formation of dimers between adjacent pyrimidine residues (fig. 13.15). This distorts one strand of the double helix and prevents base-pairing over a considerable length. The damaged area is first cut by an endonuclease and then removed one nucleotide at a time by exonucleases. Finally a new strand is laid down on the template of the complementary chain and joined on by a DNA ligase (fig. 13.16).

It seems likely that a similar process is used to repair breaks in DNA caused by ionising radiation. If two breaks are in nearby strands of DNA misrepair may take place; these may cause an increased liability to certain neoplastic diseases such as leukaemia.

Disorders of DNA repair. The best known disorder is *xeroderma pigmentosum*, a rare recessively inherited disease characterised by a greatly increased sensitivity to solar radiation. Soon after birth the patient may

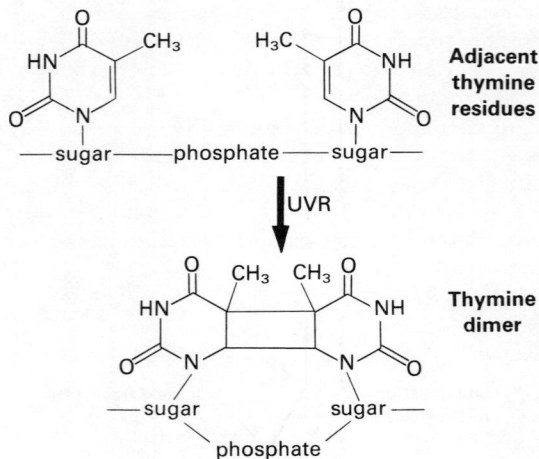

Fig. 13.15 The action of ultraviolet radiation (UVR) on DNA

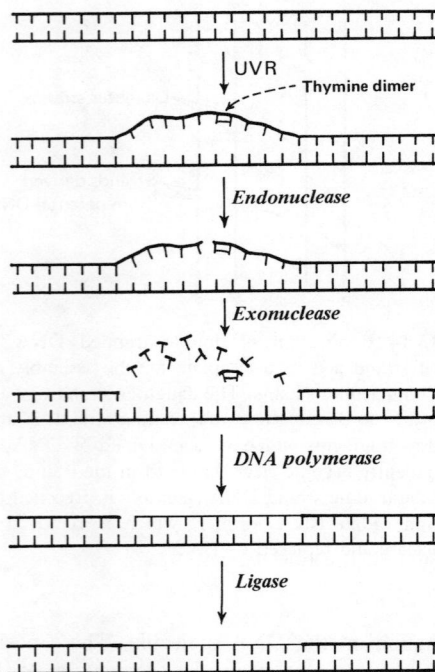

Fig. 13.16 Probable mechanism for the repair of a section of DNA in which a thymine dimer has formed as a result of the action of ultraviolet radiation. (After Lambert, B. and Ringborg, U. (1976) *Acta Medica Scandinavica* **200,** 433)

develop reddening of the skin, oedema and blisters. Freckles appear in large numbers on exposed areas of the skin and later benign and malignant skin neoplasms may develop.

Ataxia telangiectasia is a rare disorder caused by defective repair of DNA after exposure to ionising radiation. Among the features are cerebellar atrophy and an increased likelihood of malignancy. Patients have severe and sometimes fatal reactions to radiotherapy.

Genetic engineering

The practice of medicine is likely to be greatly altered by recent developments in the technology of gene manipulation. It is now possible to insert a section of DNA from one species into the cells of another, usually the bacterium *Escherichia coli*, so that the host produces the protein for which the section of DNA codes. This process has already been used to prepare large quantities of human proteins such as insulin, growth hormone and interferon.

The DNA can be prepared in two ways. In the first the required section of DNA can be excised from the DNA of the genome of the donor with great precision with endonucleases which attach DNA molecules at very specific base sequences. These enzymes, known as *restriction endonucleases* are found in bacteria where their normal function is to destroy foreign DNA entering the cell as viruses. The specificity ensures that the bacterial DNA is not affected. A large number of such enzymes are now available with a wide range of specificities.

The second technique for preparing the DNA involves first isolating the mRNA for the particular protein. By using the enzyme *reverse transcriptase*, which is isolated from cells of certain virus-induced tumours, a complementary strand of DNA (cDNA) can be synthesised.

In either case the next step is the insertion of the fragments of DNA into a *vector* which can be used to carry it into the host bacterial cell. The most usual vectors are *plasmids* which are self-replicating rings of double-stranded DNA found in bacteria in addition to the normal chromosomes. A plasmid may be split by restriction endonucleases and the DNA fragments can then be spliced in when the plasmid and the fragments are incubated together in the presence of a DNA ligase, which also reforms the ring.

The plasmid, now containing the fragment of donor DNA, can be used to infect bacteria so that the production of many copies of the required DNA takes place. Various techniques can be used to select the bacteria which contain the genes required. When this DNA is transcribed in the bacterium mRNA is produced and, in turn, the bacteria produce the required protein in large quantities.

Apart from the value of the new technology for protein production, the new techniques have proved useful for the identification of particular lengths of nucleotide sequence. Individual genes can be isolated and small differences in sequence can be shown up. The cDNA of a particular protein should be complementary to one strand of the native DNA and can therefore be used to determine the amount of the gene present in cells. This technique has demonstrated, for example, that in some forms of thalassaemia (p. 206) the genes for the α-chains of globin are absent.

Nucleotide breakdown

Various nucleotidases and nucleosidases break down free nucleotides to pentose, phosphate and the pyrimidine or purine bases.

Pyrimidine catabolism

The pathways of pyrimidine breakdown are shown in figure 13.17. They lead to the production of β-alanine, used for the formation of coenzyme A and some other compounds, and β-aminoisobutyric acid which can be converted to methylmalonyl CoA (p. 121) or excreted in the urine.

Purine catabolism

Free purines may either be reconstituted into nucleotides as described earlier (p. 149) or broken down to give urate (fig. 13.18). A normal man excretes in his urine about 600 mg (36 mmol) urate daily on a purine free diet. In addition some 200 mg urate is excreted into the intestine where it is degraded by bacteria. This extrarenal route for urate disposal is of considerable importance when diseases of the kidneys impair the renal excretion of urate.

The plasma normally contains 0.18 to 0.40 mmol/l (3.0 to 6.7 mg/dl) urate. Men tend to have higher

Cytosine

Methylcytosine

Uracil

Thymine

NADPH + H$^+$

NADP$^+$

NADPH + H$^+$

NADP$^+$

Dihydrouracil

Dihydrothymine

H$_2$O

H$_2$O

NH$_2$ COO$^-$
CO CH$_2$
NH—CH$_2$

β-ureidopropionate

NH$_2$ COO$^-$
CO HC—CH$_3$
NH—CH$_2$

β-ureido*iso*butyrate

H$_2$O

NH$_3$ + CO$_2$

H$_2$O

NH$_3$ + CO$_2$

NH$_2$CH$_2$CH$_2$COO$^-$
β-alanine

CH$_3$
NH$_2$CH$_2$CH—COO$^-$
β-amino*iso*butyrate

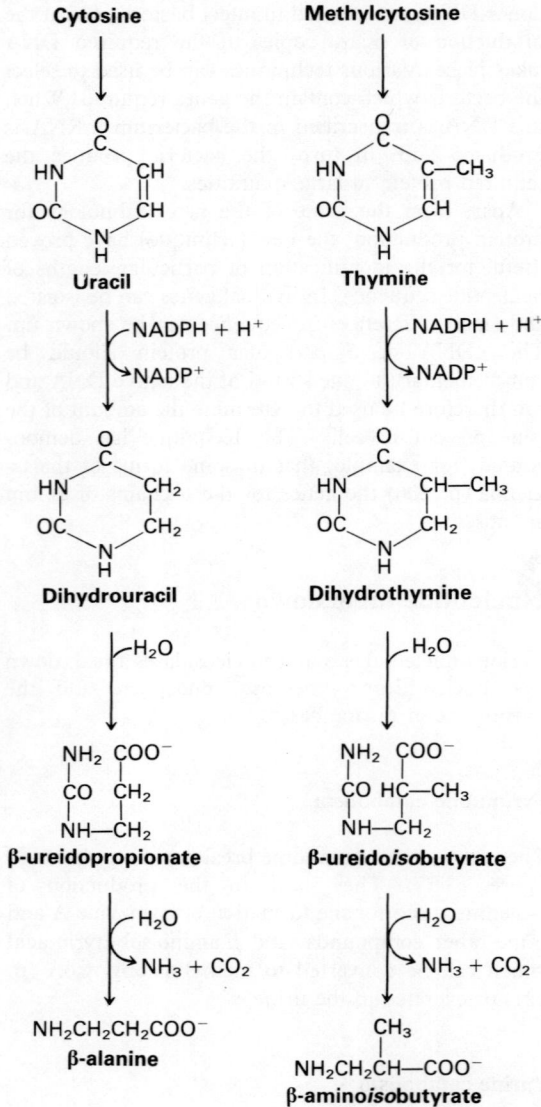

Fig. 13.17 Pathways for pyrimidine catabolism

H$_2$O   NH$_3$

AMP ⇌ **IMP**   **GMP**

APRT   *HGPRT*   *HGPRT*

Adenine   **Hypoxanthine**   **Guanine**

Xanthine oxidase   H$_2$O

Guanine deaminase

NH$_3$

Xanthine

Xanthine oxidase

Uric acid

Fig. 13.18 Pathways for the breakdown of purine bases to uric acid in man. The importance of adenine phosphoribosyl transferase (APRT) and hypoxanthine phosphoribosyl transferase (HGPRT) in the 'salvage' of purine bases was described earlier (p. 149)

high plasma uric acid itself contributes to the increase in heart disease or simply reflects other risk factors.

Urate is filtered in the glomerulus and almost completely reabsorbed in the proximal tubule. Further urate is secreted into the lumen in the distal part of the proximal tubule. On a typical western diet the daily urinary excretion of urate is about 800 mg (48 mmol) of which one-quarter is derived from nucleic acids in the diet and three-quarters from the breakdown of tissue nucleic acids. For an accurate assessment of urate output the urine collection should be made after five days on a purine-free diet.

Experiments with isotopically labelled nucleic acids have demonstrated that the purines of dietary nucleic acids are almost entirely converted to urate or degraded by intestinal bacteria. Little is used for nucleic acid synthesis.

values than women (fig. 13.19). Apart from gout, higher values tend to be found in people who are overweight, have high plasma cholesterol levels or have high blood pressure. It is not surprising therefore that people with high urate levels in the plasma also have a higher than average risk of coronary heart disease. It is not yet clear whether the

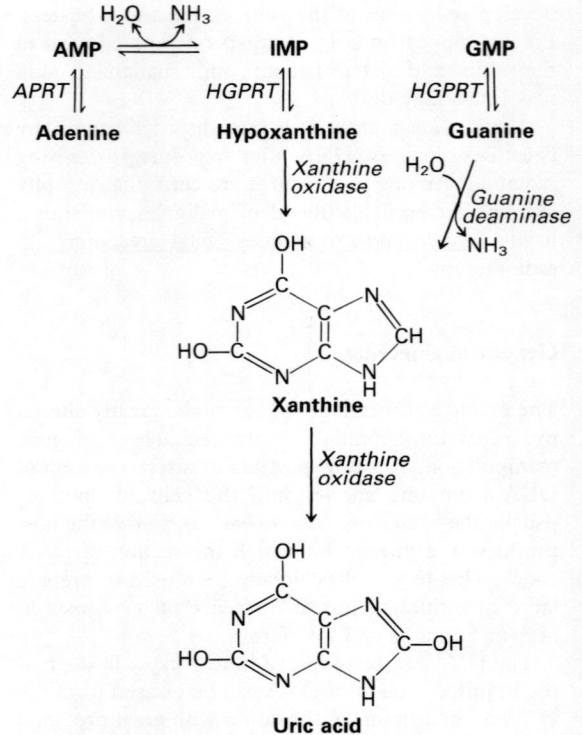

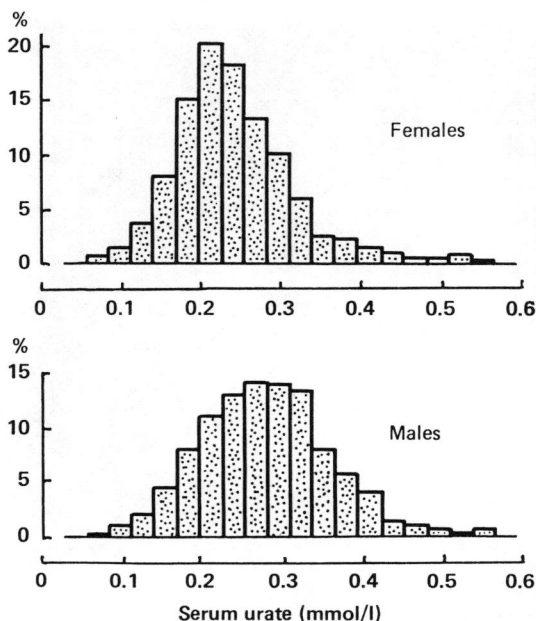

Fig. 13.19 Distribution of serum uric acid levels in normal people (2987 males and 3013 females) in Tecumseh, Michigan. In each case the height of the column represents the number of subjects with the value stated as a percentage of the total number of subjects of that sex. (After Mikkelsen, W. M. *et al.* (1965) *American Journal of Medicine* **39,** 243)

Gout. High plasma levels of urate may cause gout, a very painful acute arthritis caused by the crystallisation of monosodium urate in a synovial cavity. In about half of the cases the first joint to be affected is the metatarso-phalangeal joint of the great toe. Recurrent attacks of gout may lead to a chronic gouty arthritis affecting principally the fingers and toes (figs. 13.20 and 13.21). In addition such patients may have *tophi*, deposits of urates under the skin (fig. 13.22). Longstanding hyperuricaemia causes deposition of urate in the kidneys and chronic renal impairment (*urate nephropathy*).

In most cases the fundamental abnormality in gout is an increase in purine synthesis but this has no single cause. Some of the abnormalities which have been identified are indicated in figure 13.23. Together these known disorders account only for a minority of patients with gout.

Hyperuricaemia and gout can also result from increased nucleic acid turnover due to increased formation and breakdown of cells in disorders such as leukaemia and polycythaemia. Gout can result from the action of drugs, such as the thiazide diuretics, which impair urate secretion in the renal tubule.

Hyperuricaemia can be treated either by reducing urate production with *allopurinol*, which is a competitive inhibitor of xanthine oxidase (fig. 13.24), or by the *uricosuric drugs* such as probenecid, which increases urinary losses of urate by inhibiting its tubular reabsorption.

Xanthinuria. This is a rare disorder, probably inherited as an autosomal recessive, caused by deficiency of the enzyme xanthine oxidase. Blood and urine levels of urate are very low while levels of xanthine and hypoxanthine are raised. Some patients develop urinary calculi composed of xanthine and others muscle weakness because of the deposition of xanthine and hypoxanthine in skeletal muscle.

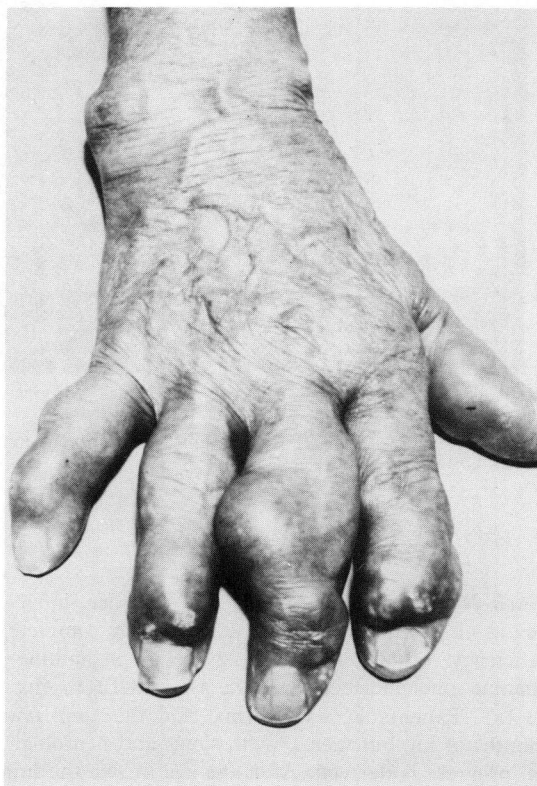

Fig. 13.20 Severe gouty changes in the fingers. (Courtesy of G. Nuki)

159

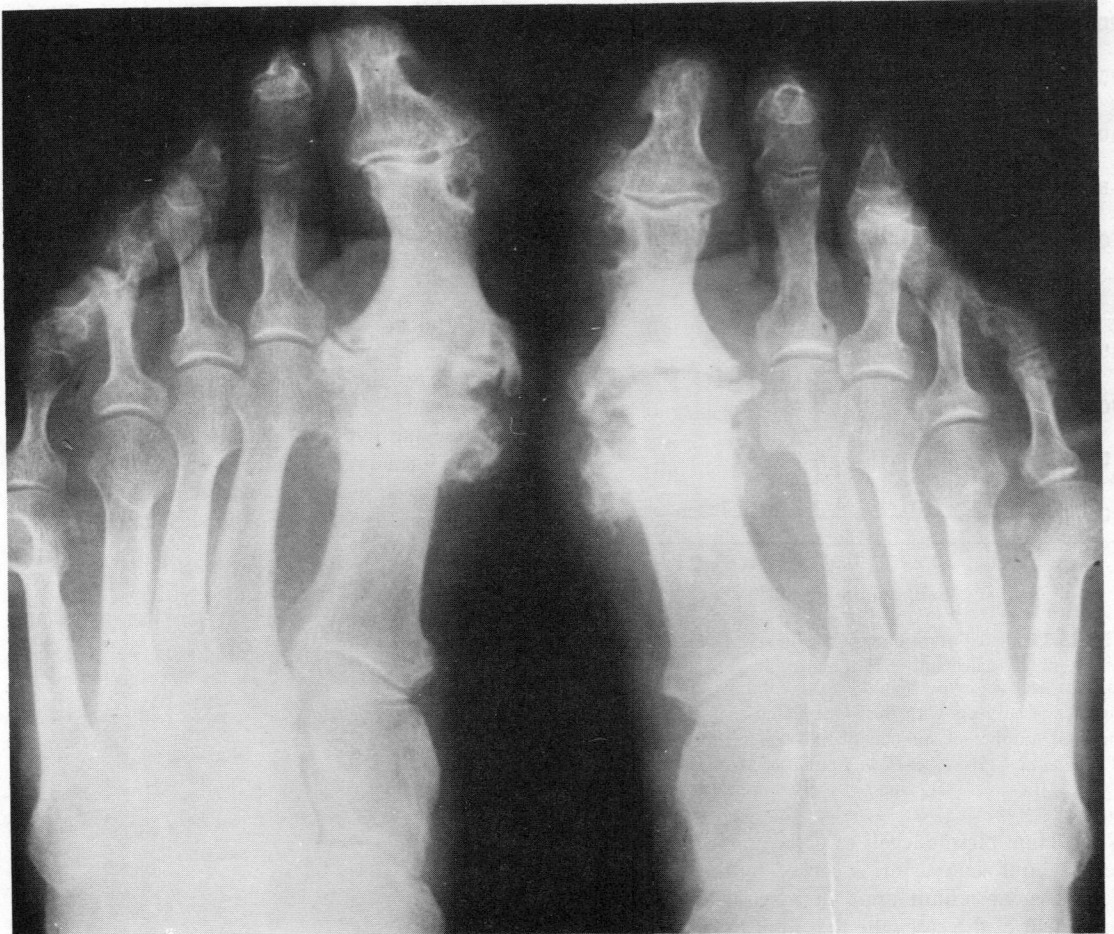

Fig. 13.21 Radiograph to show severe gouty changes in the toes. (Courtesy of F. D. Hart)

Lesch–Nyhan syndrome. This rare disorder, inherited as an X-linked recessive, is caused by complete deficiency of the salvage enzyme hypoxanthine-guanine phosphoribosyl transferase (HGPRT) (fig. 13.18). Patients appear normal for the first few months of life but then growth slows and neurological progress is delayed. After the age of two the full syndrome is apparent with spasticity, involuntary movements, mental retardation, gout, uric acid calculi and a compulsion for self-mutilation (figs. 13.25

and 13.26).

The plasma and urine levels of urate are greatly raised. There is no effective treatment for the major neurological symptoms but allopurinol protects the patient from gout and urinary calculi.

Uric acid calculi. Between 5 and 10 per cent of patients who form recurrent urinary calculi pass stones composed mainly of uric acid or monosodium urate. While some of these patients have gout and a

160

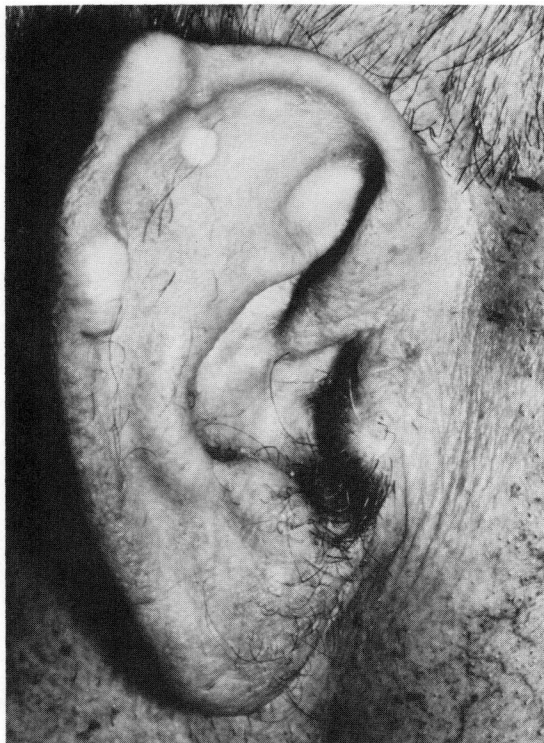

Fig. 13.22 Tophi, deposits of urate under the skin, in a patient with gout. (Courtesy of W. W. Buchanan)

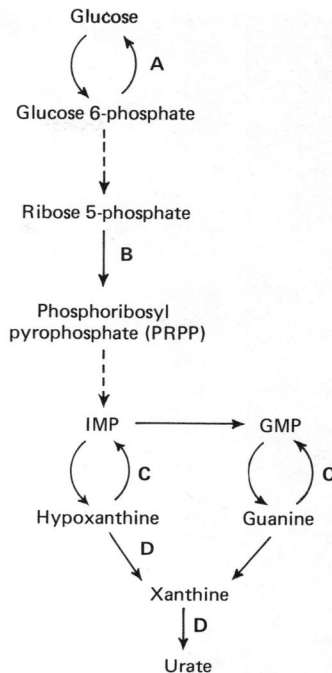

Fig. 13.23 Four points at which abnormalities lead to an increase in urate production. A: Glucose-6-phosphatase deficiency (p. 107). B: Increased activity of PRPP synthetase thought to be due to a defect in the ability of the enzyme to 'switch-off' in response to high nucleotide levels (p. 147). C: Partial deficiency of hypoxanthine-guanine phosphoribosyl transferase, leading to defects in the salvage pathway (p. 149). D: Xanthine oxidase activity increased. It is not yet clear whether this abnormality is a primary disorder or secondary to other metabolic defects

high urinary concentration of urate, the majority show no such abnormality. The principal factor in these cases is a low urinary pH of unknown cause. The disorder can be treated by giving the patient a high fluid intake and alkalis to ensure that the urine has a high volume and pH.

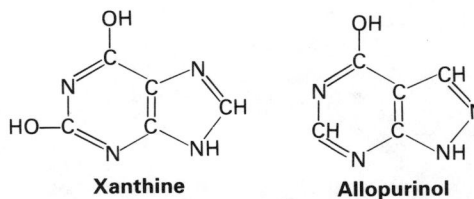

Fig. 13.24 Structures of xanthine and allopurinol

161

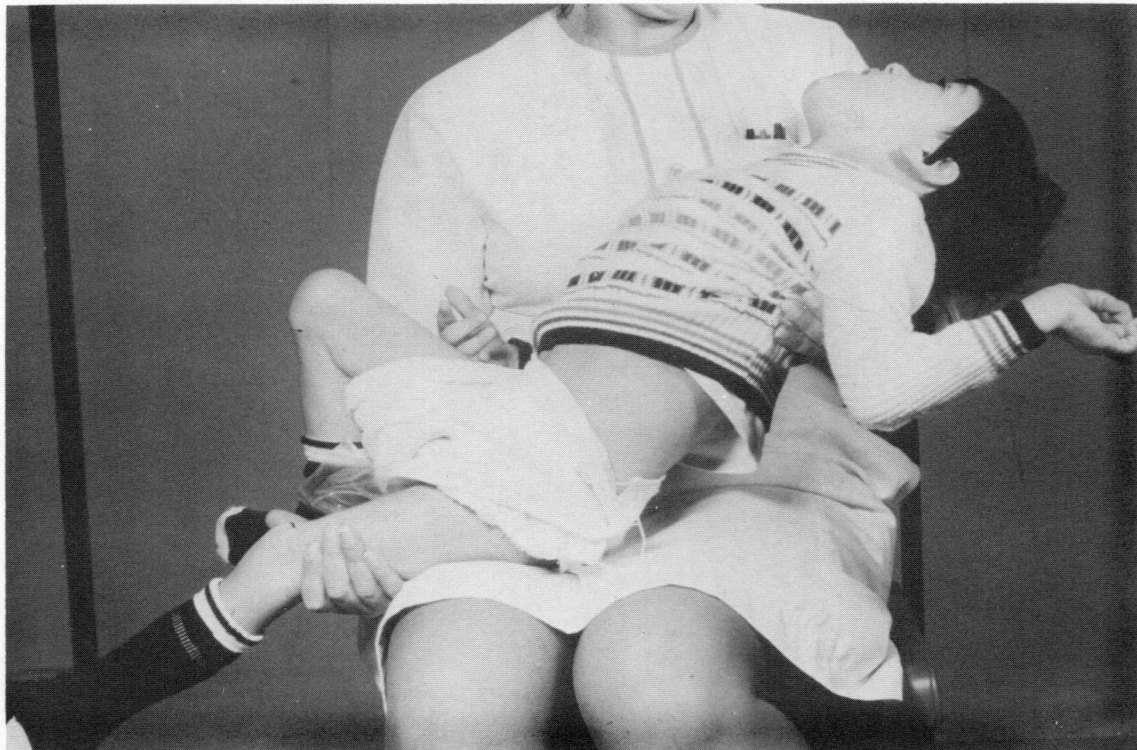

Fig. 13.25 Child with Lesch–Nyhan syndrome, mentally retarded and subject to frequent involuntary movements. (Courtesy of R. W. E. Watts)

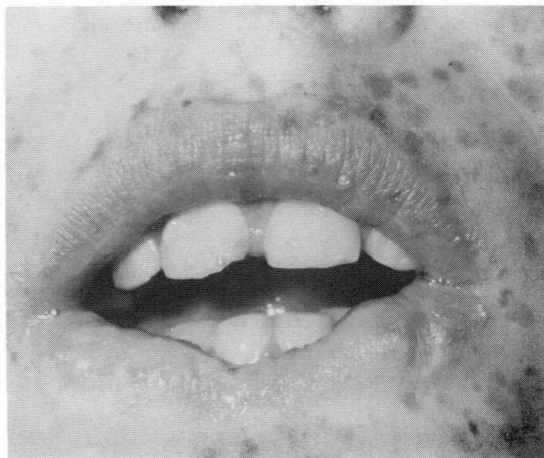

Fig. 13.26 Lips of a child with Lesch–Nyhan syndrome to show effect of biting, especially on the lower lip. (Courtesy of R. W. E. Watts)

Further reading

Adams, R. L. P., Burdon, R. H., Campbell, A. M., Leader, D. P. and Smellie, R. M. S. (1981) *The Biochemistry of the Nucleic Acids*, 9th edn. London: Chapman and Hall

Coe, F. L. and Parks, J. H. (1981) Hyperuricosuria and calcium nephrolithiasis. *Urologic Clinics of North America* **8**, 227–244

Cohen, S. N. (1980) The transplantation and manipulation of genes in microorganisms. *Harvey Lectures* **74**, 173–204

Emery, A. E. H. (1984) *An Introduction to Recombinant DNA*. Chichester: Wiley

Hoffbrand, A. V. and Wickramasinghe, S. N. (1982) Megaloblastic anaemia. In *Recent Advances in Haematology 3*, pp. 25–44. Edinburgh: Churchill Livingstone

Kelley, W. N., Wyngaarden, J. B. (1983) Clinical syndromes associated with hypoxanthine-guanine phosphoribosyl-transferase deficiency. In *Metabolic Basis of Inherited Disease*, 5th edn, eds Stanbury, J. B.,

Wyngaarden, J. B., Fredrickson, D. S., Goldstein, J. L., Brown, M. S. pp. 1115–1143. New York: McGraw-Hill

Kornberg, A. (1980) *DNA Replication.* San Francisco: Freeman

Lambert, B. and Ringborg, U. (1976) DNA repair and human disease. *Acta Medica Scandinavica* **200,** 433–439

McKeran, R. O. and Watts, R. W. E. (1978) Purine metabolism and cell physiology. In *Recent Advances in Endocrinology and Metabolism 1*, ed. O'Riordan, J. L. H., pp. 219–252. Edinburgh: Churchill Livingstone

Mainwaring, W. I. P., Parish, J. H. and Pickering, J. D. (1982) *Nucleic Acid Biochemistry and Molecular Biology.* Oxford: Blackwell

Old, R. W. and Primrose, S. B. (1980) *Principles of Gene Manipulation.* Oxford: Blackwell

Seegmiller, J. E. (1982) Disorders of purine and pyrimidine metabolism. In *Contemporary Metabolism 2*, ed. Freinkel, N., pp. 343–409. New York: Plenum

Watson, J. D. (1983) *Recombinant DNA.* San Francisco: Freeman

Weatherall, D. J. (1982) *The New Genetics and Clinical Practice.* London: Nuffield Provincial Hospitals Trust

Williamson, R. W. (ed.) (1981) *Genetic Engineering.* London: Academic Press

Wyngaarden, J. B., Kelley, W. N. (1983) Gout. In *Metabolic Basis of Inherited Disease*, 5th edn, eds Stanbury, J. B., Wyngaarden, J. B., Fredrickson, D. S., Goldstein, J. L., Brown, M. S. pp. 1043–1114. New York: McGraw-Hill

C R P   G A J G

14 Protein synthesis & metabolism

THE proteins of the tissues are continually being broken down and resynthesised. The amount of a particular protein present at any time therefore depends on the balance between its rate of degradation and its rate of formation; both processes may be influenced by factors such as diet and levels of hormones in the blood. Even under normal conditions the rates of turnover of individual proteins may vary widely; Table 14.1 shows the half-lives of several proteins in the liver of the rat. It can be seen that among the proteins with the most rapid turnover are the enzymes which regulate metabolic processes. The levels of these enzymes can clearly be altered quickly by changes either in the rate of synthesis or the rate of breakdown. Even proteins which are structural components of cell organelles are subject to continuous turnover. Thus the maintenance of both the structural and functional integrity of a tissue demands precise control of the processes involved in protein metabolism.

Protein synthesis requires energy; formation of one gram of tissue protein requires about 5.9 kJ. A 70 kg man synthesises daily some 200 to 300 g protein and therefore uses 1.2 to 1.7 MJ for this purpose. Since protein breakdown also requires energy it is clear that a significant proportion of the basal energy consumption is used for the turnover of tissue protein.

Protein synthesis

The mechanism of protein synthesis to be described in this chapter is that which is thought to take place in mammalian cells. It differs in several respects from that in bacteria; these differences are important in medicine since some antibiotics inhibit bacterial protein synthesis without affecting that of the patient.

The reaction of the carboxyl group of one amino acid with the amino group of another requires energy and an activation step is needed. The carboxyl group of an amino acid reacts first with ATP and then with a transfer RNA molecule. The first step is

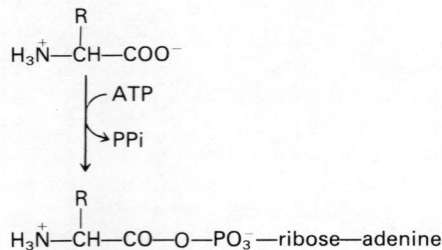

$$\overset{R}{\underset{|}{H_3\overset{+}{N}-CH-COO^-}}$$

$$\downarrow ATP$$

$$\downarrow PPi$$

$$\overset{R}{\underset{|}{H_3\overset{+}{N}-CH-CO-O-PO_3^--ribose-adenine}}$$

The aminoacyl-adenylate formed initially reacts with a transfer RNA (tRNA) specific for the amino acid to form an aminoacyl-tRNA derivative:

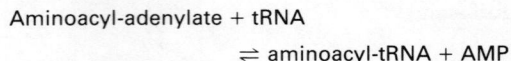

Aminoacyl-adenylate + tRNA

$$\rightleftharpoons aminoacyl\text{-}tRNA + AMP$$

Both stages take place on the same enzyme, known as an *amino acid activating enzyme* or an *aminoacyl-tRNA synthetase*. Each synthetase is specific for one particular amino acid and its acceptor tRNA. The amino acid is attached to its tRNA by an ester bond with the adenosine residue (fig. 14.1).

Table 14.1 Half-lives of some proteins in rat liver

	Half-life
Enzymes	
Ornithine decarboxylase	10 minutes
δ-Aminolaevulinate synthase	1 hour
Tyrosine aminotransferase	1.5 hours
Tryptophan pyrrolase	2.5 hours
Glucokinase	1.25 days
β-Glucuronidase (lysosomal)	15 days
Proteins of organelles	
Average lysosomal	1 day
Average plasma membrane	1.5 days
Average mitochondrial	4 days
Average ribosomal	4.5 days

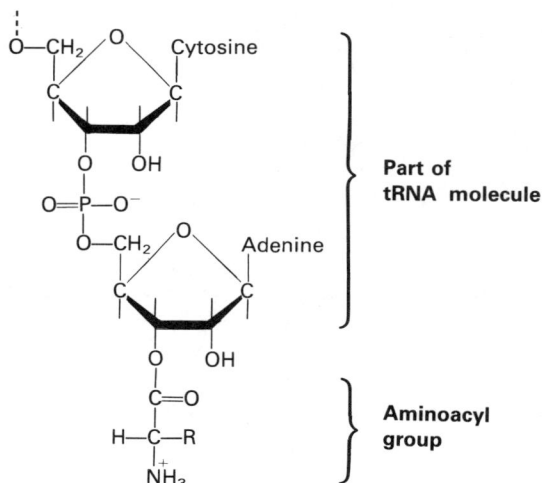

Fig. 14.1 Mode of attachment of amino acid to tRNA

Fig. 14.2 Drawing of eukaryote ribosome subunits based on electron micrographs. (After Aoubik, M. and Hellman, W. (1978) *Proceedings of the National Academy of Science* **75**, 2829)

The overall reaction catalysed by an aminoacyl-tRNA synthetase is:

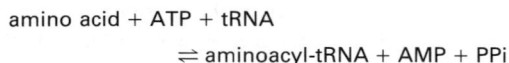

amino acid + ATP + tRNA

$$\rightleftharpoons \text{aminoacyl-tRNA} + \text{AMP} + \text{PPi}$$

The free energy of hydrolysis of the ester bond in an aminoacyl-tRNA is similar to that for the hydrolysis of ATP to AMP and pyrophosphate; $\Delta G^{0\prime}$ for the overall reaction is therefore almost zero. The reaction proceeds because the pyrophosphate formed is immediately hydrolysed to two molecules of inorganic phosphate.

Transfer RNA molecules combined with their specific amino acids are described as *charged* or *loaded* and are conventionally designated:

tRNA^{leu}: uncharged tRNA specific for leucine

leu-tRNA^{leu}: charged leucyl-tRNA

Ribosomes

The ribosome is the organelle in which the actual formation of the peptide bond takes place. Ribosomes may be free in the cytoplasm or attached to the endoplasmic reticulum. A small number of ribosomes are found within mitochondria; these are smaller than ribosomes elsewhere and have a different RNA and protein content.

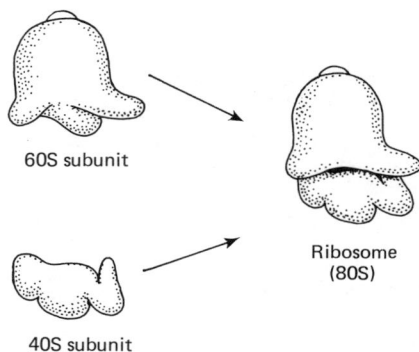

The ribosome in the mammalian cell cytoplasm has a sedimentation (p. 59) of 80 S and in conditions of low magnesium concentration dissociates into subunits with sedimentation coefficients 40 S and 60 S (fig. 14.2). Associated with the smaller subunit is a molecule of 18 S RNA and some 30 different proteins. The larger subunit comprises one molecule each of 28 S RNA, 5 S and 5.8 S RNA and about 45 different proteins.

Messenger RNA, the template for protein synthesis

The ordered sequence of amino acids in newly synthesised proteins depends upon the interaction of the ribosome with messenger RNA (mRNA) (p. 57). The process by which the sequence of bases on an mRNA molecule is interpreted by the ribosome as a sequence of amino acids is known as *translation*. Electron micrographs of cells show that ribosomes occur in clusters held together by a continuous strand which can be shown to be an mRNA molecule. These clusters of ribosomes are known as *polysomes* (fig. 14.3).

Many genes in the mammalian cell contain sequences of nucleotides in addition to those required to specify the sequence of nucleotides in the mRNA. Such additional sequences are termed *introns* and the coding sequences themselves are termed *exons*. For example the gene for the β-chain of haemoglobin has three exons and two introns. Such a gene in which

165

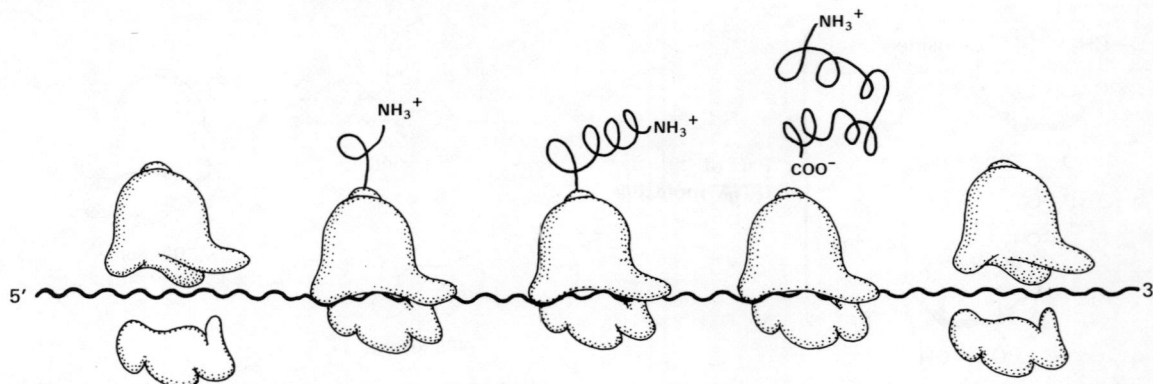

Fig. 14.3 Structure of a polysome (polyribosome). The ribosomes move along the messenger RNA in a 5′ to 3′ direction

the exons are interspersed with introns is called a *split gene*.

In the formation of the mRNA the entire gene, introns and exons, is transcribed into a primary transcript. The regions corresponding to the introns are then excised and the coding sequences joined by a specific splicing enzyme (fig. 14.4).

The information transcribed from DNA, carried on mRNA and finally translated into a sequence of amino acids on the ribosome is known as the *genetic code*. As there are some 20 different amino acids in proteins the code must allow for at least this number. DNA and RNA contain only four main bases adenine, guanine, cytosine, uracil (RNA) and thymine (DNA). A one base–one amino acid code could specify only four different amino acids and a two base–one amino acid code would allow for only $4 \times 4 = 16$ different amino acids. In fact the bases are read in threes which allows $4 \times 4 \times 4 = 64$ different arrangements. A sequence of three consecutive bases on an mRNA molecule is known as a *codon*; the various codons found on *mRNA* and the amino acids to which these have been assigned are shown in Table 14.2.

It will be seen that 61 of the possible 64 codons can be assigned to amino acids. The remaining three codons play an important part in the termination process to be described later and are known as *termination codons*. One codon, AUG, merits particular mention. This triplet codes for methionine in the synthesis of a polypeptide but it also serves as an *initiator codon*. The start of the genetic message is not at the beginning of an mRNA molecule, and in some instances the information for the synthesis of more

than one protein may be contained in a single mRNA molecule. Certain AUG codons therefore serve to tell the ribosome where to start translation. These initiation codons are recognised by virtue of the accompanying base sequences in the mRNA.

It is important to recognise that the genetic code is universal. Cytoplasmic ribosomes from bacterial, plant or animal sources in the presence of a specific mRNA synthesise identical proteins. With mitochondria, however, there are certain differences from the codon assignations given in Table 14.2; UGA specifies tryptophan and not a termination codon and AUA codes for methionine and not isoleucine. These differences may be due to structural features of mitochondrial ribosomes or tRNA.

The nature of the genetic code also indicates how a mutation at a single base on the DNA, corresponding to a particular gene, gives rise to a polypeptide with a single amino acid replacement. In *sickle-cell anaemia*, haemoglobin (Chap. 16) contains normal α-chains but the β-chains contain a valine residue at a position where the β-chain of normal haemoglobin contains a glutamic acid residue. That this change could be due to a single base change in the gene is apparent when the codons for glutamic acid and valine are compared;

Glutamic acid: GAA   Valine: GUA
GAG   GUG

Thus a mutation which involves a change of a thymine residue on the β-globin gene to an adenine residue would explain this phenomenon. The substitution of one valine residue for one glutamic acid is

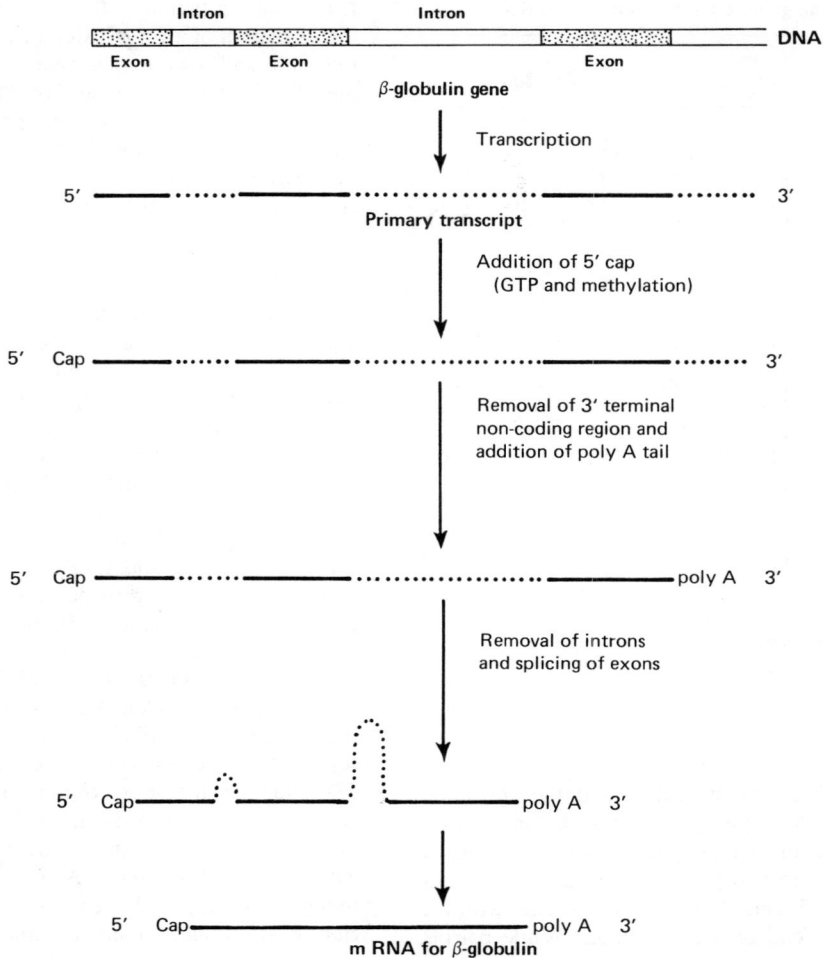

Fig. 14.4 Transcription of the β-globin gene and the processing of the transcript to produce the messenger RNA

sufficient to alter the oxygen binding of the haemoglobin such that under conditions of reduced oxygen tension the protein molecules tend to adhere and the erythrocytes become distorted into abnormal shapes (p. 208).

The thalassaemias (p. 207) are a group of inherited disorders involving defects in the synthesis of one of the globin chains of haemoglobin. In one form the β-chain has an extra 31 residues because a termination codon UAA has altered to CAA, the codon for glutamine. In another no β-chain is formed because a termination codon UAG has replaced the codon for lysine (AAG) at position 17 in the mRNA.

Before the involvement of the ribosome in the translation process is discussed, it is necessary to draw attention to the basic factor in the recognition of the correct codon during each stage in the growth of the polypeptide chain. The structure of a typical tRNA molecule was shown in figures 5.13 and 5.14 (p. 58). One of the loops has at its base a triplet of three nucleotides, the *anticodon*, which forms hydrogen bonds with the codon for a particular amino acid and with no other. Combination between the codon on the mRNA and the anticodon on the tRNA occurs when the codon on the mRNA is at a particular site on the ribosome. It is this specific

Table 14.2   The genetic code: codons on mRNA

5' terminal base	Middle base	3' terminal base			
		A	G	C	U
A	A	Lys	Lys	Asn	Asn
	G	Arg	Arg	Ser	Ser
	C	Thr	Thr	Thr	Thr
	U	Ile	Met*	Ile	Ile
G	A	Glu	Glu	Asp	Asp
	G	Gly	Gly	Gly	Gly
	C	Ala	Ala	Ala	Ala
	U	Val	Val	Val	Val
C	A	Gln	Gln	His	His
	G	Arg	Arg	Arg	Arg
	C	Pro	Pro	Pro	Pro
	U	Leu	Leu	Leu	Leu
U	A	TS	TS	Tyr	Tyr
	G	TS	Trp	Cys	Cys
	C	Ser	Ser	Ser	Ser
	U	Leu	Leu	Phe	Phe

* This codon is also the initiation signal.
TS = termination signal.

combination between the codon and the anticodon which ensures the accuracy of the translation process.

During protein synthesis the mRNA and tRNA molecules line up anti-parallel to each other so that the base at the 5' end of the codon associates with the base at the 3' end of the anticodon. While pairing between the base at the 5' end of the codon and the 3' end of the anticodon obeys the usual rules (G with C and A with U) some variation is seen in the pairing between the other two bases. For example if G is in the 5' anticodon base it can pair with C or U in the codon. The nucleoside inosine (Chap. 5) can only pair with C in a double helix but when it occurs in the 5' position of an anticodon it can pair with A, U or C. These variations from the general rules are known as the *wobble hypothesis*.

Messenger RNA translation

This process can conveniently be divided into three distinct phases—initiation, elongation and termination.

Initiation. Initiation is the process which results in the formation of a complex of a ribosome, an mRNA molecule and a particular charged tRNA molecule, met-tRNA. This phase involves the ribosomal subunits, several non-ribosomal proteins known as *initiation factors* (eIF), GTP and ATP. The details are summarised in figure 14.5. The cell normally contains a pool of 80 S ribosomes which are inactive in initiation but which can dissociate into subunits:

$$80\ S \rightleftharpoons 60\ S + 40\ S$$

The conditions in the cell are such that the equilibrium is in favour of reassociation. 40 S subunits are needed for initiation and in order to maintain a supply of these, combination with one of the initiation factors, eIF-3, is required. 40 S subunits combined with eIF-3 can no longer reassociate and react with met-tRNA$_f$ in the presence of eIF-2 and GTP.

The complex of the 40 S subunit with eIF-3, eIF-2, met-tRNA$_f$ and GTP can now combine with an mRNA molecule provided some further initiation factors eIF-1, eIF-4a and eIF-4b, are available. The precise role of these factors is not yet clear. Hydrolysis of ATP takes place at this stage.

The initiation codon AUG is not always found at the 5' end of the mRNA; there may be as many as several hundred nucleotides between the 5'-cap (p. 57) and the initiation codon. The 40 S subunit complex moves towards the 3' end of the mRNA until the anticodon of the bound met-tRNA$_f$ engages with the initiation codon AUG. ATP hydrolysis is thought to supply the energy for this. The 60 S ribosomal subunit then becomes associated, yet another initiation factor, eIF-5, being required. Lastly all the initiation factors are released from the 40 S complex; this requires the hydrolysis of the GTP bound earlier.

Elongation. Translation of mRNA proceeds in a 5'→3' direction. The growing peptide chain is synthesised from its N-terminal end at a rate of about 7 amino acids per second.

There are two sites on the ribosome where a transfer RNA molecule can bind; two molecules of tRNA may be associated with a ribosome at any time. These sites are termed the *aminoacyl site* (A) and the *peptide site* (P). Both involve several of the ribosomal proteins and parts of the ribosomal RNA. After initiation is completed the met-tRNA$_f^{met}$ is located in the P-site. The adjacent unoccupied A-site

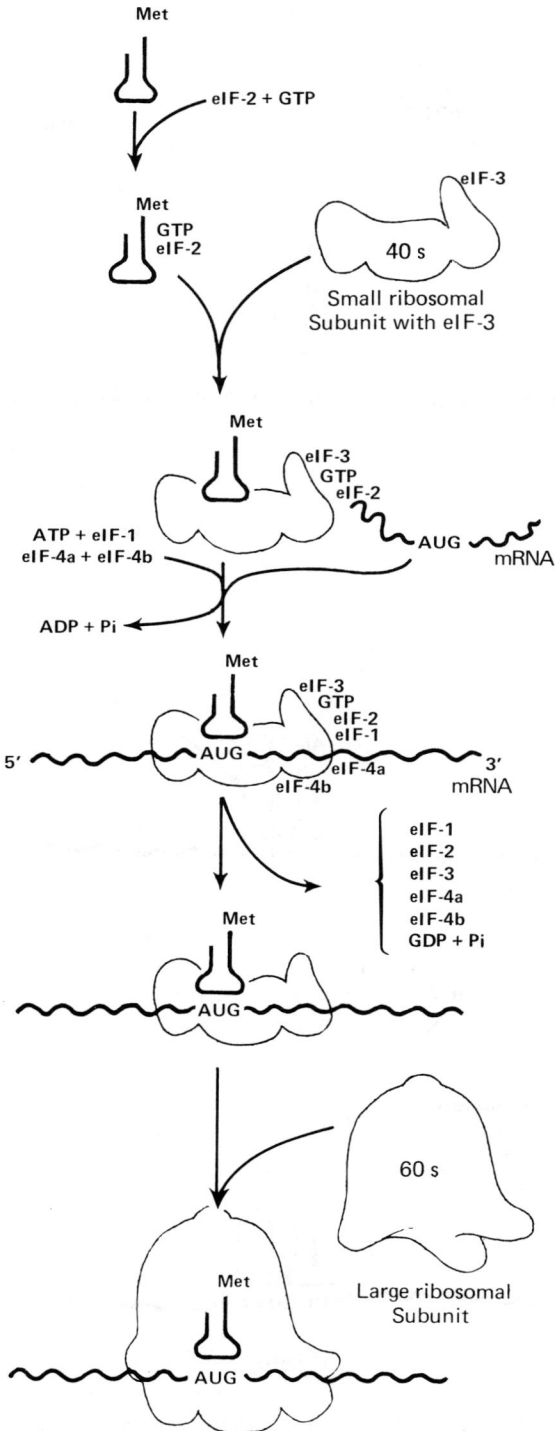

is associated with the codon representing the next amino acid to be inserted. Before the required charged tRNA can approach this site it undergoes a reaction with a cytoplasmic protein, elongation factor 1α (EF-1α) and GTP to form a complex:

EF-1α:GTP:amino acid-tRNA

This complex reacts with the A-site, GTP hydrolysis occurs, EF-1α:GDP is released from the ribosome and the charged tRNA remains bound to the A-site (fig. 14.6). The EF-1α:GDP exchanges the GDP for a molecule of GTP, a process which requires another elongation factor, EF-1β. The amino group of the newly introduced amino acid forms a peptide bond with the carboxyl group of the methionyl residue bound to the tRNA in the P-site. This reaction is catalysed by peptidyl transferase, an enzymic activity associated with the large subunit of the ribosome and involving several of its proteins. The energy required for the formation of the peptide bond is derived from that released on the breaking of the bond between methionine and the tRNA^{met}. The A-site now contains a tRNA molecule attached to a dipeptide.

Before the next charged tRNA can associate with the A-site three events must occur. The peptidyl-tRNA has to move from the A- to the P-site, the next codon must move to the A-site and the uncharged tRNA^{met} has to be discharged from the ribosome. This state is called *translocation* and requires another cytoplasmic factor, EF-2. EF-2 forms a complex with GTP before it binds to the ribosome and hydrolysis of the GTP is needed for the process to continue. The peptidyl-tRNA is now in the P-site; the A-site is unoccupied and associated with the next codon. The next charged tRNA then binds to the A-site (complexed with EF-1α and GTP) and the cycle of events repeats. The process continues; at each translocation step the ribosome moves one codon along the mRNA.

Chain termination. At the end of the informational sequences on the mRNA there is a termination codon, UAA, UAG or UGA. At this stage another cytoplasmic protein, release factor (RF), and GTP are involved (fig. 14.7). When the A-site is occupied by a termination codon, RF binds to the ribosome. It

Fig. 14.5 The initiation step in mammalian protein synthesis

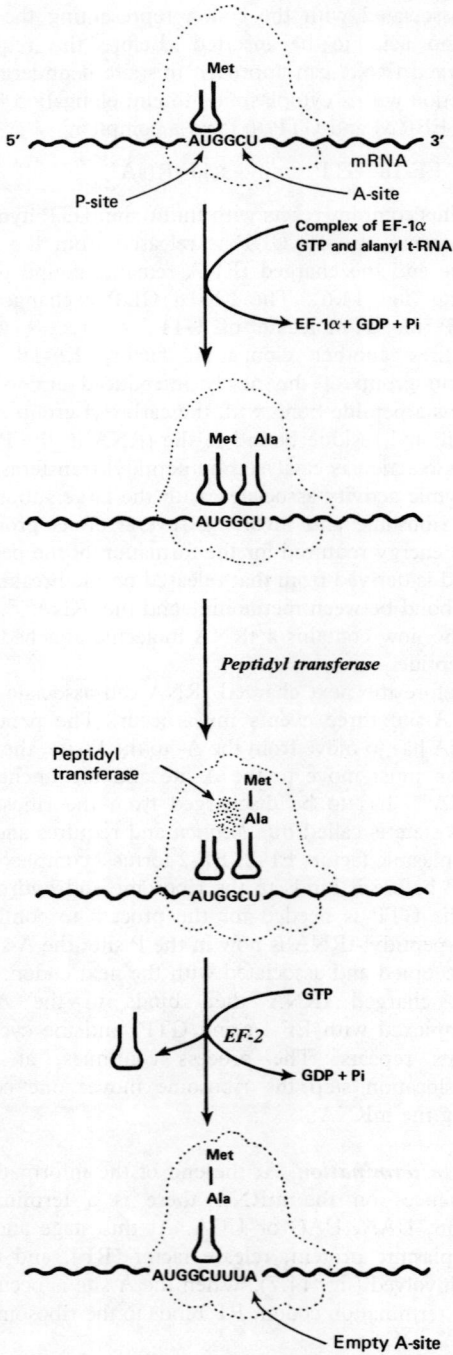

Fig. 14.6 The elongation step in protein synthesis. The elongation factors (EF) are proteins

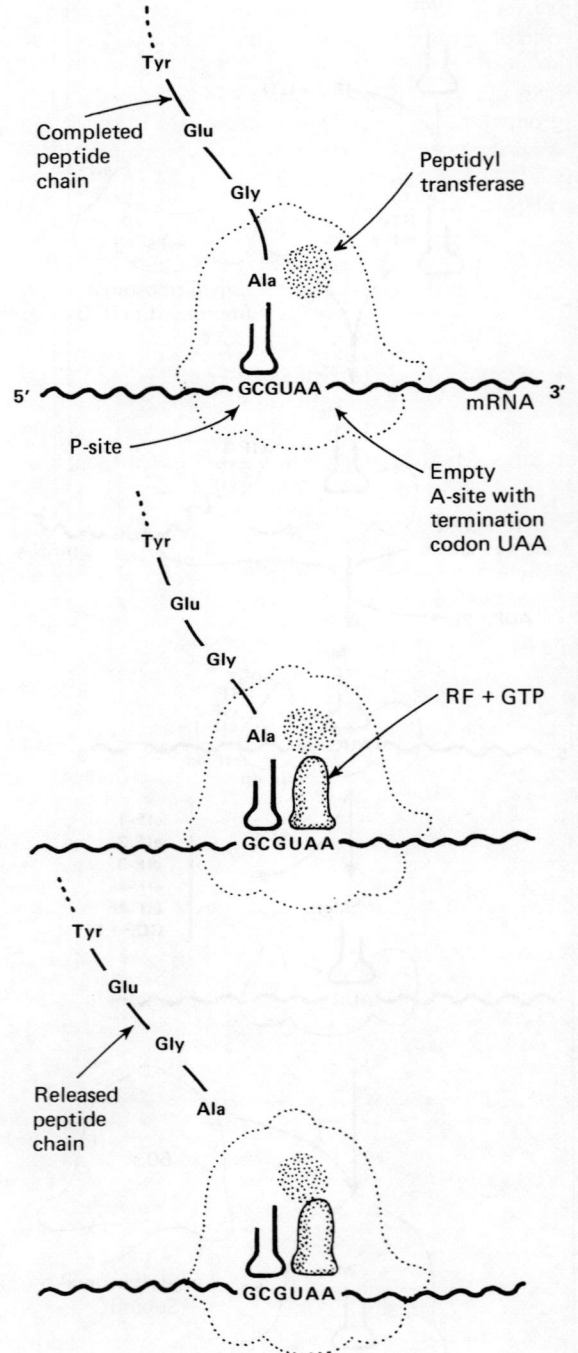

Fig. 14.7 Steps in the termination of protein synthesis

converts the peptidyl transferase activity of the ribosome to an esterase activity and the peptide attached to tRNA at the P-site is released by hydrolysis of the bond between the terminal carboxyl group and the tRNA. The free tRNA leaves the ribosome, the RF is released, with hydrolysis of ATP, and the ribosomal subunits dissociate from the mRNA.

GTP and protein synthesis. It is clear that GTP is involved in all stages of mRNA translation; it cannot be replaced by ATP or any other nucleoside triphosphate. The energy from the hydrolysis of GTP is not required for peptide bond synthesis but for the binding of various initiation, elongation and termination factors. The subsequent hydrolysis of GTP may be needed to allow the ribosome to adopt a conformation which permits the factors to be discharged from the ribosome when no longer required.

Rough endoplasmic reticulum. Electron micrographs of cells in which a considerable proportion of the protein synthesised is to be exported outside the tissue show many polysomes attached to the endoplasmic reticulum. Because of its appearance on electron microscopy this endoplasmic reticulum is called *rough endoplasmic reticulum* (p. 65). Plasma cells which produce immunoglobulins contain large amounts of rough endoplasmic reticulum as do cells of the liver and pancreas; undifferentiated tumour cells have few membrane-bound ribosomes.

Post-translational modifications

Glycoproteins. Many of the proteins secreted by mammalian cells are glycoproteins (Chap. 3). The synthesis of glycoproteins occurs on the polysomes attached to the rough endoplasmic reticulum (fig. 14.8). The polypeptide chains enter the lumen of the reticulum and the carbohydrate component is then attached by covalent bonds. The glycoproteins are packaged into vesicles (fig. 6.12, p. 67) which are formed by budding of the endoplasmic reticulum and

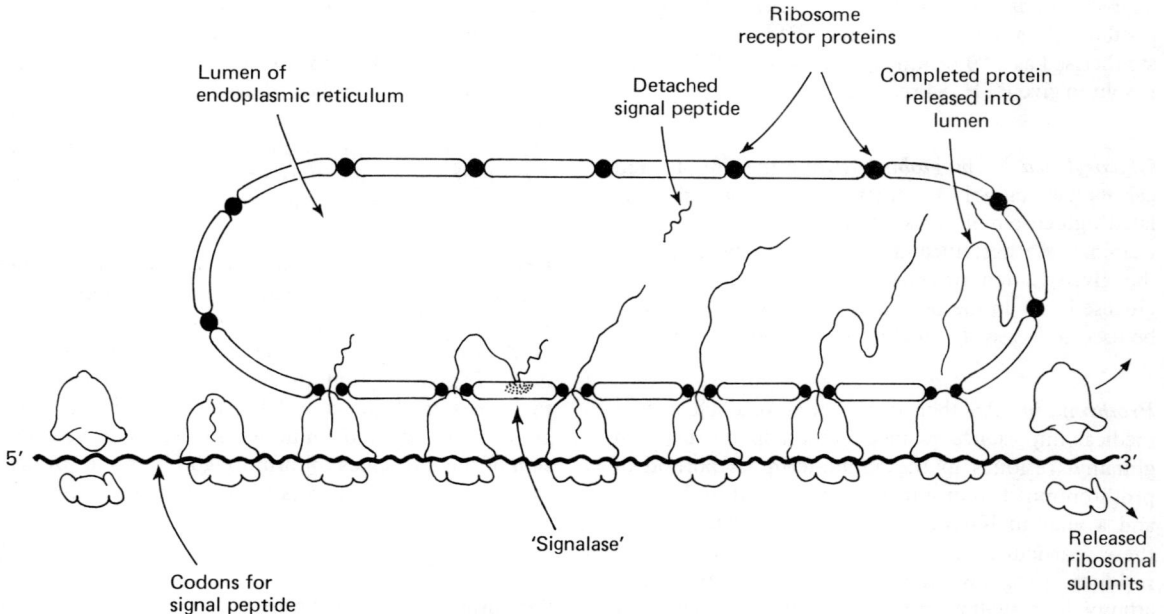

Fig. 14.8 Protein synthesis by ribosomes attached to the rough endoplasmic reticulum. The first segment of the polypeptide chain to emerge is the signal peptide consisting largely of hydrophobic amino acid residues. This is recognised by a specific receptor on the surface of the reticulum and the ribosome binds to it. The growing peptide chain then enters the lumen of the reticulum and the signal peptide is split off. Other enzymes on the inner surface of the reticulum include those which add the carbohydrate units to convert the proteins to glycoproteins. (After Campbell, P. N. and Blobel, G. (1976) *FEBS Letters* **72,** 215)

171

pass to the Golgi apparatus where further modification of the carbohydrate part may occur. The glycoproteins leave the Golgi apparatus in vesicles and are carried to sites such as lysosomes or plasma membranes.

Proteins synthesised on the polysomes attached to the endoplasmic reticulum all have a sequence of about 20 hydrophobic amino acids near the amino end. This is termed a *signal sequence*. In each case protein synthesis starts on ribosomes free in the cytoplasm. When the signal sequence is translated the large ribosomal subunit binds to the endoplasmic reticulum. The emerging peptide is threaded through a tunnel in the membrane. When the signal sequence reaches the lumen it is split off by a specific protease ('signalase').

Peptide hormones. Almost all peptide hormones are now known to be formed initially as larger precursor molecules and parts of the molecule are later split off to give the hormone. Insulin, for example, is formed first as a polypeptide chain of 84 residues, proinsulin. This is split as shown in figure 12.11 (p. 139) to give insulin. Parathyroid hormone (PTH) is initially synthesised as a 90 residue precursor pro-PTH which is split to give PTH (84 residues).

Glycosylated haemoglobin. During the life of a red cell its haemoglobin becomes steadily more 'glycosylated', glucose molecules being attached at the amino terminal. As described in Chapter 16 the extent of the glycosylation provides a measure of the blood glucose levels in the preceding two months. This can be used to assess the quality of diabetic control.

Prothrombin. Another post-translational change of medical importance is an alteration in the first ten glutamyl residues in the amino-terminal portion of prothrombin. Prothrombin is synthesised in the liver and a vitamin K-requiring enzyme system converts these particular glutamyl into γ-carboxyglutamyl residues. γ-Carboxyglutamyl residues have a strong affinity for calcium and the binding of calcium by prothrombin is essential for the association of the protein to phospholipid membranes of platelets after injury.

The role of vitamin K in prothrombin formation is antagonised by various anticoagulant drugs such as warfarin (fig. 14.9) which is also used as a rat poison.

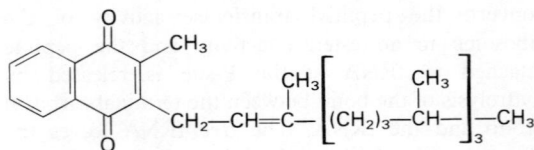

Vitamin K₁

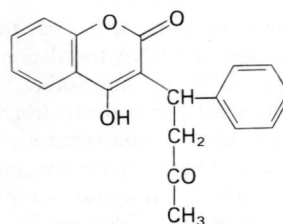

Warfarin

Fig. 14.9 Similarities between warfarin and vitamin K

Actin and myosin. Specific histidine residues in actin and myosin in skeletal muscle undergo a post-translational methylation to form 3-methylhistidine. When muscle is broken down this amino acid is released and cannot be reincorporated into muscle protein. Estimation of the urinary excretion of 3-methylhistidine therefore provides a measure of the rate of muscle breakdown.

Collagen. Collagen formation includes a large number of different post-translational modifications which are described on page 210. The modifications include (1) the hydroxylation of certain proline and lysine residues, (2) glycosylation, (3) aggregation of three chains to form a triple helix, (4) the removal of segments at both carboxy- and amino-terminals, (5) the association of the molecules to form tough fibres and (6) the formation of covalent cross-links between molecules.

Inhibitors of protein synthesis

Protein synthesis may be inhibited in various ways. Mutations in the structural gene giving an inappropriate termination codon have been mentioned earlier. In certain types of α-thalassaemia a complete gene may be missing.

172

Exposure of cells to *interferon* causes an inhibition of protein synthesis. This anti-viral protein causes the production of an enzyme known as *2-5 A synthetase* which brings about the production of an unusual oligonucleotide with three adenosine components. This nucleotide in turn activates a ribonuclease which causes splitting of ribosomal RNA and possibly also mRNA.

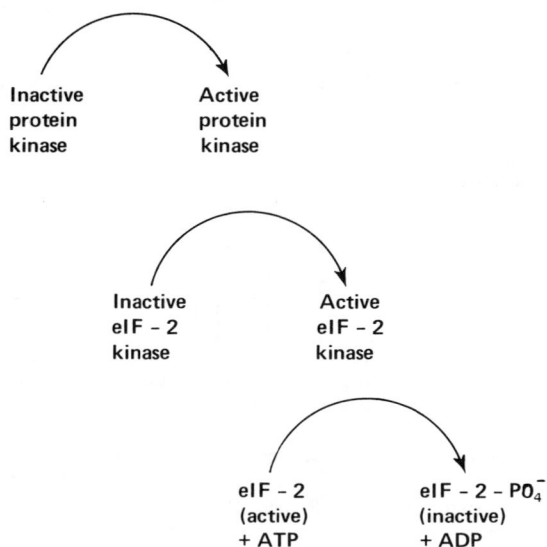

Fig. 14.10 Inactivation of initiation factor 2 (eIF-2) by phosphorylation

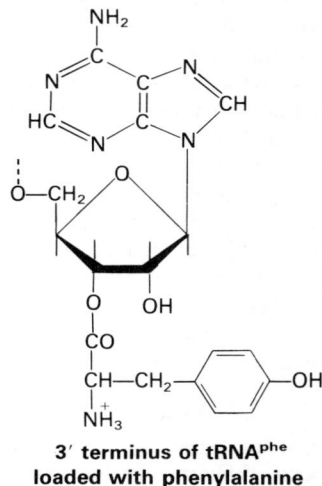

3′ terminus of tRNA^{phe}
loaded with phenylalanine

Initiation factor eIF-2 may be inactivated by phosphorylation which prevents its release from the ribosome after the initiation tRNA has been bound. This phosphorylation is carried out by a specific kinase which exists in active and inactive forms (fig. 14.10). In reticulocytes the presence of haemin prevents eIF-2 phosphorylation. It is not known whether similar inhibitory systems act on eIF-2 in other tissues. Double-stranded RNA, which may occur in cells infected with RNA viruses also activates a protein kinase which phosphorylates eIF-2.

Inhibition of the translocation step can also occur. Diphtheria toxin, produced by *Corynebacterium diphtheriae*, specifically inactivates EF-2.

Puromycin, an antibiotic produced by a strain of *Streptomyces* inhibits protein synthesis. It has a structure similar to that of the 3′ terminus of a charged tRNA (fig. 14.11) and enters the A-site of the ribosome. A peptide bond forms between its free amino group and the carboxyl end of the peptidyl residue in the P-site. The peptide with a puromycin residue at its carboxyl end dissociates from the ribosome and translation ceases.

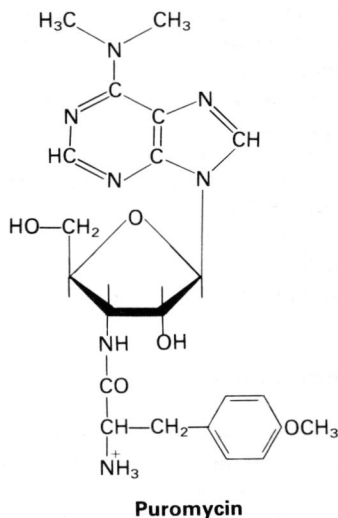

Puromycin

Fig. 14.11 Structure of puromycin to show its similarities to that of the 3′ terminus of tRNA loaded with phenylalanine

Protein breakdown

As discussed earlier the turnover rates of the proteins vary widely (Table 14.1). Those with a short half-life tend to have a high molecular mass, a low isoelectric point and a high surface hydrophobicity. Glycoproteins are generally degraded more rapidly than other proteins. Denaturation appears to increase the susceptibility of proteins to degradation.

Binding of ligands to proteins may affect their rate of turnover. Most enzymes are degraded more slowly when associated with their normal substrates or cofactors. In the case of glutamine synthetase however an increased concentration of glutamine increases the rate of breakdown of the enzyme. The effect of ligand binding on protein turnover is particularly important in the case of protein receptor molecules for specific hormones. If target cells are grown in culture in the presence of a high concentration of a specific ligand such as insulin or growth hormone the number of receptor molecules is reduced. The likely explanation is that the receptor–hormone complex enters the cell by endocytosis and is broken down in lysosomes.

Lysosomes. The lysosomes (p. 67) play an important part in protein degradation; they contain several proteases, with different specificities, which act optimally at low pH. Energy is required for intracellular protein degradation. One reason may be the need to maintain a low pH within the lysosome by a proton pump driven by ATP hydrolysis.

The differences between the turnover rates of individual proteins may be explained if lysosomes are the sites of protein degradation. It has already been noted that the rate of degradation of hydrophobic proteins is usually less than that of more hydrophilic ones. The rate of degradation could be related to a protein's ability to attach to the lysosomal membrane. Proteins which attach most readily have been shown to have short half-lives.

Factors affecting tissue protein turnover

Skeletal muscle protein constitutes about 45 per cent of the total protein of an adult and is the major protein reserve of the body. In an adult the turnover of muscle protein is about 25 per cent of the total body protein turnover. The rate of muscle protein breakdown, estimated from the urinary excretion of

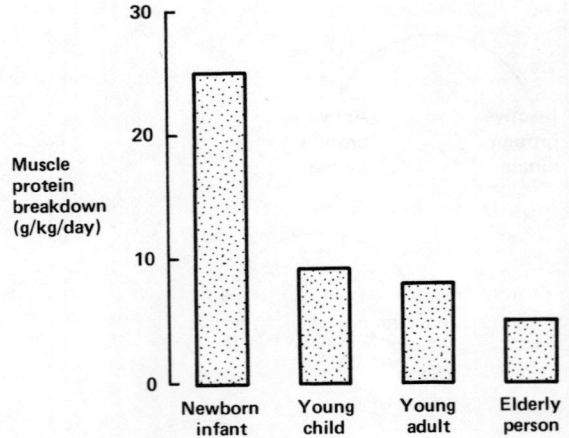

Fig. 14.12 Values for muscle protein breakdown, estimated by measurements of 3-methylhistidine excretion at different ages. (After Young, V. R. and Munro, H. N. (1980) In *Degradative Processes in Heart and Skeletal Muscle*, ed. Wildenthal, K. Amsterdam: Elsevier)

3-methylhistidine (p. 7), varies with age, values being highest in the newborn and lowest in the elderly (fig. 14.12).

Fasting causes an increased mobilisation of amino acids from muscle protein. These may provide substrates for gluconeogenesis in the liver and renal cortex (Chap. 9), and serve as sources of energy (Chap. 12). This release of amino acids from muscle is due in the first instance to a decrease in the rate of synthesis of muscle protein. An increase in protein degradation also occurs later. If fasting is prolonged loss of muscle protein diminishes possibly because ketone bodies derived from fat catabolism inhibit the degradation of muscle protein.

Protein turnover in tissues is regulated by several hormones. Insulin promotes protein synthesis in muscle, not only by increasing amino acid uptake and glucose utilisation (Chap. 12) but also by increasing the number of ribosomes present and their efficiency in the initiation step of translation. Insulin also inhibits protein degradation both in liver and muscle.

After injury or in the presence of infection muscle protein breakdown is increased. This may be due to the increase in circulating glucocorticoids which increase muscle protein breakdown and inhibit its synthesis. In the liver, on the other hand, glucocorticoids cause an increase in protein synthesis and a decrease in degradation. The synthesis of certain liver proteins, notably the enzymes of gluconeogenesis is markedly increased by glucocorticoids.

Alterations in the activity of the thyroid gland cause major changes in the turnover of body protein. Both the synthesis and the breakdown of skeletal muscle protein are reduced in thyroid insufficiency. In hyperthyroidism both are increased, and a diminished muscle mass is a characteristic finding.

Physical activity causes a decrease in muscle protein breakdown and an increase in synthesis.

Further reading

Adams, R. L. P., Burdon, R. H., Campbell, A. M., Leader, D. P. and Smellie, R. M. S. (1981) *The Biochemistry of the Nucleic Acids*, 9th edn. London: Chapman and Hall

Caskey, C. T. (1980) Peptide chain termination. *Trends in Biochemical Sciences* **5,** 234–237

Clark, B. (1981) The elongation step of protein biosynthesis. *Trends in Biochemical Sciences* **5,** 207–209

Davidson, J. M. and Berg, R. A. (1981) Post-translational events in collagen biosynthesis. *Methods in Cell Biology* **23,** 119–136

Doherty, K. and Steiner, D. F. (1982) Post-translational proteolysis in polypeptide hormone biosynthesis. *Annual Review of Physiology* **44,** 625–638

Hershko, A. and Ciechanover, A. (1982) Mechanisms of intracellular protein turnover. *Annual Review of Biochemistry* **51,** 335–364

Holzer, H. and Heinrich, P. C. (1980) Control of proteolysis. *Annual Review of Biochemistry* **49,** 635–91

Hunt, T. (1983) Phosphorylation and the control of protein synthesis. *Philosophical Transactions of the Royal Society* **B302,** 127–134

Schachter, H. (1981) Glycoprotein biosynthesis and processing. In *Lysosomes and Lysosomal Storage Diseases*, eds Callahan, J. W. and Lowden, J. A., pp. 73–93. New York: Raven Press

Temple, G. F., Dozy, A. M., Roy, K. L. and Kan, Y. W. (1982) Construction of a functional human suppressor tRNA gene: an approach to gene therapy for β-thalassaemia. *Nature* **296,** 537–540

Traugh, J. A. (1981) Regulation of protein synthesis by phosphorylation. In *Biochemical Actions of Hormones, VIII,* ed. Litwak, G., pp. 167–208. New York: Academic Press

Wildenthal, K. (1980) *Degradative Processes in Heart and Skeletal Muscle.* Amsterdam: Elsevier

Wool, I. G. (1979) Structure and function of eukaryotic ribosomes. *Annual Review of Biochemistry* **48,** 719–754

G A J G

15 Amino acid metabolism

Amino acids arising from tissue protein breakdown together with those derived from dietary protein form a pool in which there is no distinction between the amino acids from the two sources. Amino acids are constantly being removed from the pool to serve as precursors for the synthesis of proteins or other substances of biological importance, or to be broken down to yield energy.

Five tissues, gut, liver, kidney, brain and skeletal muscle are of particular importance in the metabolism of amino acids. An indication of the uptake or output of amino acids by a tissue can be obtained from a comparison of the concentrations of amino acids in its arterial and venous blood. If the concentration of an amino acid in venous blood is less than that in the arterial blood the tissue is taking up the amino acid. The opposite is true when a tissue is putting out an amino acid. Figure 15.1 shows the results of such measurements for skeletal muscle in a

man who had fasted for more than 12 hours. It is clear that there is a net release of amino acids by muscle. More than 50 per cent of the total amount is released as alanine and glutamine. Similar results have been obtained with cardiac muscle where alanine constitutes 80 per cent of the amino acids released.

Similar studies have shown that the liver is the main site of alanine uptake, while the gut in fact has a net output of alanine. Most of the glutamine uptake is by the gut. The liver has a very small capacity for the uptake of the branched-chain amino acids, valine, leucine and isoleucine, but it readily takes up serine. While the brain takes up most amino acids, the uptake of the branched-chain amino acids, notably valine, is particularly rapid. The kidney has a net uptake of glutamine and a net output of alanine and serine. These processes are summarised in figure 15.2. The important role of alanine and its interconversion with glucose, the *glucose–alanine cycle* are discussed later (p. 181).

After a protein meal there is, as might be expected, a rise in the total concentration of amino acids in the plasma. Analyses of portal vein plasma have shown that the proportions of the individual amino acids is not that which might be expected from the amino acid composition of the dietary protein. Alanine predominates while aspartate and glutamate are virtually absent. It seems clear therefore that transamination reactions (p. 177) between amino acids and pyruvate to form alanine are taking place within the mucosal cells. The liver takes up most of the amino acids coming from the gut other than the branched-chain amino acids which are taken up by the peripheral tissues, particularly muscle and brain. As a result of these processes amino acid levels in plasma fall to fasting values in about three hours.

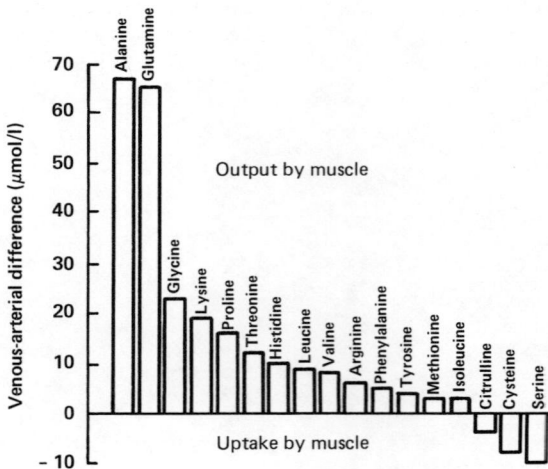

Fig. 15.1 Exchange of amino acids in muscle in normal postabsorptive man. (After Felig, P. (1975) *Annual Review of Biochemistry* **44**, 933)

Metabolic pathways involving amino acids. The reactions in which amino acids are incorporated into proteins were described in the previous chapter.

176

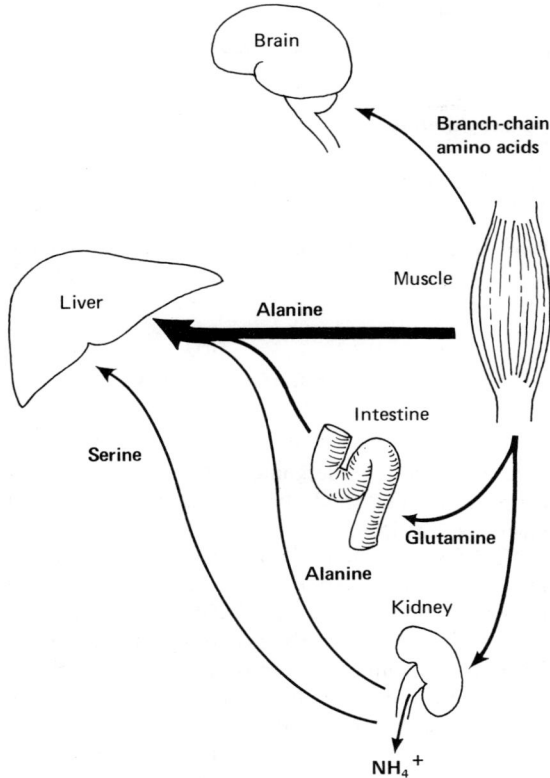

Other reactions of amino acids include the removal of the amino group (*deamination*) and the catabolism of the remaining carbon skeleton to provide energy. Amino acids are also utilised for the formation of specific compounds of physiological importance, such as haem, serotonin, thyroxine and the catecholamines.

Deamination of amino acids

Transamination. An important step in the breakdown of an amino acid is the transfer of its amino group to α-oxoglutarate or pyruvate, a reaction catalysed by enzymes known as *aminotransferases* or *transaminases*. These reactions result in the formation of glutamate or alanine and an α-oxo-acid (fig. 15.3). The aminotransferases show some specificity; the enzyme catalysing the reaction between alanine and α-oxoglutarate is known as *alanine aminotransferase* while that involved in the reaction between aspartate and α-oxoglutarate is termed *aspartate aminotransferase*.

Oxidative deamination. Amino groups on glutamate can be converted to ammonium ions by a reaction catalysed by the enzyme *glutamate dehydrogenase* (fig. 15.4). The activity of this enzyme is affected by nucleoside tri- and diphosphates. ATP and GTP act as allosteric inhibitors while ADP and GDP are allosteric activators. The activity of the enzyme thus increases when the available energy in the cell is low, in other words, when increased oxidation of amino acids is needed. This reaction is of particular

Fig. 15.2 Principal exchanges of amino acids between tissues. The key role of alanine output from muscle and gut and its uptake by the liver is emphasised. (After Felig, P. (1975) *Annual Review of Biochemistry* **44**, 933)

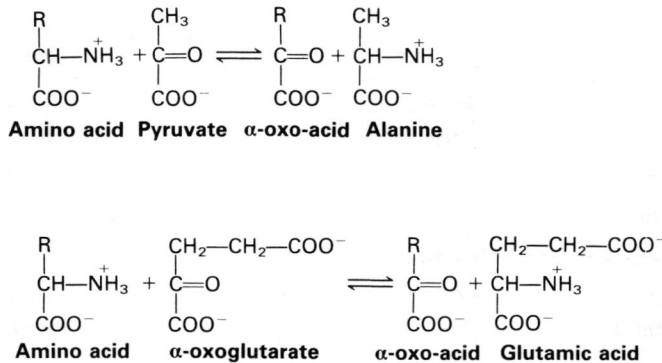

Fig. 15.3 Transamination reactions catalysed by aminotransferases

177

$$CH_2-CH_2-COO^-$$
$$CH-\overset{+}{N}H_3 \quad + NAD^+ + H_2O \rightleftharpoons NH_4^+ + \quad C=O \quad + H^+ + NADH$$
$$COO^- \qquad\qquad\qquad\qquad\qquad\qquad COO^-$$

Glutamate — **α-oxoglutarate**

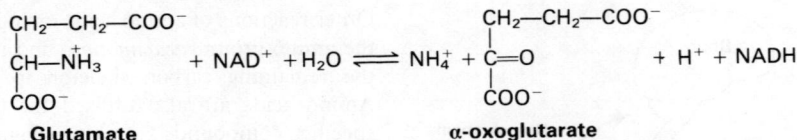

Fig. 15.4 Oxidative deamination of glutamate to yield α-oxoglutarate and ammonium ions

importance in the kidney where it is a source of urinary ammonia.

Other deamination reactions. The deamination of serine, threonine and cysteine occur by non-oxidative mechanisms which start with the elimination of H_2O or H_2S (fig. 15.5). Pyridoxal phosphate is the coenzyme for both the dehydratase and the desulphyridase. The dehydratase, which is found mainly in liver, increases in starvation, diabetes and after a high intake of dietary protein. Since all these situations call for an increase in gluconeogenesis, the increased enzyme activity may act to provide more pyruvate for this purpose.

Formation of urea

Deamination of amino acids gives rise to NH_4^+ ions which are toxic to tissue cells. The level of NH_4^+ ions in mammals is maintained at low levels by the formation and excretion of urea. Urea formation occurs almost exclusively in the liver. The overall reaction is

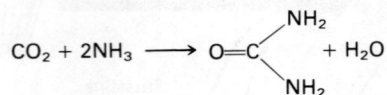

$$CO_2 + 2NH_3 \longrightarrow O=C\begin{matrix} NH_2 \\ \\ NH_2 \end{matrix} + H_2O$$

This reaction is endergonic ($\Delta G^{0'} = +59$ J); it is coupled to ATP hydrolysis in a cyclic series of reactions (fig. 15.6).

NH_4^+ ions combine with CO_2 in a reaction involving 2 molecules of ATP with the formation of carbamyl phosphate; N-acetyl glutamate is required as a co-factor. The carbamyl group is then transferred to the amino acid ornithine to form citrulline. Citrulline combines with aspartate to yield argininosuccinate, a reaction requiring the hydrolysis of a further molecule of ATP to AMP. Argininosuccinate is hydrolysed to arginine and fumarate. Finally, as a result of the action of arginase on arginine, arginine

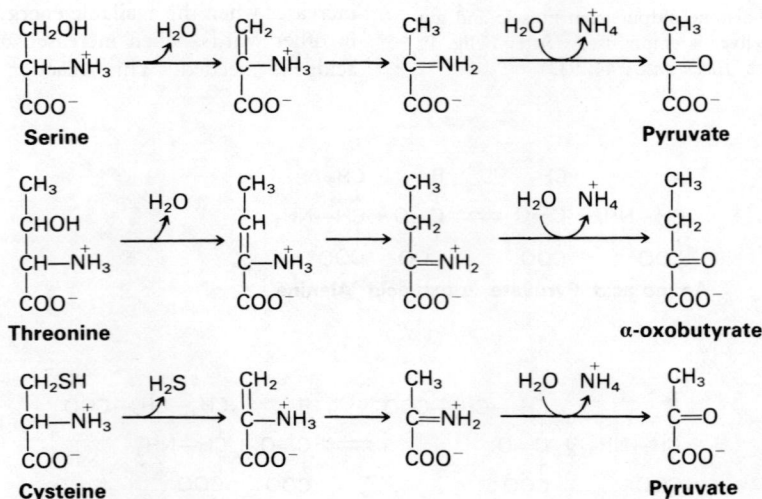

Fig. 15.5 The non-oxidative mechanisms for the deamination of serine, threonine and cysteine. The enzyme involved in the first step in serine and threonine metabolism is the same, serine/threonine dehydratase

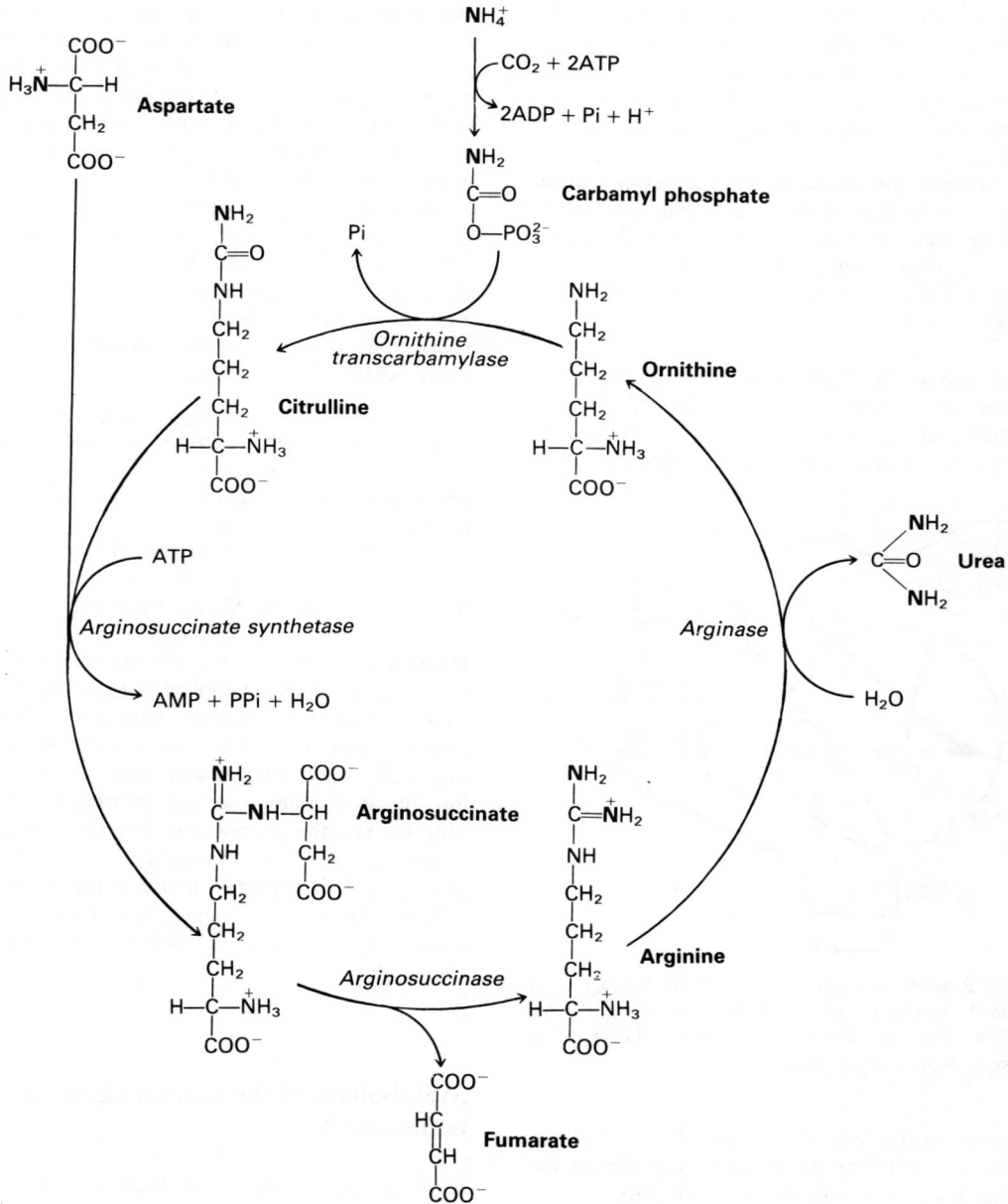

Fig. 15.6 The urea cycle. Note that of the two nitrogen atoms excreted in a urea molecule (shown in bold letters), one is derived from ammonium ions while the other is derived from aspartate formed as a result of transamination reactions. The formation of carbamyl phosphate and the conversion of ornithine to citrulline take place in mitochondria. The other reactions take place in the cytosol

is broken down to yield urea and ornithine. The production of NH_4^+ ions by the action of glutamate dehydrogenase on glutamate, the formation of carbamyl phosphate and citrulline all occur in the mitochondrial matrix; the other steps take place in the cytosol.

The fumarate produced by the urea cycle can be broken down further by the tricarboxylic acid cycle. One of the N atoms of urea is derived from the amino group of aspartate but it can also be derived from other amino acids by the transamination reactions described earlier.

Colonic ammonia production. The liver has to deal not only with NH_4^+ ions formed by the deamination of amino acids but also with NH_4^+ ions produced by bacteria in the large intestine (fig. 15.7)

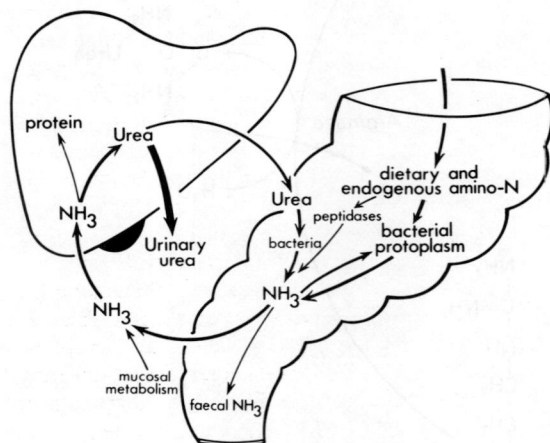

Fig. 15.7 Role of the colonic bacteria in the metabolism of urea and the production of ammonia. (From Wrong, O. M. et al. (1981) *The Large Intestine*, Lancaster: MTP Press, by courtesy of authors and publisher)

which pass to the liver via the portal vein. Colonic bacteria can hydrolyse proteins and peptides in the intestinal lumen to amino acids and break these down to yield ammonia. In addition ammonia is formed by the action of bacterial urease on urea which has been secreted into the gut.

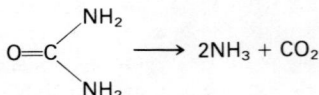

$$O=C \begin{matrix} NH_2 \\ \\ NH_2 \end{matrix} \longrightarrow 2NH_3 + CO_2$$

Blood ammonia levels. The capacity of the liver to convert ammonium ions to urea is such that the blood ammonium concentration does not normally exceed 50 μmol/l. High values are found in patients with severe liver disease partly because of defective liver cell function and partly because of the shunting of blood from the portal to the systemic circulation. In liver failure blood ammonia levels may rise as high as 250 μmol/l; patients become confused, develop coma (*hepatic encephalopathy*) and die. It is thought that the high blood levels of ammonia play a major part in the effect of liver failure on the brain, possibly by disturbing the concentrations of key intermediates. For example:

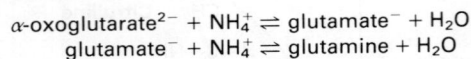

$$\alpha\text{-oxoglutarate}^{2-} + NH_4^+ \rightleftharpoons \text{glutamate}^- + H_2O$$
$$\text{glutamate}^- + NH_4^+ \rightleftharpoons \text{glutamine} + H_2O$$

Disorders of the urea cycle. Inherited defects of each of the enzymes of the urea cycle have been recognised. All are characterised by high blood ammonia levels. All are rare; some details of the three best known disorders are given in Table 15.1.

Blood urea levels. Normal plasma urea levels are in the range 2.5 to 6.0 mmol/l (15 to 40 mg/dl). High values are found in patients with a low glomerular filtration rate due either to renal disease or to impaired renal perfusion as a result, for example, of low blood pressure. At any particular glomerular filtration rate the plasma urea depends on the rate of synthesis (fig. 15.8); in turn this is determined principally by the protein intake in the diet.

Low plasma urea values are found in severe hepatic disease because of defective function of the urea cycle.

Metabolism of the carbon skeletons of amino acids

After an amino acid has been deaminated, the product can enter one of the pathways of oxidative metabolism. Some such as the pyruvate formed from alanine and the α-oxoglutarate formed from glutamate can enter directly. Other substances need preliminary metabolic modifications. The principal pathways for the metabolism of the carbon skeletons are shown in figure 12.1 (p. 131). It can be seen that deamination products of all amino acids can be

Table 15.1 Principal inherited disorders of the enzymes of urea synthesis

Disorder	Blood ammonia levels (µmol/l)	Clinical findings
Ornithine transcarbamylase deficiency	100–1000	Vomiting, lethargy, coma, death at the age of 10 days in severe cases. Some milder cases can survive well on a low protein diet
Citrullinaemia (arginosuccinic acid synthetase deficiency)	100–500	Vomiting, irritability, convulsions first seen at about 9 months of age. Permanent mental retardation. Enormous urinary excretion of citrulline
Arginosuccinicaciduria (arginosuccinase deficiency)	100–200	Convulsions, ataxia and severe mental retardation. Very high urinary excretion of arginosuccinic acid

catabolised to CO_2 and H_2O by the tricarboxylic acid cycle. It is also clear that certain amino acids can be used for the production of glucose. For example valine is converted to succinyl coenzyme A and hence oxaloacetate and thus glucose. Leucine on the other hand cannot give rise to glucose. The extent to which the carbon skeleton of an amino acid is oxidised or is converted into glucose in a particular tissue depends on the enzymic complement of the tissue and on various hormonal stimuli (Chap. 12).

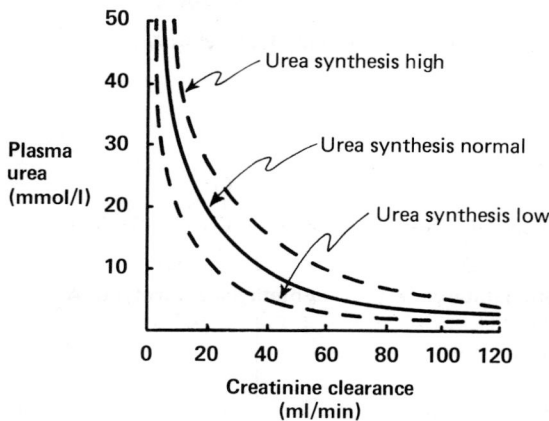

Fig. 15.8 Relationship between plasma urea and glomerular filtration rate as measured by creatinine clearance. A high urea synthesis occurs with a high protein diet, with bleeding into the gut or tissues and with infections, burns or trauma. A low urea synthesis occurs with a low protein intake and also in liver disease. (After Morrison R. B. I. (1979) In *Renal Disease*, ed. Black, D. A. K. and Jones, N. F., 4th edn, p. 305. Oxford: Blackwell)

Oxidation of branched-chain amino acids

While the liver is the principal site of the metabolism of most amino acids, the branched-chain amino acids are degraded mainly in skeletal muscle. The pathways of catabolism of valine, leucine and isoleucine are shown in figure 15.9. The first step is a transamination reaction involving α-oxoglutarate to form glutamate; the same aminotransferase is involved in each case. The other product of the reaction is a branched-chain α-oxo-acid. The three α-oxo-acids undergo an oxidative decarboxylation which requires a mitochondrial enzyme complex known as *branched-chain α-oxo-acid dehydrogenase*. This enzyme requires NAD^+, thiamine pyrophosphate and coenzyme A as co-factors; it is inhibited by ATP. The succeeding steps lead to the formation of succinyl CoA, acetoacetyl CoA or acetyl CoA.

The glucose–alanine cycle

The amount of alanine released by muscle (fig. 15.1) is much greater than can be accounted for by the breakdown of muscle protein; it must arise largely from pyruvate by a transamination involving glutamate and alanine aminotransferase (p. 177). Much of the glutamate needed is that formed in the catabolism of the branched-chain amino acids. Thus the release of alanine by muscle provides a means for the disposal of the nitrogen from these amino acids. The alanine is ultimately catabolised in liver and its nitrogen excreted as urea.

One source of the pyruvate required for alanine formation is glycolysis. Muscle and liver are therefore

181

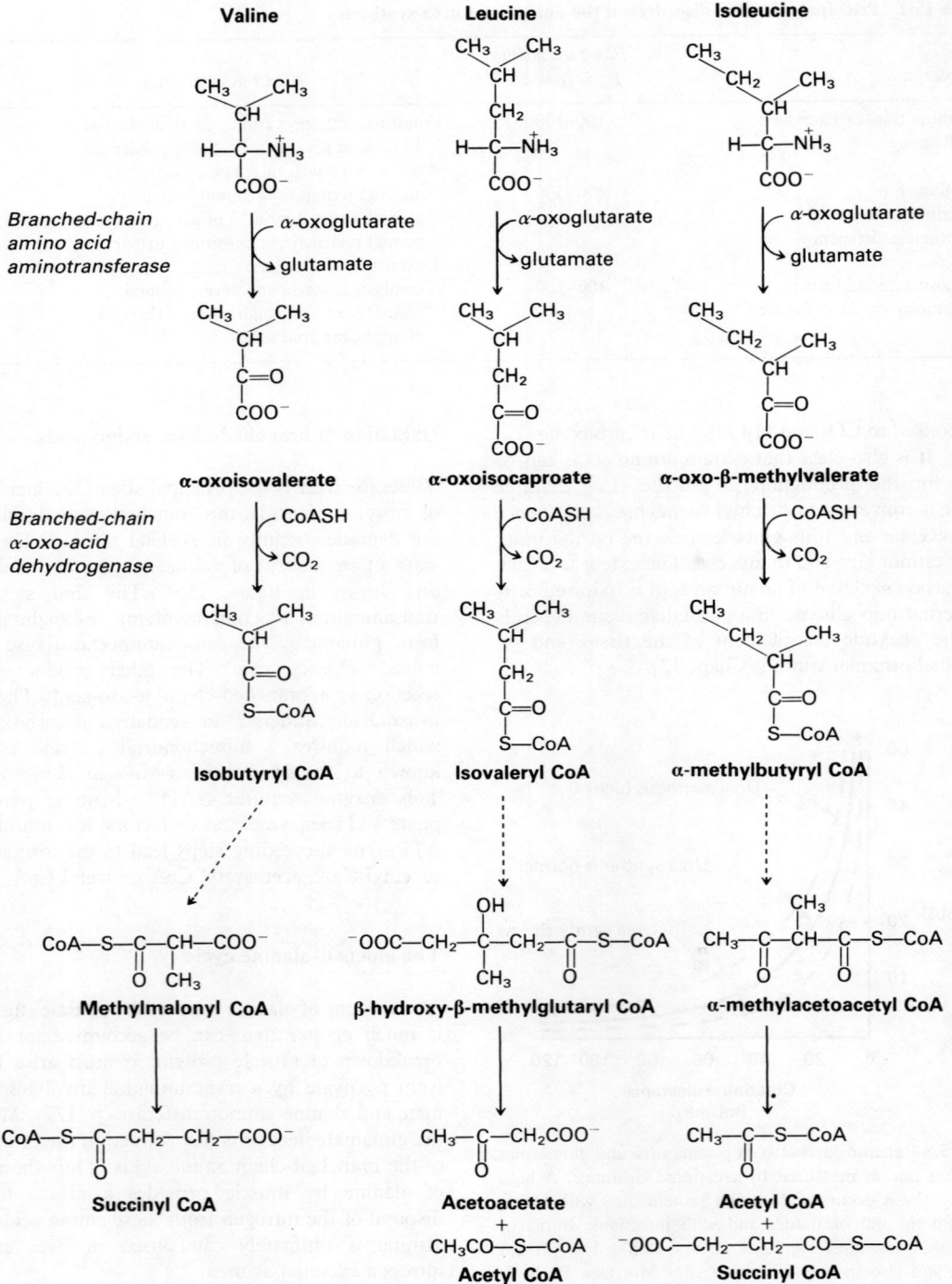

Fig. 15.9 Metabolism of the branched-chain amino acids

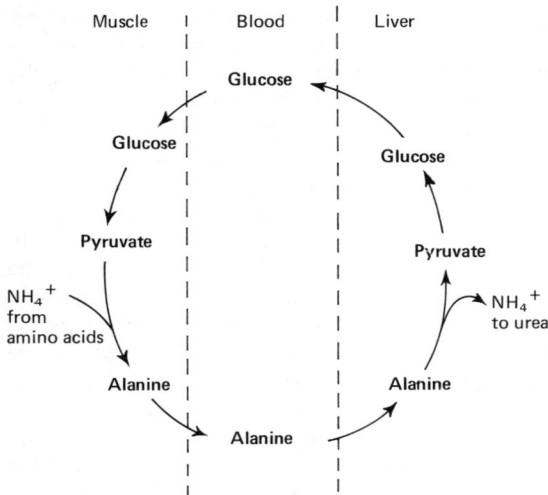

Fig. 15.10 The glucose–alanine cycle

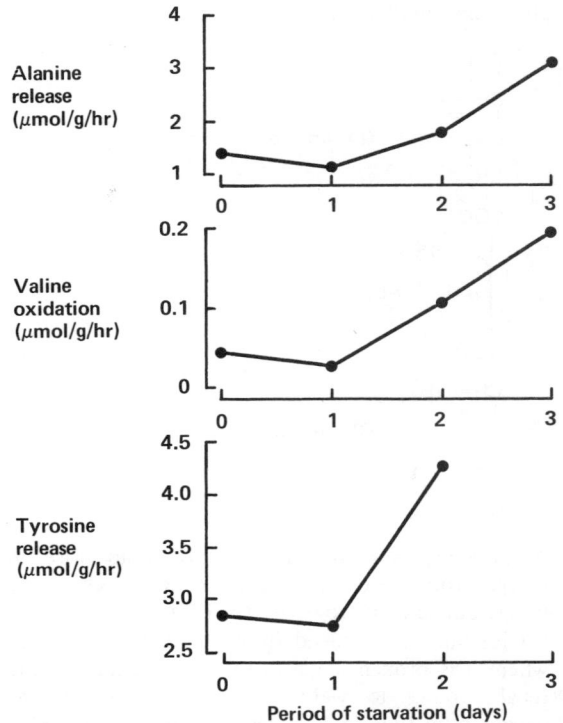

Fig. 15.11 Effect of progressive starvation on alanine release, valine oxidation and protein breakdown by rat muscle. (After Snell, K. (1980) *Biochemical Society Transactions* **8,** 209)

linked in a cyclic process, the *glucose–alanine cycle*. Alanine formed from pyruvate in the muscle is transported to the liver where it is converted to glucose by reverse glycolysis (p. 95). Glucose is taken up by muscle where it is broken down by glycolysis to form pyruvate for alanine formation (fig. 15.10). The glucose–alanine cycle is thus similar to the Cori cycle (p. 96), alanine serving as an alternative to lactate for gluconeogenesis.

Alanine formation in muscle has one important advantage over lactate formation. High levels of lactate within the cell affect the $NAD^+/NADH$ ratio and consequently various redox systems. The existence of the glucose–alanine cycle prevents an excessive rise in lactate concentration. Although alanine is a major substrate for gluconeogenesis in the liver, no net synthesis of glucose takes place if the alanine is formed from pyruvate derived from glycolysis. The main purpose of the glucose–alanine cycle therefore appears to be the removal of NH_2 groups formed from amino acid catabolism to the liver.

Pyruvate is formed in muscle, not only by glycolysis but also from the carbon skeletons of amino acids. This process is particularly important after a protein meal, during fasting and in diabetes mellitus. In the last two cases the amino acids become available as a result of increased breakdown of muscle protein. When muscle from rats starved for different periods is incubated a good correlation is observed between protein breakdown, valine oxidation and alanine release (fig. 15.11). Since many of the amino acids can give rise to glucose the operation of the glucose–alanine cycle in starvation results in the net formation of glucose.

Glutamine

Glutamine is another amino acid released in large amounts by muscle. The carbon skeleton of this amide can be formed in muscle from aspartate, asparagine, glutamate, valine or isoleucine. Work with rats has suggested that 50 per cent of the carbon chains of these amino acids entering the tricarboxylic acid cycle in muscle are converted to glutamine.

The formation of glutamine involves a reaction between NH_4^+ and ATP, catalysed by an enzyme,

183

glutamine synthetase:

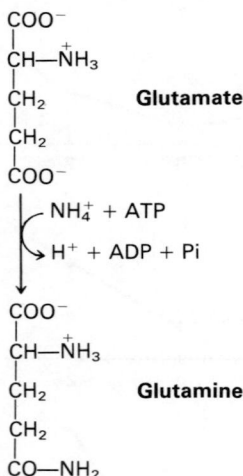

$$
\begin{array}{l}
\text{COO}^- \\
| \\
\text{CH}-\overset{+}{\text{NH}_3} \\
| \\
\text{CH}_2 \qquad \textbf{Glutamate} \\
| \\
\text{CH}_2 \\
| \\
\text{COO}^-
\end{array}
$$

$$\text{NH}_4^+ + \text{ATP} \searrow$$
$$\text{H}^+ + \text{ADP} + \text{Pi} \searrow$$

$$
\begin{array}{l}
\text{COO}^- \\
| \\
\text{CH}-\overset{+}{\text{NH}_3} \\
| \\
\text{CH}_2 \qquad \textbf{Glutamine} \\
| \\
\text{CH}_2 \\
| \\
\text{CO}-\text{NH}_2
\end{array}
$$

Ammonium ions for this process could arise in muscle from the oxidative deamination of glutamate by glutamate dehydrogenase (p. 177).

Glutamine is removed from blood by the kidney where it is broken down in the mitochondria of the renal tubules to yield ammonium ions and α-oxoglutarate (fig. 15.12). The ammonium ions play an important role in the regulation of acid–base balance and the conservation of cations (Chap. 19).

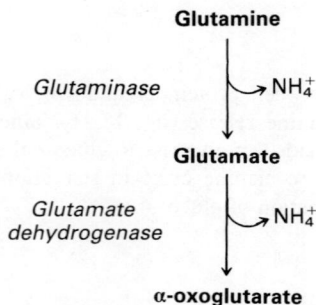

Glutamine

Glutaminase $\quad\searrow \text{NH}_4^+$

Glutamate

Glutamate
dehydrogenase $\quad\searrow \text{NH}_4^+$

α-oxoglutarate

Fig. 15.12 The actions of glutaminase and glutamine dehydrogenase in the production of ammonium ions in the renal tubule

Essential and non-essential amino acids

An amino acid is classified as essential if it must be present in the diet to ensure either normal growth in a child or the maintenance of nitrogen balance in an adult. Leucine, isoleucine, valine, lysine, methionine, phenylalanine, tryptophan, threonine, histidine and arginine are considered *essential* for adults. The other amino acids, the *non-essential amino acids*, can be synthesised in amounts sufficient for the body's needs. The nature of the carbon skeleton of a particular amino acid determines whether or not it is essential. The requirement for the branched-chain amino acids, for example, can be met if the corresponding α-oxo-acids are included in the diet; transamination reactions convert these to the corresponding amino acid. This can be used in the treatment of patients with chronic renal failure. Such patients can be maintained in nitrogen balance on a low-protein diet supplemented with branched chain α-oxo-acids. The accumulation of urea and other metabolites in toxic amounts is thus limited. Tyrosine is not an essential amino acid because it is readily formed from phenylalanine in the liver (p. 191). However, if tyrosine is present in the diet the requirement for phenylalanine is reduced. Similarly cysteine has a sparing effect on methionine because cysteine can be synthesised from methionine (p. 188).

Glutamic acid derived from α-oxoglutarate can supply the carbon chain for the other non-essential amino acids, proline (fig. 15.13) and glutamine (p. 183). The carbon skeleton of serine arises from glucose via 3-phosphoglycerate (fig. 15.14). Glycine can be produced from serine by a reaction involving tetrahydrofolic acid (fig. 15.15).

Requirements for the essential amino acids. An adult requires approximately 1 g of each of the essential amino acids daily. The amino acids are present in approximately the ideal proportions in animal foods such as meat, fish and eggs, which are therefore sometimes called 'first class proteins'. Certain plants, notably groundnuts, peas, beans and maize, are sources of protein but the amino acid proportions are not ideal and a larger amount is needed to provide the equivalent of first class protein. Peas and beans for instance have only a small proportion of methionine and cereals such as maize contain little lysine. These deficiencies are of limited significance in practice since, in a diet consisting of a mixture of plant proteins, the deficiency of amino acids in one source is compensated by their relative abundance in another. This is important in many developing countries where meat, eggs and milk may

COO⁻
|
CH₂
|
CH₂   **Glutamate**
|
CH—$\overset{+}{N}$H₃
|
COO⁻

NADH + ATP
NAD⁺ + ADP + Pi

CHO
|
CH₂
|
CH₂
|
CH—$\overset{+}{N}$H₃
|
COO⁻

CH₂——CH₂
| |
CH   CH—COO⁻
  \\ /
   $\overset{+}{N}$H

NADPH + H⁺
NADP⁺

CH₂——CH₂
| |
CH₂   CH—COO⁻   **Proline**
   \\ /
   $\overset{+}{N}$H₂

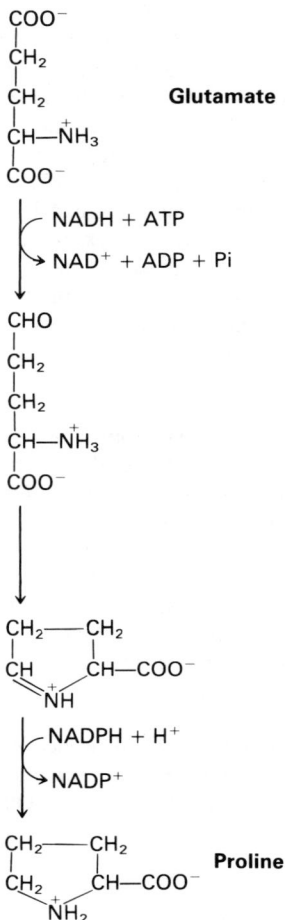

Fig. 15.13 Synthesis of proline from glutamate

Glucose

COO⁻
|
CH—OH   **3-phosphoglycerate**
|
CH₂—O—PO₃²⁻

NAD⁺
NADH + H⁺

COO⁻
| **3-phospho-**
C=O   **hydroxypyruvate**
|
CH₂—O—PO₃²⁻

glutamate
α-oxoglutarate

COO⁻
|
CH—$\overset{+}{N}$H₃   **3-phosphoserine**
|
CH₂—O—PO₃²⁻

H₂O
Pi

COO⁻
|
CH—$\overset{+}{N}$H₃   **Serine**
|
CH₂OH

Fig. 15.14 Synthesis of serine from glucose. The first step is the production of 3-phosphoglycerate by glycolysis (Chap. 9)

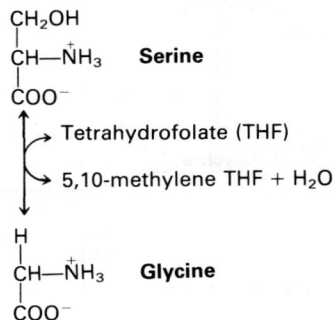

be extremely expensive. Protein-energy malnutrition can be prevented cheaply by the use of appropriate vegetable proteins.

Decarboxylation of amino acids

L-Amino acid decarboxylases catalyse the decarboxylation of amino acids to form amines some of which are of physiological importance. *Histamine* is formed by the decarboxylation of histidine and 5-hydroxytryptophan gives rise to *5-hydroxytryptamine*. Decarboxylation of serine yields ethanolamine (p. 39) which is a precursor for the synthesis of choline and

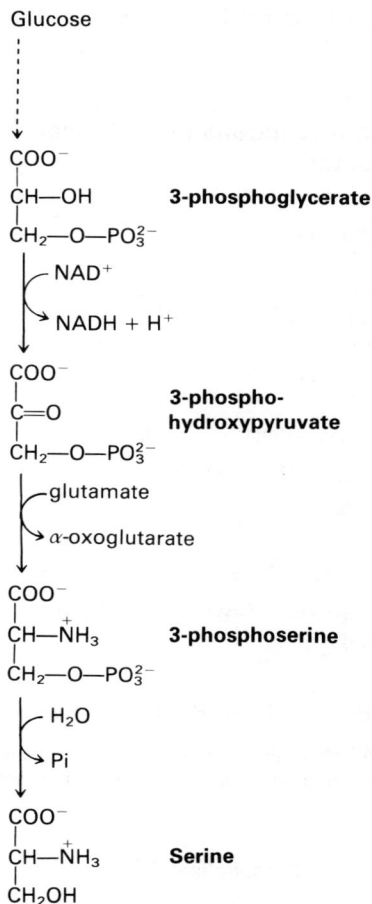

CH₂OH
|
CH—$\overset{+}{N}$H₃   **Serine**
|
COO⁻

Tetrahydrofolate (THF)
5,10-methylene THF + H₂O

H
|
CH—$\overset{+}{N}$H₃   **Glycine**
|
COO⁻

Fig. 15.15 Role of tetrahydrofolate (p. 152) in the interconversion of serine and glycine

185

the phospholipids (Chap. 12). The decarboxylases all require pyridoxal phosphate (p. 87) as a coenzyme.

Important reactions of particular amino acids

Glycine and serine

Glycine can react with 5,10-methylene tetrahydrofolate to yield serine (fig. 15.15). After incorporation into serine the carbon atoms of glycine can enter the general pathways of oxidation as pyruvate (fig. 12.1).

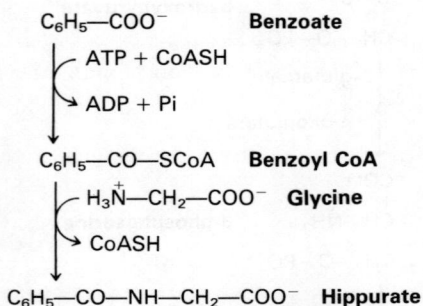

$$C_6H_5—COO^-\quad \textbf{Benzoate}$$

ATP + CoASH

ADP + Pi

$$C_6H_5—CO—SCoA\quad \textbf{Benzoyl CoA}$$

$H_3\overset{+}{N}—CH_2—COO^-\quad \textbf{Glycine}$

CoASH

$$C_6H_5—CO—NH—CH_2—COO^-\quad \textbf{Hippurate}$$

Fig. 15.16 Detoxication of benzoic acid by conjugation with glycine to give hippuric acid which is then excreted in the urine

Glycine is also a precursor for the synthesis of purines (Chap. 13) and porphyrins (Chap. 16).

Glycine also acts as a conjugating agent to render toxic metabolites or foreign chemicals (*xenobiotics*) more soluble and thus facilitate their excretion. An example is the formation of hippuric acid in the liver after the ingestion of benzoic acid (fig. 15.16). In this reaction a peptide bond is formed by a mechanism involving coenzyme A, a process quite different from that in protein synthesis (Chap. 14). Another example of this role of glycine is the formation of the bile acid glycocholic acid by a reaction between the coenzyme A derivative of cholic acid (p. 44) and glycine.

Glyoxylate metabolism. Approximately 1 per cent of glycine metabolism is by another pathway involving the key intermediate glyoxylate. Glyoxylate also provides the principal pathway for the breakdown of glycolic acid in vegetable foodstuffs and of glycolaldehyde and thus ethanolamine (p. 39) (fig. 15.17).

Glyoxylate is oxidised in a cycle of reactions involving α-oxoglutarate and the enzyme carboligase. Lack of this enzyme is one cause of *hyperoxaluria* (oxalosis).

Oxalate metabolism. Oxalate occurs in the diet notably in rhubarb, spinach, tea and cocoa. Only about 5 per cent is absorbed. Oxalate is also

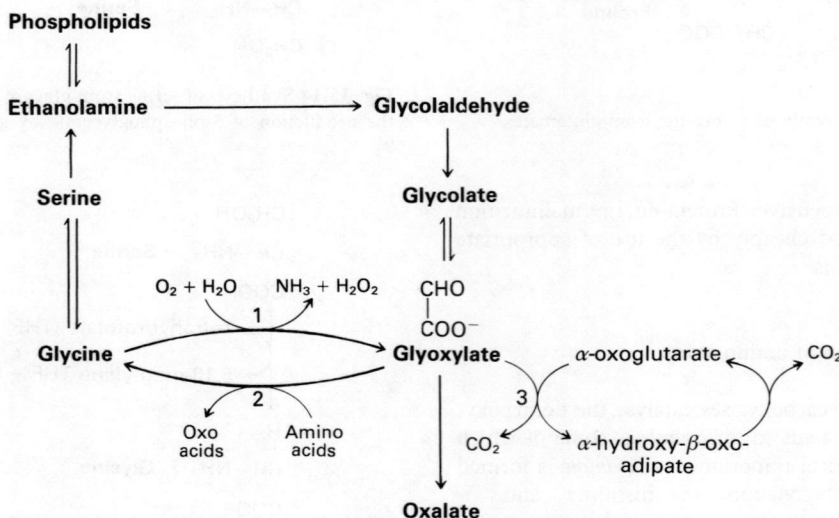

Fig. 15.17 Some of the metabolic pathways involving glyoxylate. Reaction 1 involves glycine oxidase and FAD. Reaction 2 involves the aminotransferases (fig. 15.3) and pyridoxal phosphate. Reaction 3 involves the enzyme carboligase

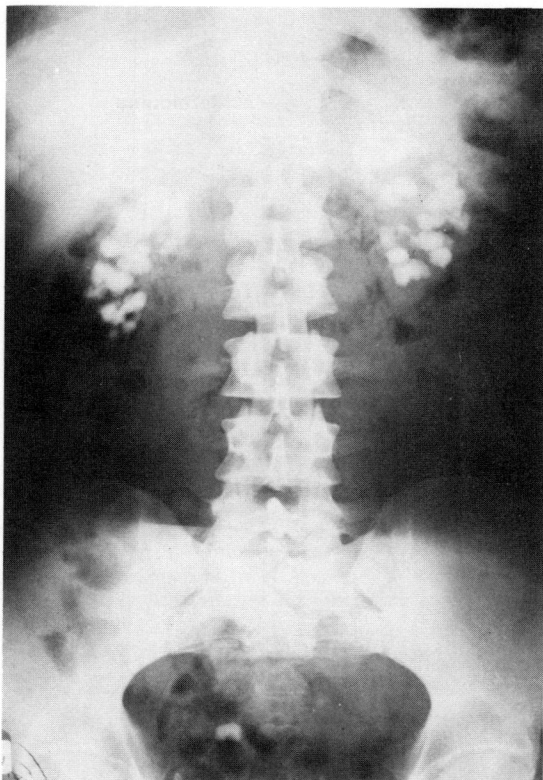

Fig. 15.18 Extensive deposits of calcium oxalate in the kidneys in a patient with hyperoxaluria. (From Rose, G. A. (1981) *Urinary Stones: Clinical and Laboratory Aspects.* Lancaster: MTP Press, Baltimore: University Park Press, by courtesy of author and publisher)

oxalate stone formation is also recognised in patients who have treated themselves with very large doses of ascorbate (vitamin C).

Creatine and creatinine. Glycine is a precursor for the synthesis of creatine phosphate which is found in muscle and is important as a reserve energy source. The synthetic pathway is shown in figure 15.19. Creatine produced by the liver circulates in the plasma and is picked up by muscle where, as creatine phosphate, it forms about 0.5 per cent of the muscle weight.

Each day approximately 2 per cent of the creatine in muscle is broken down to creatinine which diffuses into the plasma and is excreted by the kidney. The creatinine excretion is a measure of the total creatine stores and therefore of muscle mass (fig. 15.20). It is little affected by dietary protein and is remarkably constant from day to day.

Since creatinine is filtered by the glomerulus and passes down the tubule without reabsorption (but it is however secreted to a small extent) the creatinine clearance provides a valuable measure of the glomerular filtration rate.

In disorders characterised by excessive muscle damage such as fever of any cause, starvation, thyroid overactivity and muscle diseases such as muscular dystrophies, the excretion of creatine may be greatly increased. Muscle damage and even severe exercise also cause the release of large amounts of creatine kinase into the plasma (p. 91).

produced by the action of various enzymes such as lactate dehydrogenase on glyoxylate and as the product of ascorbate metabolism. Oxalate is an end-product of metabolism; a normal adult has a urinary output of 0.25 to 0.6 mmol oxalate in 24 hours.

High blood and urine levels of oxalate occur in two rare inherited conditions, primary *hyperoxaluria* I and II. Hyperoxaluria I is caused by lack of the carboligase involved in glyoxylate metabolism. Both conditions cause the formation in the kidney of calculi containing calcium oxalate (fig. 15.18). Eventually renal failure ensues and death occurs usually before the age of 20. At post mortem deposits of calcium oxalate are found in many tissues including the heart. Excessive oxalate excretion and calcium

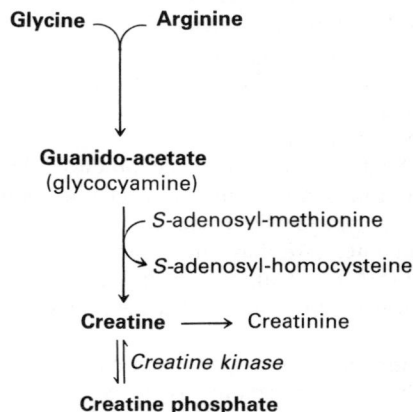

Fig. 15.19 Pathway for the synthesis of creatine phosphate

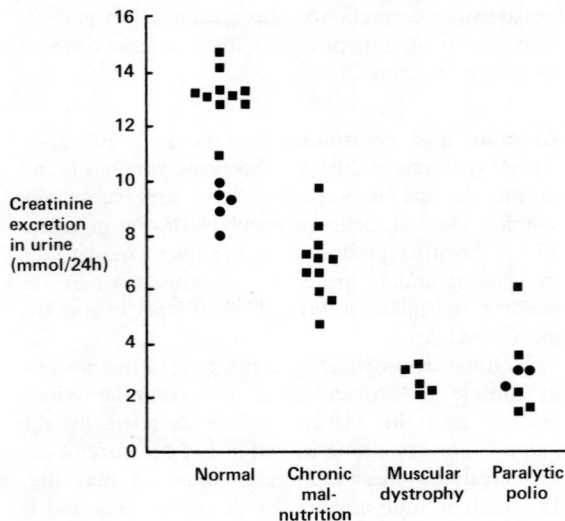

Fig. 15.20 Excretion of creatinine in the urine in normal subjects and patients with greatly reduced muscle mass for various reasons. ■ = males, ● = females

The sulphur-containing amino acids

Of the three sulphur-containing amino acids methionine, cysteine and cystine, only methionine is essential for man. Methionine is particularly important as a source of the methyl groups required in the synthesis of important substances such as adrenaline (epinephrine) (p. 260), choline (p. 39) and creatine (fig. 15.19). The methyl group of methionine is activated as a result of a reaction between methionine and ATP to form a S-adenosyl-methionine (SAM) (fig. 15.21). The linkage between the sulphur and the methyl group in SAM is readily broken and the compound serves as a *methyl donor* in various reactions (fig. 15.21).

The S-adenosyl-homocysteine produced by such reactions can undergo further metabolic steps (fig. 15.22) leading to the formation of cysteine and α-oxobutyrate. α-oxobutyrate is further degraded to succinyl CoA by a pathway which is also used for the breakdown of threonine, isoleucine and valine (fig. 15.23).

Cysteine can be converted to pyruvate as described earlier (p. 178); the sulphur is excreted in the urine as sulphate. Cysteine can also undergo oxidation to cysteine sulphinate which can serve as a source of the

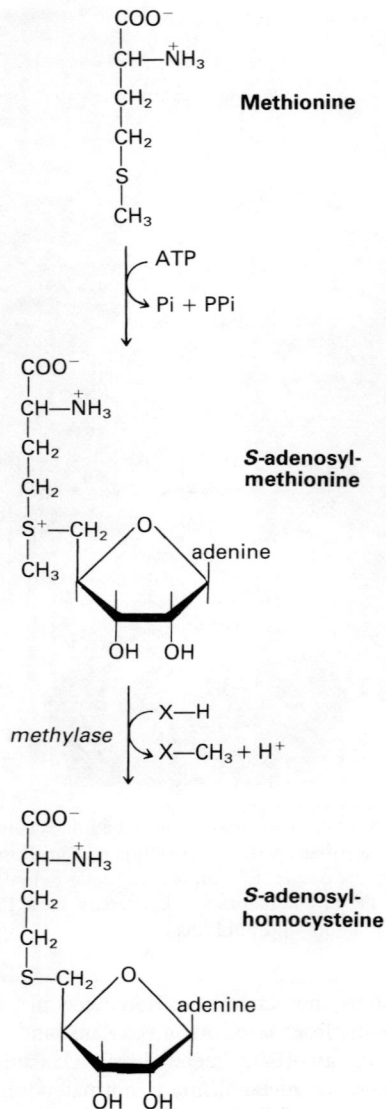

Fig. 15.21 Formation of S-adenosyl-methionine and its role in methylation reactions

sulphate needed for the synthesis of the proteoglycans, heparin and glucuronic acid (Chap. 3). Cysteine sulphinate can also be decarboxylated to yield taurine which is found in high concentration in muscle and is conjugated with cholic acid in the liver to form the bile salt, taurocholate (Chap. 4).

adenosine $\overset{+}{N}H_3$
H—S—CH$_2$—CH$_2$—CH—COO$^-$

S-adenosyl-homocysteine

↓ H_2O
↘ adenosine

$\overset{+}{N}H_3$
H—S—CH$_2$—CH$_2$—CH—COO$_3^-$

Homocysteine

Cystathionine synthetase   serine

$\overset{+}{N}H_3$
S—CH$_2$—CH$_2$—C—COO$^-$
CH$_2$—CH—COO$^-$
$\overset{+}{N}H_3$

Cystathionine

Cystathionase

$\overset{+}{N}H_3$
HO—CH$_2$—CH$_2$—CH—COO$^-$ ←   → CH$_2$SH—CH—COO$^-$
Homoserine $\overset{+}{N}H_3$
Cysteine

CH$_3$—CH$_2$—CO—COO$^-$
α-oxobutyrate

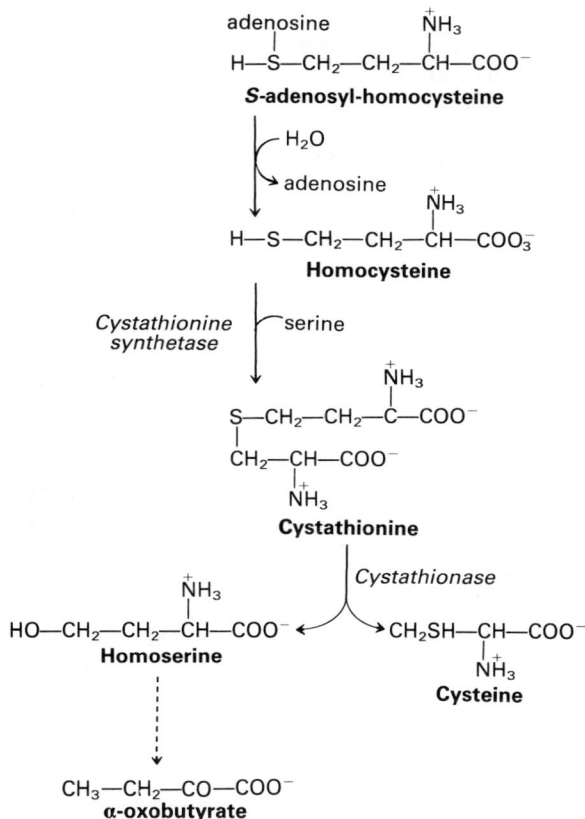

Fig. 15.22 Principal steps in the metabolism of *S*-adenosyl-homocysteine to give cysteine and α-oxobutyrate

Methionine   **Threonine**

↓ ↘

CH$_3$—CH$_2$—CO—COO$^-$ **α-oxobutyrate**

↓ CoA.SH
↘ CO_2 **Isoleucine**

CH$_3$—CH$_2$—CO—SCoA   **Propionyl CoA**

↓ CO_2
Propionyl CoA carboxylase   **Valine**

COO$^-$ **Methyl malonyl CoA**
CH$_3$—CH—CO—SCoA

↓ *Methyl malonyl CoA mutase*

CH$_2$—COO$^-$ **Succinyl CoA**
CH$_2$—CO—SCoA

Fig. 15.23 Pathway for the metabolism of α-oxobutyrate derived from methionine, and also for the metabolism of the carbon skeletons of threonine, isoleucine and valine. Propionyl CoA carboxylase contains biotin and methyl malonyl CoA mutase contains cobalamin (vitamin B_{12}) (Chap. 13)

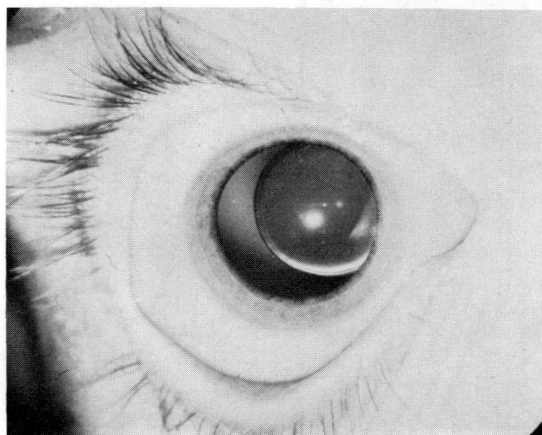

Homocystinuria. This is the most common disorder of methionine metabolism and is caused by deficiency of the enzyme cystathionine synthetase. It has an incidence of about 1 in 25 000 births. The patients have greatly increased blood and urine levels of homocysteine and methionine, and also of the compound formed by the combination of two homocysteine molecules, *homocystine*. Patients have a variety of symptoms and signs. The lenses of the eyes become dislocated (fig. 15.24) because of disorganisation of the collagen fibres which normally keep them in place. The bones may be excessively fragile and the joints deformed (fig. 15.25), both probably because of collagen abnormality. About 50 per cent of patients are mentally retarded. The collagen defects may be related to abnormalities of –S–S– cross-links resulting from replacement of cysteine by

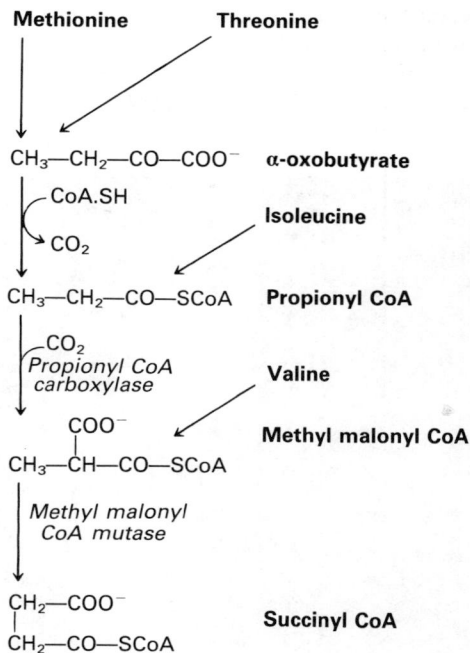

Fig. 15.24 Dislocated lens in homocystinuria. (Courtesy of Nina Carson)

189

homocysteine. In addition homocysteine may compete with cysteine for active transport.

There appear to be two variants of homocystinuria. In one the patients respond biochemically to large doses of pyridoxine (vitamin B_6); the defect may be in

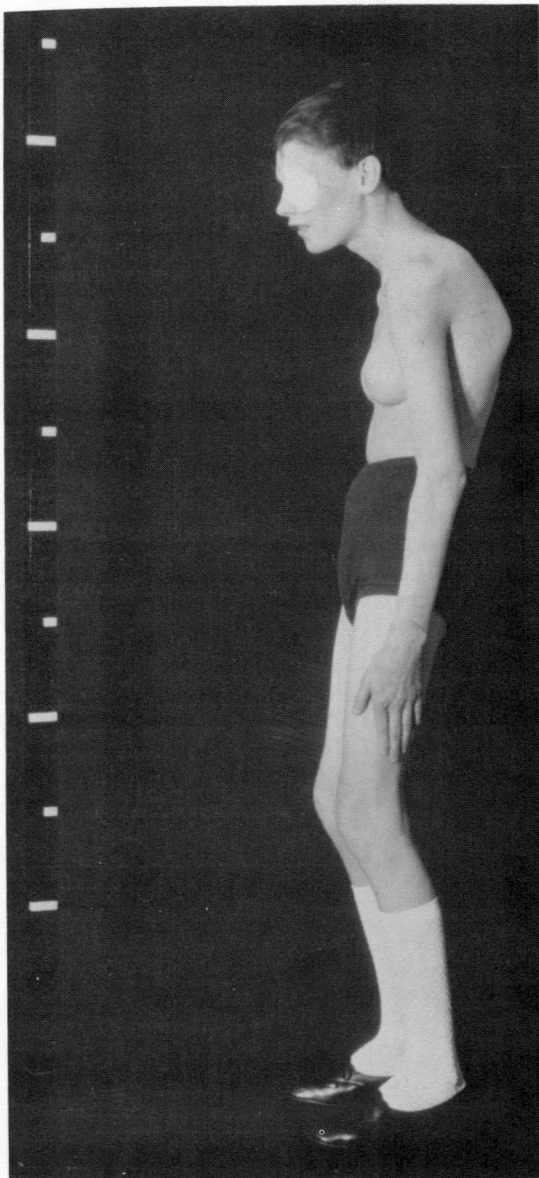

Fig. 15.25 Severe kyphosis in a 12-year-old girl with homocystinuria. (Courtesy of Nina Carson)

the enzyme's binding site for the coenzyme. In both variants patients improve if they are given a low methionine diet.

Phenylalanine and tyrosine

Of the two aromatic amino acids, phenylalanine is essential but tyrosine is not. Phenylalanine can be converted to tyrosine in the smooth endoplasmic reticulum of the liver but this reaction is not reversible (fig. 15.26). Tyrosine can be broken down to fumarate and acetoacetate as shown; it is also the precursor for the synthesis of several important compounds such as the thyroid hormones (p. 262), the neurotransmitters dopamine, adrenaline (epinephrine) and noradrenaline (norepinephrine) (p. 260) and the pigment melanin. Phenylalanine can be metabolised to a small extent to derivatives such as phenylpyruvate; this pathway is important in phenylketonuria.

Phenylketonuria. This is the most common inborn error of metabolism causing mental retardation and accounts for up to 3 per cent of patients in institutions for the mentally handicapped. It is caused by deficiency of phenylalanine hydroxylase. The incidence of the disorder in the United States and in many western countries is about 1 in 18 000 births. It is very rare in Finns, Jews and Africans. A high incidence (1 in 4000 births) has been found in Eire.

In affected individuals phenylalanine and its metabolites such as phenylpyruvate, phenylacetate and phenyllactate accumulate in the plasma and the tissues and can be detected in the urine, which has a mousy smell. Patients appear normal at birth but within a few weeks become irritable, have convulsions and vomit often. Mental retardation becomes evident by four to six months and at this stage is irreversible. The link between the metabolic disorder and the brain disturbance is not yet understood. Older patients are usually grossly retarded in the 'imbecile' range with purposeless movements, convulsions and frequent tantrums.

This disastrous outcome can be prevented by early treatment with a low phenylalanine diet which includes tyrosine. Some phenylalanine must be included for protein synthesis and several suitable diets are now available commercially. Figure 15.27 shows the effect of delay in treatment on the eventual

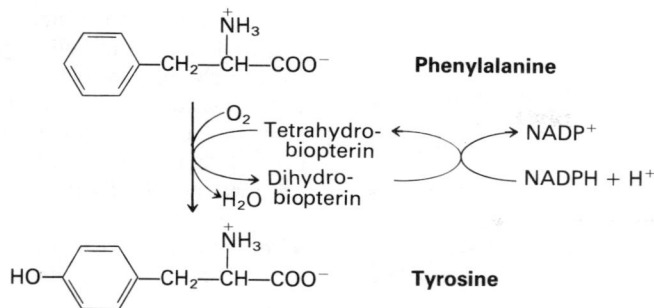

Fig. 15.26 Conversion of phenylalanine to tyrosine by phenylalanine hydroxylase. This is a 'mixed function' oxidase and uses molecular oxygen, NADPH and a pteridine cofactor, tetrahydrobiopterin

intelligence. Early diagnosis of phenylketonuria is so important that in many countries screening tests are carried out on all newborn infants.

Phenylketonuria is one of the inborn errors in which it is possible to identify the heterozygotes. Although they are symptomless and of normal intelligence, they usually have plasma phenylalanine levels above the normal range (fig. 15.28) and have an abnormal response to a test dose of phenylalanine (fig. 15.29).

Albinism. This is a group of hereditary disorders of the synthesis of melanin from tyrosine by melanocytes. Two major variants have been identified: a severe form caused by lack of the enzyme tyrosinase and a slightly less severe form in which tyrosinase is present. In both disorders the hair, skin and eyes lack pigment (fig. 15.30); the eyes are exceptionally sensitive to bright light (*photophobia*).

This condition was the first in which it was demonstrated that the same clinical abnormality could be caused by two distinct genetic disorders (*heterogeneity*). Part of the evidence for this was that in two families, in which both parents were albinos, all the children were normal. Since albinism is inherited as an autosomal recessive, all the children would have been expected to be affected in such a family if the parents had the same disorder. Further study of the parents in each family demonstrated that one parent in each couple was tyrosinase positive and the other tyrosinase negative.

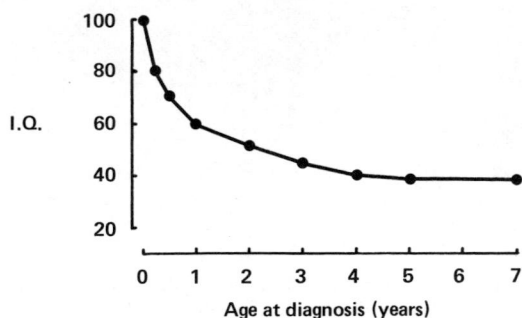

Fig. 15.27 The effect of early diagnosis and treatment on the eventual I.Q. of patients with phenylketonuria. (Data collected by C. Clow and C. R. Scriver in Rosenberg, L. E. and Scriver, C. R. (1974) In *Duncan's Diseases of Metabolism*, eds Bondy, P. K. and Rosenberg, L. E., 7th edn, p. 465. Philadelphia: Saunders)

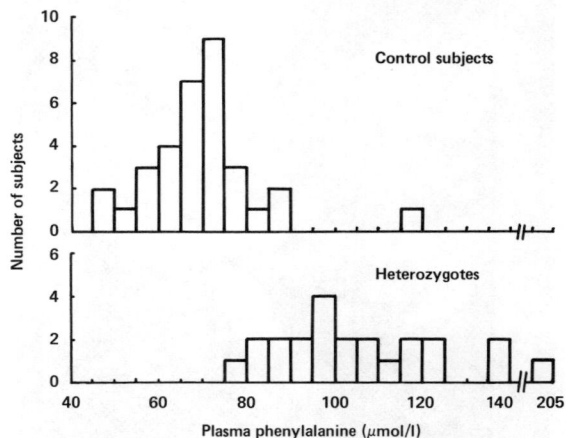

Fig. 15.28 Plasma phenylalanine levels in 33 control subjects and 23 heterozygotes for phenylketonuria (parents of affected children). (Data of Knox, W. E. (1958) *American Journal of Human Genetics* **10**, 53)

191

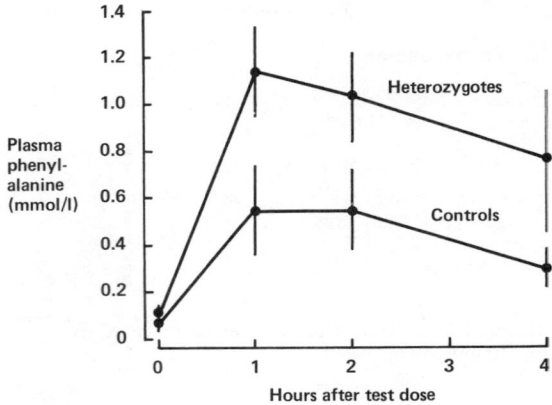

Fig. 15.29 Plasma phenylalanine levels in normal subjects and heterozygotes for phenylketonuria after oral doses of phenylalanine 100 mg/kg body weight. The vertical bars represent 1 standard deviation. (Data of Hsia, D. Y. *et al.* (1956) *Nature* **178,** 1239)

Tyrosinase negative albinism has an incidence of about 1 in 35 000 in both negroes and caucasians. A particularly high incidence is found in Ireland (1 in 15 000). Tyrosinase positive albinism is more common in negroes than in caucasians. It is also particularly common in native americans such as the San Blas Indians of Panama (1 in 150).

Alcaptonuria. This is a rare disorder of particular interest since it was the first in which a genetic defect leading to the lack of a specific enzyme was postulated (by Sir Archibald Garrod in 1902). Alcaptonuria has an incidence of about 1 in 300 000 and is characterised by the excretion of homogentisic acid in the urine. If the urine is alkaline it darkens on standing. Alcaptonuria causes no symptoms in childhood but, in later life, deposition of alcapton, a metabolite of homogentisic acid, causes black discolouration of cartilage and other tissues (fig. 15.31), and also arthritis and spondylitis (fig. 15.32).

Fig. 15.31 Eye in an elderly patient with alcaptonuria to show pigmentation due to the deposition of alcapton. (Originally published as a colour illustration in Boyle, A. C. (1980) *A Colour Atlas of Rheumatology*. London: Wolfe, reprinted by courtesy of author and publisher)

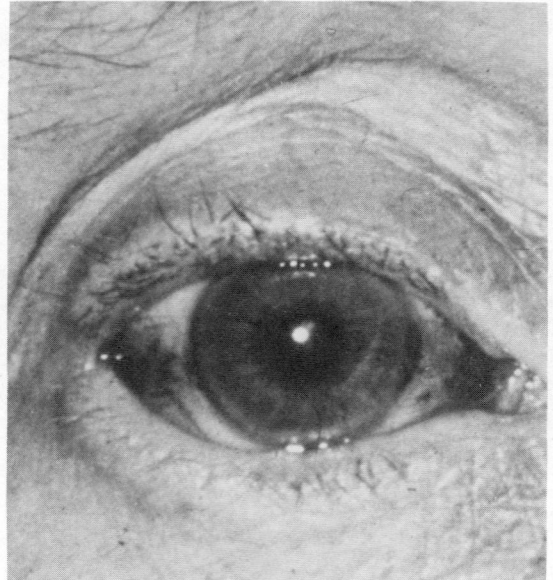

Fig. 15.30 African twins, one of whom had albinism. (Courtesy of I. W. E. Hunter)

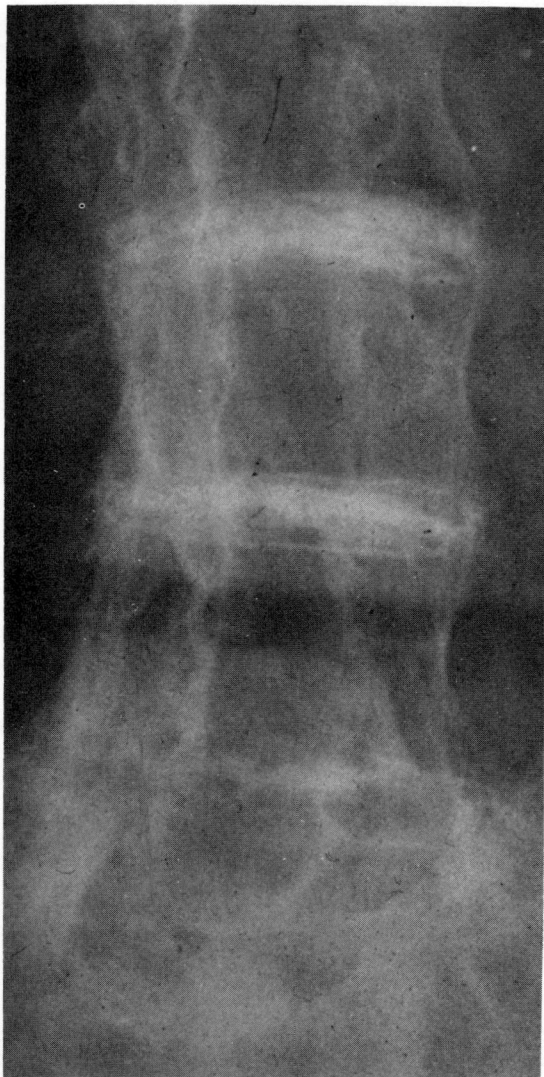

Fig. 15.32 Fusion of the vertebral bodies in a patient with alcaptonuria. (From Boyle, A. C. (1980) *A Colour Atlas of Rheumatology*. London: Wolfe, by courtesy of author and publisher)

Tryptophan

This is an essential amino acid in man and its metabolism yields several important substances.

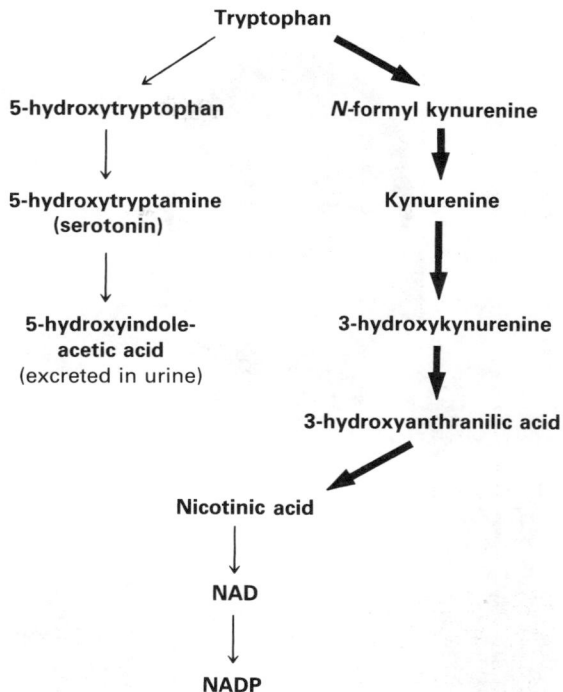

Fig. 15.33 Pathways of tryptophan metabolism

Hydroxylation yields 5-hydroxytryptophan, the precursor of 5-hydroxytryptamine (*serotonin*). The major pathway of tryptophan metabolism in the liver yields 3-hydroxyanthranilic acid, a precursor of nicotinic acid and so of NAD (fig. 15.33). This is an important source of nicotinic acid in man; about 60 mg tryptophan yields 1 mg nicotinic acid.

A low dietary tryptophan intake together with deficiency of nicotinic acid in the diet causes *pellagra*. This was particularly common in the southern United States between 1900 and 1940 at a time when many poor people lived on an inadequate diet consisting mainly of maize and molasses. Although pellagra is now uncommon it still occurs in communities subsisting mainly on maize and as part of a multiple deficiency syndrome in chronic alcoholics. Pellagra is characterised by skin changes, particularly in exposed areas (fig. 15.34), diarrhoea, loss of weight and eventually dementia. The symptoms can be relieved readily by administration of nicotinic acid or tryptophan.

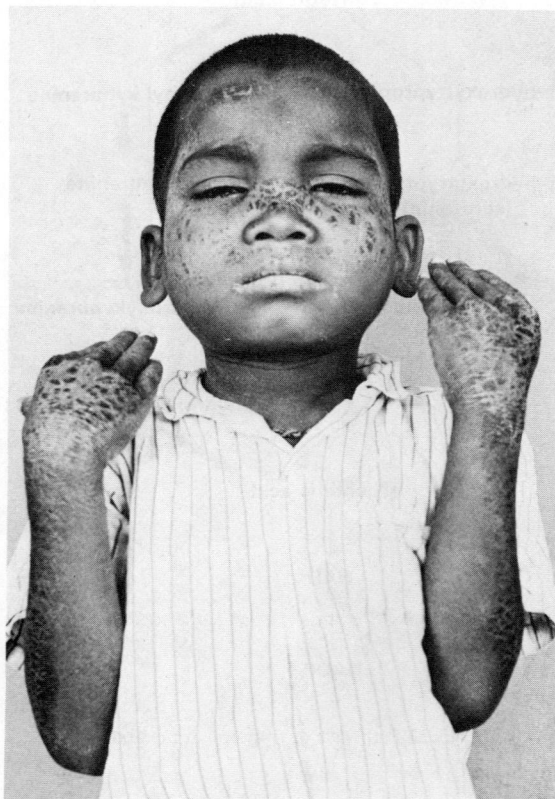

Fig. 15.34 Pellagra. (Courtesy of M. Mohan Ram.) A further case is shown in figure 5.3

Further reading

Bender, D. A. (1975) *Amino Acid Metabolism*. New York: Wiley

Felig, P. (1975) Amino acid metabolism in man. *Annual Review of Biochemistry* **44,** 933–955

Harris, H. (1963) *Garrod's Inborn Errors of Metabolism*. London: Oxford University Press

Harris, H. (1980) *Principles of Human Biochemical Genetics*, 3rd edn. Amsterdam: Elsevier

Leonard, J. V. (1984) Hyperammonaemia in childhood. In *Chemical Pathology and the Sick Child*, eds Clayton, B. E. and Round, J. M., pp. 245–264. Oxford: Blackwell

Mamune, P. (1980) Neo-natal screening tests. *Paediatric Clinics of North America* **27,** 733–751

Menon, M. and Mahle, C. J. (1982) Oxalate metabolism and renal calculi. *Journal of Urology* **127,** 148–151

Tourian, A., Sidbury, J. B. (1983) Phenylketonuria and hyperphenylalaninemia. In *Metabolic Basis of Inherited Disease*, 5th edn, eds Stanbury, J. B., Wyngaarden, J. B., Fredrickson, D. S., Goldstein, J. L., Brown, M. S. pp. 270–286. New York: McGraw-Hill

Walser, M. (1983) Urea cycle disorders and other hereditary hyperammonemic syndromes. In *Metabolic Basis of Inherited Disease*, 5th edn, eds Stanbury, J. B., Wyngaarden, J. B., Fredrickson, D. S., Goldstein, J. L., Brown, M. S. pp. 402–438. New York: McGraw-Hill

Wellner, D. and Meister, A. (1981) A survey of inborn errors of amino acid metabolism and transport in man. *Annual Review of Biochemistry* **50,** 911–968

Williams, H. E., Smith, L. H. (1983) Primary hyperoxaluria. In *Metabolic Basis of Inherited Disease*, 5th edn, eds Stanbury, J. B., Wyngaarden, J. B., Fredrickson, D. S., Goldstein, J. L., Brown, M. S. pp. 204–228. New York: McGraw-Hill

Witkop, C. J., Quevedo, W. C., Fitzpatrick T. B. (1983) Albinism and other disorders of pigment metabolism. In *Metabolic Basis of Inherited Disease*, 5th edn, eds Stanbury, J. B., Wyngaarden, J. B., Fredrickson, D. S., Goldstein, J. L., Brown, M. S. pp. 301–346. New York: McGraw-Hill

G A J G   C R P

16 Haemoglobin

HAEMOGLOBIN plays an essential role in the carriage of oxygen and carbon dioxide between the lungs and the tissues. Its unique properties, to be described in this chapter, allow it to do this very efficiently. Haemoglobin, the most abundant protein in the blood, also plays a major part in the buffering of hydrogen ions (Chap. 1).

The haemoglobin molecule consists of four sub-units each of which contains protein (*globin*) and the iron-containing substance *haem*. Each subunit contains one atom of iron and has a molecular weight of about 16 750. The components of haem and of haemoglobin are illustrated in figure 16.1.

Haem

Haem (heme) consists of a porphyrin ring and one atom of iron. The porphyrin is composed of four pyrrole nuclei joined together as shown in figure 16.2.

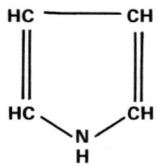

Pyrrole ring

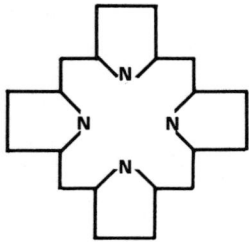

Porphyrin ring

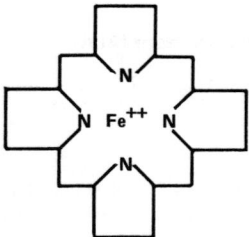

Haem

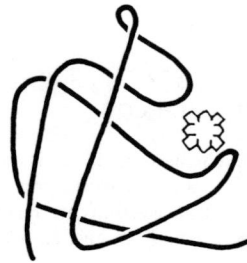

Haem-globin complex as in myoglobin

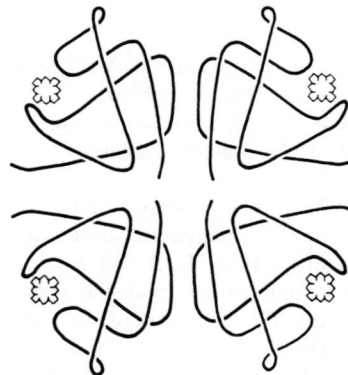

Haemoglobin

Fig. 16.1 The chemical composition of the haemoglobin molecule (After Lehmann, H. and Huntsman, R. G. (1974) *Man's Haemoglobins*, 2nd edn. Amsterdam: North Holland)

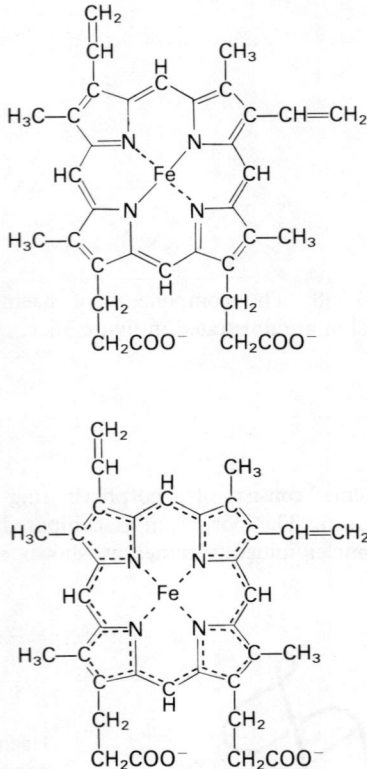

Fig. 16.2 Structure of haem. The haem molecule is conventionally described, as shown at the top, with a mixture of single and double bonds. In fact the electrons are shared, as shown below, by resonance (Chap. 7) to give an 'electron cloud'

Haem synthesis

The porphyrin ring is synthesised in the body from glycine and succinyl coenzyme A produced in the tricarboxylic acid cycle. The first two steps, summarised in figure 16.3, yield δ-amino-laevulinic acid (ALA).

Two molecules of ALA then condense under the influence of the enzyme ALA dehydrase to give the pyrrole *porphobilinogen* (fig. 16.4). Four molecules of porphobilinogen now combine to yield the cyclic derivative uroporphyrin III which is decarboxylated to coproporphyrin III. A further decarboxylation and then a dehydrogenation step lead to the formation of protoporphyrin IX. Under the influence of the enzyme haem synthetase one

atom of iron is introduced to yield haem. These steps are summarized in figure 16.5.

Side-reactions lead to the production of uroporphyrin III from uroporphyrinogen III and of coproporphyrin III from coproporphyrinogen III. The urine normally contains 10 to 200 μg of porphyrin per day and the faeces 150 to 300 μg per day.

Control of haem synthesis. ALA synthetase, the rate-controlling enzyme at the beginning of the pathway, is stimulated by erythropoietin and inhibited by haem, a good example of a negative feed-back mechanism. Normally this means that the

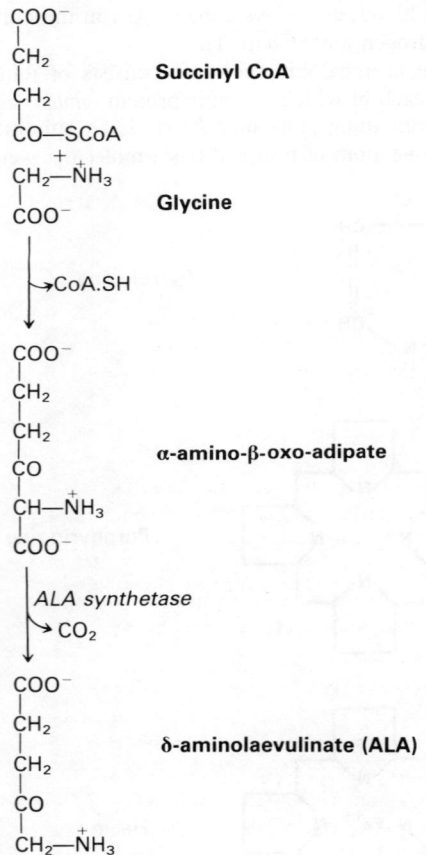

Fig. 16.3 The earliest steps in the synthesis of haem. The step catalysed by ALA synthetase is the rate limiting one and abnormalities in its amount lead to recognisable disorders. Pyridoxal phosphate (vitamin B_6) is a coenzyme for this step and deficiency leads to anaemia

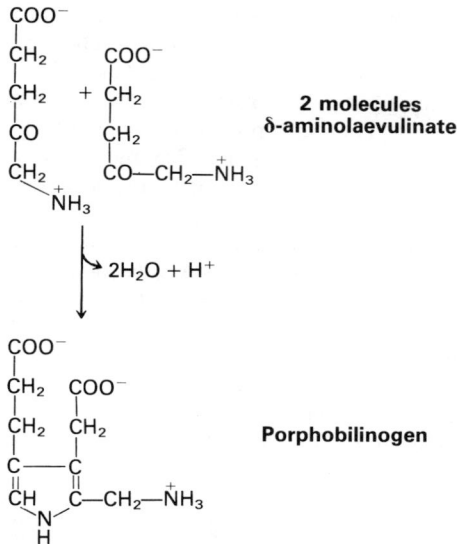

Fig. 16.4 The formation of porphobilinogen

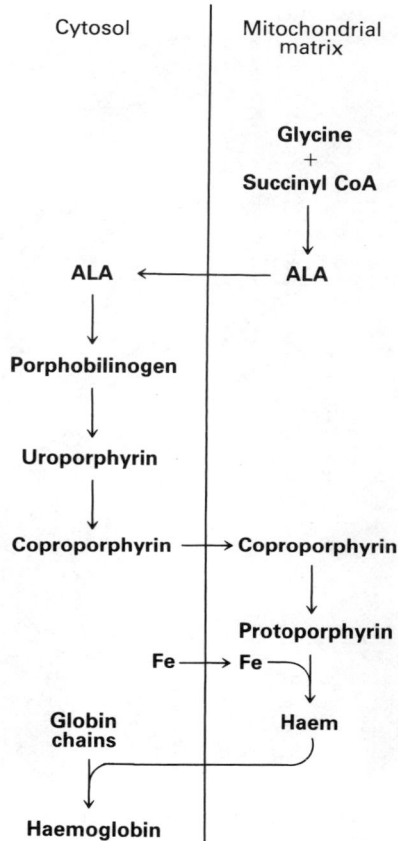

Fig. 16.5 The complete pathway for the synthesis of haem and haemoglobin. Note that the steps involve enzymes in the cytosol and in the mitochondrial matrix

synthetic pathway proceeds economically with little accumulation or excretion of intermediates. However a group of disorders, the *porphyrias*, are characterised by the excessive accumulation and excretion of porphyrins, porphobilinogen or ALA; many probably result from disorders of the regulation of haem synthesis.

The porphyrias. Several clinical disorders affecting haem synthesis are known and are classified according to whether the principal disorder is of the haem synthesis in the bone marrow required for haemoglobin formation (*erythropoietic porphyrias*), or in the hepatic haem synthesis required for the production of haem proteins such as cytochrome P_{450} (*hepatic porphyrias*).

Congenital (erythropoietic) porphyria is a very rare disorder inherited as an autosomal recessive and characterised by the excretion of large quantities of abnormal (type I) porphyrins in the faeces and in the urine, which is often coloured red. The main symptoms are due to the deposition of porphyrins in the skin which is very sensitive to sunlight (fig. 16.6).

Another disorder associated with photosensitivity is porphyria cutanea tarda (fig. 16.7), a hepatic porphyria in which there appears to be a defect at a late stage in haem synthesis; this causes clinical problems only in patients who also have a defect in

the hepatic excretion of porphyrins, as occurs for example in the liver disease of chronic alcoholics.

Another fairly common hepatic porphyria is acute intermittent porphyria in which skin disease does not occur but patients develop defects of peripheral nerve function and episodic psychiatric disturbances. They may have recurrent abdominal pain. Acute attacks are precipitated by certain drugs notably barbiturates and sulphonamides. This disorder is inherited as an autosomal dominant and is characterised by an excessive urinary excretion of ALA and porphobilinogen. It is associated with an increase in the activity of the hepatic ALA-synthetase, the rate-controlling enzyme.

197

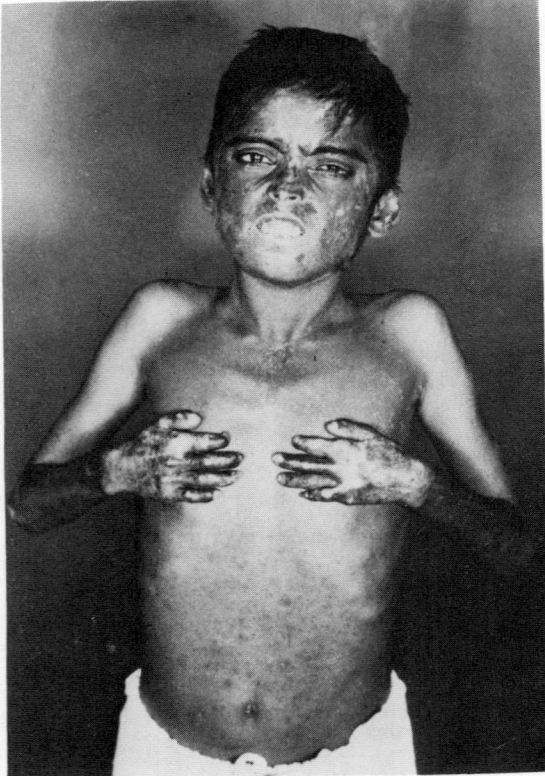

Fig. 16.6 14-year-old boy with congenital porphyria (Gunther's disease) to show severe damage to areas of the skin exposed to light. (Courtesy of L. K. Bhutani and I. Magnus)

Porphyrin metabolism may also be affected in liver disease, in various haematological disorders and in lead poisoning which affects, among other enzymes, ALA dehydrase.

Haem catabolism

Some 7 g of haemoglobin are broken down each day following the destruction of defective red cells in the spleen. The first step is the conversion of haem to biliverdin by the microsomal enzyme haem oxygenase (fig. 16.8). The details of this reaction are not yet understood but the carbon atom eliminated from the ring is excreted as carbon monoxide in the lungs. The oxygen used for this and for forming the two

new hydroxyl groups are all derived from molecular oxygen. The iron is returned to the metabolic pool in the plasma and reutilised or stored (Chap. 22).

In the next step biliverdin is converted to bilirubin (fig. 16.8). The daily production of bilirubin from red cells is about 385 μmol (225 mg) per day. An additional 85 μmol (50 mg) of bilirubin comes from haem that has not been incorporated into haemoglobin, or from haemoglobin of defective new red cells quickly destroyed, or from the breakdown of other proteins containing haem, such as the cytochromes.

Bilirubin conjugation. Bilirubin, which is insoluble in water, is carried in the plasma to the liver bound to albumin. Liver cells are efficient at removing bilirubin from plasma; the half-life of plasma

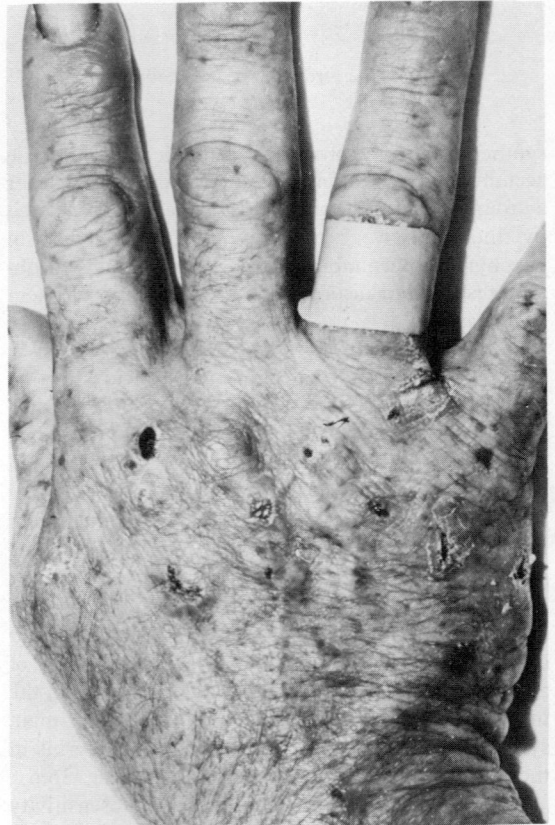

Fig. 16.7 Fragile pigmented skin in an exposed area of a patient with porphyria cutanea tarda. (Courtesy of W. Frain-Bell)

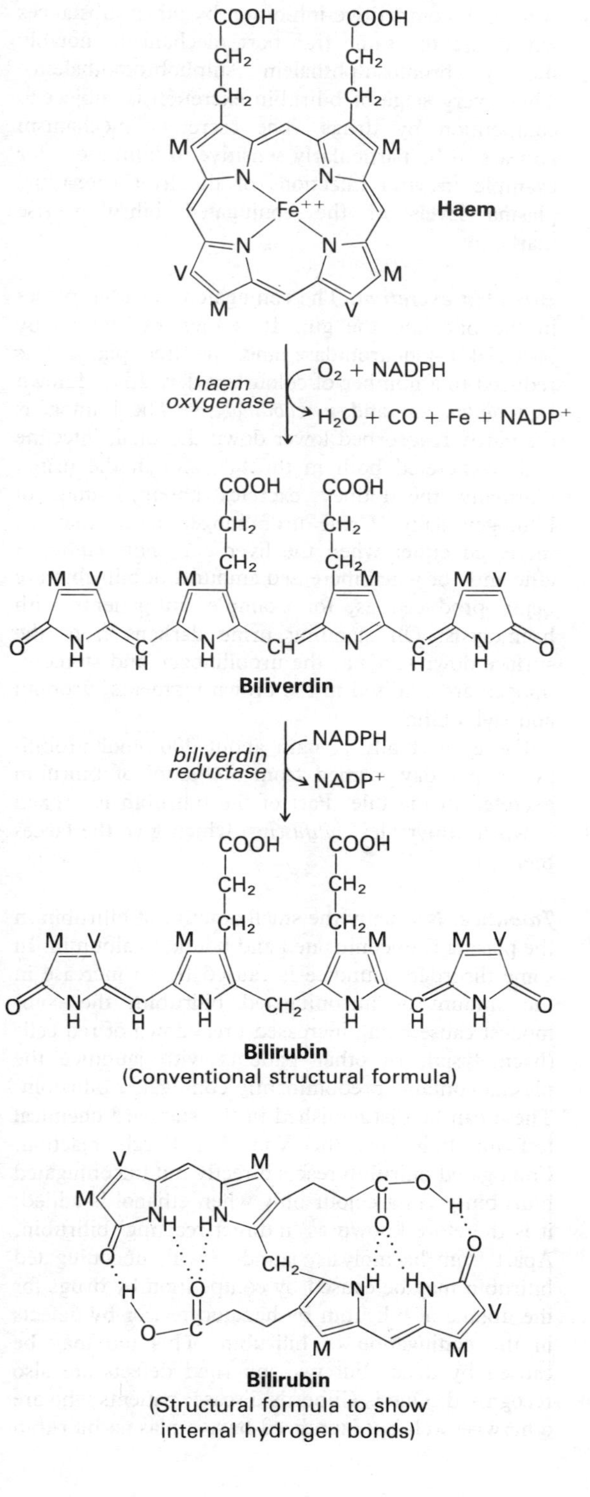

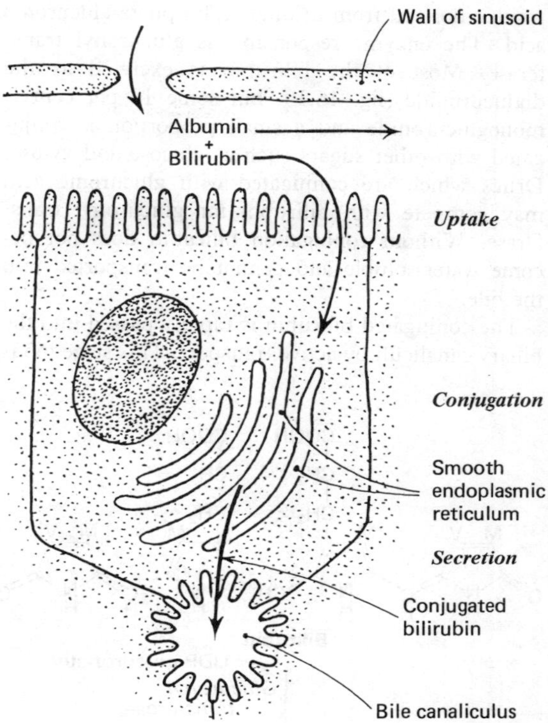

Fig. 16.9 Stages in the passage of bilirubin from the blood to the biliary canaliculus. (After Lathe, G. H. (1972) *Essays in Biochemistry* **8,** 107)

bilirubin is only 18 minutes. The liver sinusoids have large pores in their walls; these ensure that hepatocytes have easy access to plasma and, in addition, absorption is aided by the numerous microvilli on the sinusoidal surface of the hepatocytes (fig. 16.9).

The uptake of bilirubin is probably an active process, dependent on ATP and used also by certain drugs such as the antibiotic rifampicin. Once within the hepatic cell bilirubin is bound to a protein, *ligandin* (Y protein), and transported to the microsomes where it is conjugated with a glucuronic acid

Fig. 16.8 Degradation of haem to bilirubin. Haem oxygenase is in the endoplasmic reticulum while biliverdin reductase is in the cytosol. The insolubility of bilirubin in water is explained by the fact that the potential sites of hydrogen bonds are all used up in intramolecular bonds shown by dotted lines in the lower formula. M represents a methyl group and V represents a vinyl ($-CH=CH_2$) group. (After Bonnett, R. *et al.* (1976) *Nature* **262,** 326)

residue derived from uridine-diphosphate-glucuronic acid. The enzyme responsible is glucuronyl transferase. Most of the bilirubin is excreted as the diglucuronide (fig. 16.10) but some 10 per cent is monoglucuronide and a small proportion is conjugated with other sugars such as glucose and xylose. Drugs which are conjugated with glucuronic acid may compete with bilirubin for glucuronyl transferase. Without conjugation bilirubin does not become water-soluble and cannot be transported into the bile.

The conjugated bilirubin is rapidly passed into the biliary canaliculi by active transport. This stage too is

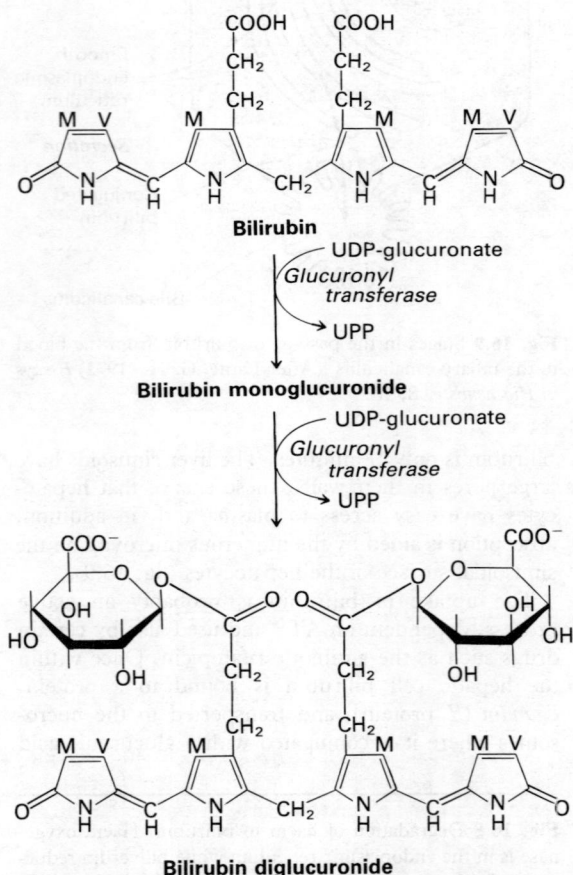

Bilirubin

— UDP-glucuronate
Glucuronyl transferase
→ UPP

Bilirubin monoglucuronide

— UDP-glucuronate
Glucuronyl transferase
→ UPP

Bilirubin diglucuronide

Fig. 16.10 Conjugation of bilirubin by the microsomal enzyme glucuronyl transferase. It is likely that two different glucuronyl transferases are involved: one at each step. M represents a methyl group and V a vinyl group. UPP = uridine pyrophosphate

subject to competitive inhibition by other substances which use the same transport mechanism, notably the dye bromosulphthalein (sulphobromothalein). Thus every stage of bilirubin excretion is subject to competition by drugs. The secretory mechanism appears to be particularly sensitive to inhibition, for example in viral infections of the liver (hepatitis) plasma levels of the conjugated bilirubin rise markedly.

Bilirubin excretion. The conjugated bilirubin passes in the bile into the gut. It is then hydrolysed by bacterial β-glucuronidase and the free pigment is reduced to a number of colourless derivatives known as *urobilinogens* and *stercobilinogens*. The bilinogens are partly reabsorbed lower down the small intestine and re-excreted both in the bile and in the urine. Normally the kidney excretes about 1 μmol of bilinogen daily. Urine urobilinogen levels may be increased either when the liver does not remove it efficiently or when increased amounts of bilirubin are being produced as, for example, in patients with haemolysis. On standing urine darkens from the surface downwards as the urobilinogen and stercobilinogen are oxidised to the brown pigments, urobilin and stercobilin.

Faeces normally contain about 200 μmol urobilinogen per day formed from 500 μmol of bilirubin excreted in the bile. Part of the bilirubin is broken down to dipyrroles, *bilifuscins*, which give the faeces their colour.

Jaundice. Normally the small amount of bilirubin in the plasma is unconjugated and bound to albumin. In some disorders jaundice is caused by an increase in the amount of unconjugated bilirubin, the commonest cause being increased breakdown of red cells (haemolysis). In other patients with jaundice the plasma contains predominantly conjugated bilirubin. These can be distinguished in the standard chemical test for bilirubin, the Van den Bergh reaction. Conjugated bilirubin reacts directly but unconjugated bilirubin gives a colour only when ethanol is added; it is therefore known as 'indirect reacting' bilirubin. Apart from haemolysis, jaundice with unconjugated bilirubin may be caused by competition by drugs for the uptake of bilirubin by hepatocytes, or by defects in the conjugation of bilirubin. This too may be caused by drugs but two inherited defects are also recognised. One is Gilbert's disease; patients who are otherwise well but jaundiced have a plasma bilirubin in the range 20 to 70 μmol/l. These patients have in their liver about 20 per cent of the normal amount of glucuronyl transferase. A much more severe disorder is the Crigler–Najjar disease in which glucuronyl transferase appears to be absent altogether and bilirubin levels are generally in the range 250 to 500 μmol/l. In this uncommon condition the basal nuclei of the brain may be damaged by the deposition of bilirubin (*kernicterus*). This can also occur when the plasma unconjugated bilirubin is raised for any other reason, for example in severe haemolytic disease in newborn infants, generally due to rhesus incompatibility between maternal and fetal red cells. Many normal newborn infants develop a mild jaundice with unconjugated bilirubin because at this age the level of glucuronyl transferase is low.

Conjugated bilirubin is found in the plasma of patients who have mechanical obstruction to the outflow of bile, for example by gall-stones or a neoplasm, or because of widespread obstruction of the small biliary canaliculi within the liver which occurs in certain chronic liver diseases. Such jaundice is sometimes known as *cholestatic jaundice*. Conjugated bilirubin may also be found in the plasma in uncommon disorders in which the secretion of conjugated bilirubin by the hepatocyte is defective, notably because of competition by drugs, viral diseases of the liver and a rare inherited defect, the Dubin–Johnson syndrome.

Jaundice is due to pigmentation of the skin by bilirubin. Because unconjugated bilirubin is firmly attached to albumin, jaundice is not usually clinically obvious until its level in the plasma reaches 70 μmol/l. The corresponding figure for conjugated bilirubin is about 40 μmol/l. Table 16.1 summarises the causes of jaundice.

Haemoglobin

Globin. The various globin molecules resemble each other in consisting of a number of helical sections connected by non-helical chains (fig. 16.11). They differ in the details of their amino acid sequences. For example the globin of myoglobin contains 153 amino acid residues whereas the α-chain of haemoglobin contains 141 residues and the β, γ and δ-chains 146 residues. The important structural feature of the globin molecule is the pocket in which the haem molecule is almost entirely buried. Globin provides a

Table 16.1 Types of jaundice

Defect	Type of plasma pigment
Increased production of bilirubin haemolytic disease	Unconjugated bilirubin
Reduced liver uptake of bilirubin drug competition	Unconjugated bilirubin
Reduced conjugation of bilirubin development defect drug competition inherited enzyme defects Gilbert's disease Crigler–Najjar disease	Unconjugated bilirubin
Decreased secretion of conjugated bilirubin drug competition inherited defects Dubin–Johnson syndrome Rotor's syndrome	Mainly conjugated bilirubin
Obstruction of biliary tree (cholestasis) within the liver primary biliary cirrhosis drugs side effects outside the liver gall-stones neoplasms	Mainly conjugated bilirubin

means of rendering soluble the otherwise non-polar molecule of haem. The polar side-chains of globin are on the exterior while the non-polar side-chains remain in the interior of the molecule and also line the haem pocket. The stability of the combined molecule is ensured by van der Waals forces which link the non-polar groups of the haem to the non-polar amino acids which line the haem pocket. There is also one bond between the iron and the histidine at position F8 of the globin.

The normal function of haemoglobin and myoglobin depends on the precise amino acid composition of the globin molecule particularly in the region of the haem pocket. Abnormalities in globin structure may be responsible for major disturbances of haemoglobin function, the *haemoglobinopathies* (p. 206).

The haemoglobin molecule consists of four globin subunits each containing a single haem molecule. The various forms of human haemoglobin all possess one pair of α-polypeptide chains but differ in the

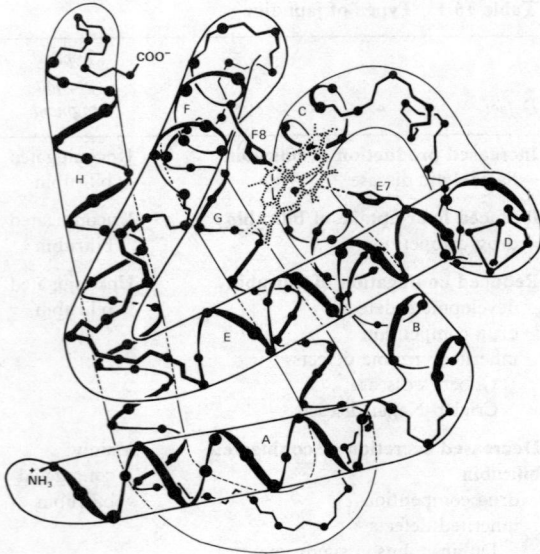

Fig. 16.11 Structure of myoglobin, typical of all the globins. Some 80 per cent of the molecule is in a helical form. In order to facilitate comparisons between globin molecules, each helical section is identified by a letter (A–H as shown). The two histidine residues involved in the haem pocket are also shown; F8 is known as the 'proximal histidine' and E7 as the 'distal histidine'. (After Carrell, R. W. and Lehmann, H. (1979) In *The Chemical Diagnosis of Disease*, eds Brown, S. S. *et al.* Amsterdam: Elsevier)

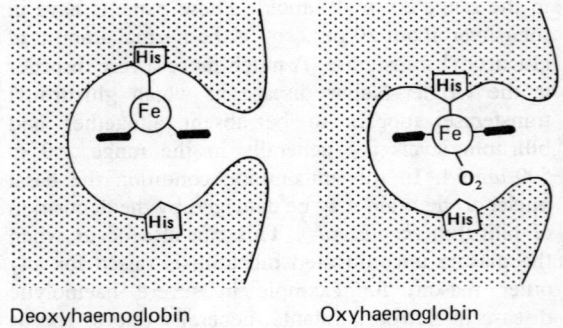

Deoxyhaemoglobin   Oxyhaemoglobin

Fig. 16.12 Schematic representation of the haem pockets of deoxy- and oxyhaemoglobin. The porphyrin ring is shown 'edge-on'. In deoxyhaemoglobin the iron atom is outside the plane of the ring attached to the histidine residue at F8 (the proximal histidine). When oxygen is taken up it is attached to the iron near the histidine residue at E7 (the distal histidine) to which it is linked by a hydrogen bond. (After Carrell, R. W. and Lehmann, H. (1979) In *Chemical Diagnosis of Disease*, eds Brown, S. S. *et al.* Amsterdam: Elsevier)

Oxygen carriage

The uptake of oxygen by haemoglobin is accompanied by a change in the shape of the haem molecule and of the haem pocket of the globin (fig. 16.12).

The oxygen dissociation curve of myoglobin has a hyperbolic shape but that of haemoglobin is sigmoid (fig. 16.13). Myoglobin gives appreciable release of

other pair. Haemoglobin A (96 to 98 per cent of the haemoglobin in adults) contains two β-chains. Haemoglobin A_2 (1.5 to 3.2 per cent of adult haemoglobin) contains two δ-chains. Haemoglobin F forms up to 80 per cent of the haemoglobin in the fetus and contains a pair of γ-chains. It diminishes gradually in early childhood and constitutes only 0.5 to 0.8 per cent of adult haemoglobin. Adult haemoglobin may be designated $\alpha_2\beta_2$, the A_2 form is $\alpha_2\delta_2$ and haemoglobin F is $\alpha_2\gamma_2$.

Myoglobin. This single haem–globin complex with a molecular weight of approximately 17 200 is found in striated muscle. In sea-diving mammals it serves as an oxygen reservoir to meet requirements during prolonged diving. In man it assists in the diffusion of oxygen within the muscle mass.

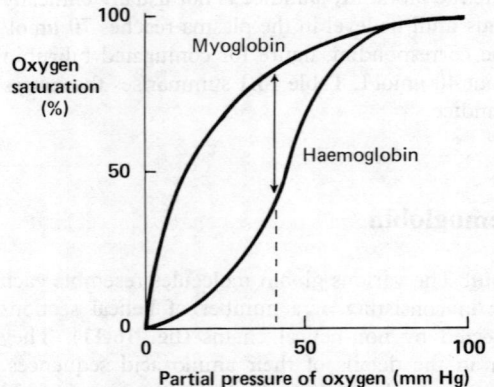

Fig. 16.13 Comparison of the (hyperbolic) oxygen dissociation curve of myoglobin with the (sigmoid) curve for haemoglobin. Note the increased efficiency of oxygen release at the mean venous partial pressure of 40 mmHg. (After Lehmann, H. and Huntsman, R. G. (1974) *Man's Haemoglobins*. Amsterdam: North Holland)

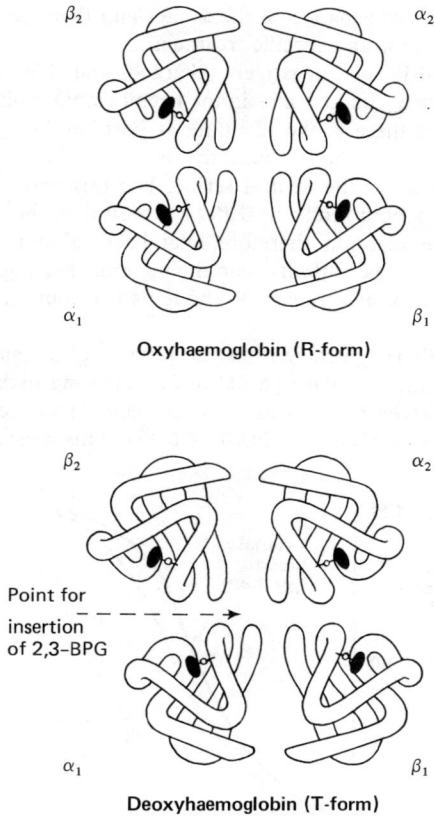

Oxyhaemoglobin (R-form)

β_2 α_2

α_1 β_1

Point for insertion of 2,3–BPG

β_2 α_2

α_1 β_1

Deoxyhaemoglobin (T-form)

Fig. 16.14 Change in the quaternary structure of oxy-haemoglobin as oxygen is given up, with the movement of one pair of subunits relative to the others and the insertion of one molecule of 2,3-bisphosphoglycerate (2,3-BPG)

oxygen only at very low partial pressures but haemoglobin releases oxygen at the higher oxygen partial pressures found in normal tissues. The affinity of haemoglobin for oxygen varies with the state of the haemoglobin. Haemoglobin behaves in this way because of changes in the relationships of the four subunits. Figure 16.14 indicates the symmetrical quaternary structure of oxyhaemoglobin sometimes described as the R form (relaxed form). When oxygen is given up, one subunit moves in relation to the other three subunits and a T (tense) structure is adopted; this has a low affinity for oxygen and therefore readily gives up further oxygen.

2,3-Bisphosphoglycerate. The changes in the subunit structure which occur on deoxygenation are accom-

panied by the insertion into the molecule of 2,3-bisphosphoglycerate (2,3-BPG). This substance is produced by a side-reaction of glycolysis in a number of tissues, particularly in the red cell, as shown in figure 16.15. 2,3-BPG is present in high concentration within red cells and reduces the oxygen affinity by making deoxyhaemoglobin more stable (fig. 16.16).

Control of the concentrations of 2,3-BPG in red cells is of great importance both physiologically and in disease. For example anoxia stimulates the synthesis both of haemoglobin and of 2,3-BPG, but the increase in 2,3-BPG levels occurs within 8 to 12 hours, compared with several weeks for the increase in haemoglobin concentration. Thus a climber moving to a high altitude increases his 2,3-BPG levels after a few hours and so is able to deliver to his tissues more oxygen than would otherwise have been possible. The synthesis of 2,3-BPG is promoted by alkalosis and inhibited by acidosis. This serves to

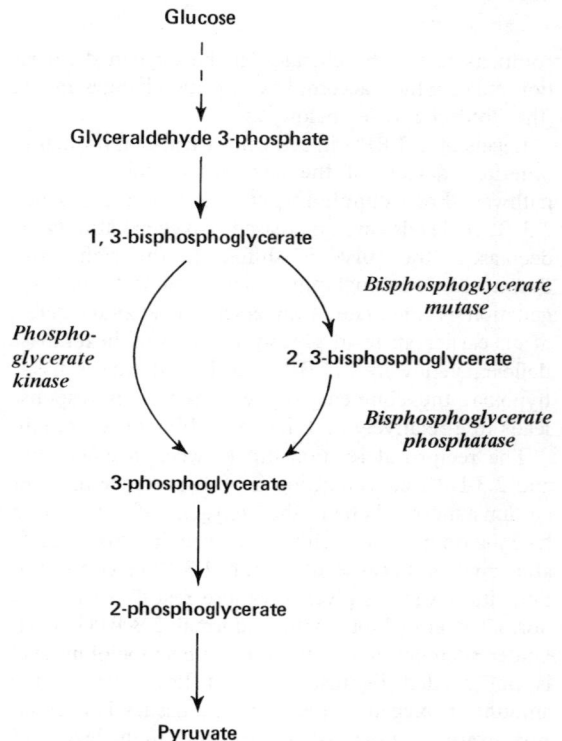

Glucose

Glyceraldehyde 3-phosphate

1, 3-bisphosphoglycerate

Phospho-glycerate kinase

Bisphosphoglycerate mutase

2, 3-bisphosphoglycerate

Bisphosphoglycerate phosphatase

3-phosphoglycerate

2-phosphoglycerate

Pyruvate

Fig. 16.15 Pathway for the production and disposal of 2,3-bisphosphoglycerate. The details of the remainder of the glycolytic pathway are given in Chapter 9

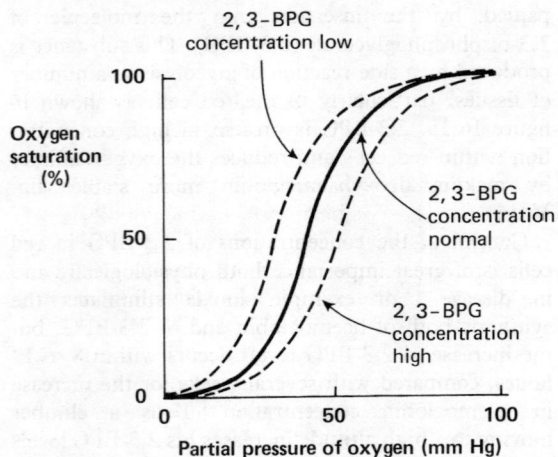

Fig. 16.16 Effect of changing 2,3-BPG levels on the oxygen dissociation curves. A small shift in the curve to the right may mean a substantial increase in the amount of oxygen released from haemoglobin at the mixed venous PO_2 of 40 mmHg

compensate for the changes in the oxygen dissociation curve which accompany chronic changes in pH (the Bohr effect, see below).

Levels of 2,3-BPG in red cells are also abnormal in inherited defects of the enzymes of the glycolytic pathway. For example in pyruvate kinase deficiency 2,3-BPG levels are increased, oxygen affinity is decreased (the curve is shifted to the right) and despite low haemoglobin concentrations tissue oxygenation is adequate. Conversely an enzyme defect at an earlier stage in glycolysis, such as hexokinase deficiency, lowers 2,3-BPG levels and causes tissue hypoxia; the compensatory erythropoietin response leads to a relatively high haemoglobin concentration.

The reciprocal relationship between haemoglobin and 2,3-BPG concentration is of importance in other circumstances. When the oxygen affinity of a haemoglobin is low either because its structure is abnormal or because of a high 2,3-BPG concentration, its oxygen is given up more readily than from normal haemoglobin with a normal 2,3-BPG level. Under such conditions the normal haemoglobin level is not needed because, even at lower levels, the amount of oxygen delivered to the tissues is normal. For example in renal failure the high levels of phosphate and sulphate stimulate 2,3-BPG synthesis and therefore contribute to the anaemia characteristic of this condition. This anaemia is accompanied by

normal oxygenation of the tissues and therefore does not call for any specific treatment.

2,3-BPG is negatively charged and bound by positively charged sites in the haemoglobin molecule. One of these is the histidine at position 143 in the β-chain. In the γ-chain the corresponding site is occupied by the neutral serine. For this reason fetal haemoglobin binds 2,3-BPG less well than the adult. In the placenta therefore fetal haemoglobin has a higher oxygen affinity than the maternal haemoglobin so that oxygen transfer to the fetus is favoured.

The Bohr effect. Within the physiological range of pH, indeed down to a pH of 6, increasing hydrogen ion concentration leads to a decrease in the oxygen affinity of haemoglobin (fig. 16.17). This means that

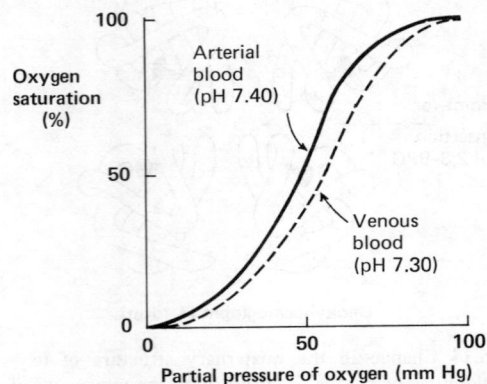

Fig. 16.17 The effect of changes in pH on oxygen dissociation curves (the alkaline Bohr effect)

in the tissues, where the hydrogen ion concentration is higher than in the arterial blood, oxygen is more readily released.

Since the Bohr effect is associated with the uptake of hydrogen ions, it has the effect of buffering the acid produced in tissues. Haemoglobin therefore acts as a carrier of hydrogen ions to the lungs where they combine with bicarbonate.

Special forms of haemoglobin

Carboxyhaemoglobin. Haemoglobin can combine with carbon monoxide to form carboxyhaemoglobin. This compound is more stable than oxyhaemoglobin but the affinity of haemoglobin for carbon monoxide is some 250 times greater than that for oxygen. The breathing of carbon monoxide, for example from coal gas or from motor car exhaust fumes, is therefore dangerous because the carboxyhaemoglobin deprives the blood of its ability to transport oxygen and may cause death from anoxia. The carbon monoxide may be displaced gradually from haemoglobin by making the patient breathe pure oxygen. Non-smokers normally have about 1 per cent of their haemoglobin as carboxyhaemoglobin. This carbon monoxide is endogenous and is derived from the breakdown of haem (p. 199). Tobacco smokers may have 5 per cent or more of their haemoglobin as carboxyhaemoglobin and this may be a factor contributing to hypoxia in the fetus of a mother who smokes.

Methaemoglobin. Methaemoglobin is formed by the oxidation of the ferrous iron of haemoglobin to ferric iron. About 3 per cent of the circulating haemoglobin in normal people is methaemoglobin. Whereas oxyhaemoglobin can become deoxyhaemoglobin simply by lowering the oxygen tension, methaemoglobin can only be reduced by the enzyme methaemoglobin reductase. Deficiency of this enzyme is a cause of a hereditary methaemoglobinaemia. Methaemoglobinaemia also occurs after the administration of certain drugs such as phenacetin.

Haemichromes. Haemichrome I is a denaturation product of haemoglobin resulting from the formation of a bond between the iron and the distal histidine within the haem pocket of globin (fig. 16.18). Small amounts of haemichromes are present in normal blood.

Glycosylated haemoglobins. Between 6 and 7 per cent of adult haemoglobin is a fraction (haemoglobin A_1) which elutes ahead of the rest of the haemoglobin on cation exchange chromatography. In 1968 it was noted that the proportion of haemoglobin A_1 was greatly increased in diabetic patients (fig. 16.19). The concentration of haemoglobin A_1 related closely to the average glucose concentrations over the preceding 60 days.

It is now clear that haemoglobin A_1 is made up of at least three components, haemoglobins A_{1a}, A_{1b} and A_{1c}, which represent respectively about 1.6 per cent, 0.8 per cent and 4 per cent of adult haemoglobin. Haemoglobin A_{1c} has a glucose molecule attached to the amino end of the globin chain. It appears that during the lifespan of a red cell there is a progressive increase in the proportion of glycosylated haemoglobin. This increase is considerably more rapid in patients with high blood glucose levels. In diabetic patients the concentration of haemoglobin A_{1c} provides a valuable measure of the quality of diabetic control over the previous two months.

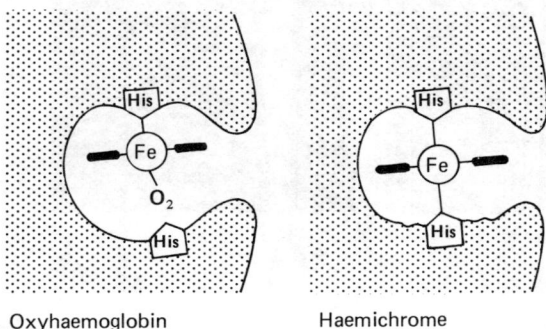

Oxyhaemoglobin   Haemichrome

Fig. 16.18 Haem and the haem pocket in oxyhaemoglobin and haemichrome. The formation of a bond between the iron and the distal histidine leads to distortion of the globin. (After Carrell, R. W. and Lehmann, H. (1979) In *Chemical Diagnosis of Disease*, eds Brown, S. S. *et al.* Amsterdam: Elsevier)

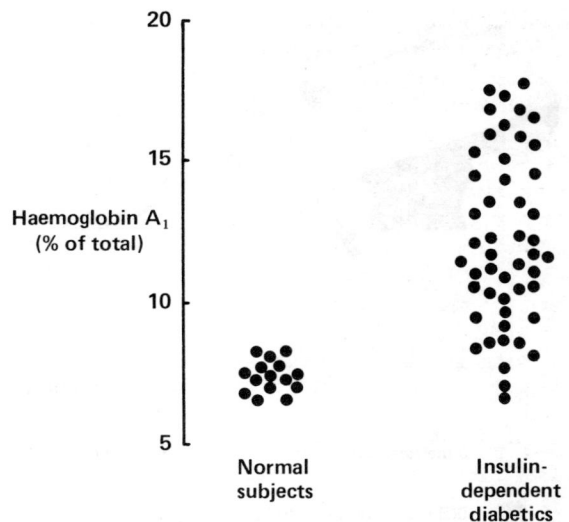

Fig. 16.19 Haemoglobin A_1 levels as a percentage of the total haemoglobin in blood samples from 15 normal subjects and 50 patients with insulin-requiring diabetes mellitus. (Courtesy of P. Caiger)

The haemoglobinopathies

The effect of small variations in the molecule of haemoglobin illustrates the close relationship between molecular biology and medicine. The two main types of abnormality are the thalassaemias, with defective synthesis of one or other globin chain, and the diseases caused by variation in the primary structure of a globin. Such changes are almost always caused by a single amino acid substitution most commonly in the β-chain. The effect may be so benign that the patient is clinically and haematologically normal; other substitutions cause major pathological changes such as sickle-cell anaemia. Figure 16.20 shows the geographical distribution of the most important haemoglobinopathies.

The thalassaemias. In normal adults and in children outside the newborn period, fetal haemoglobin (haemoglobin F) forms only a small proportion (less than 0.5 per cent) of the total haemoglobin. In β-thalassaemia major (the homozygous state) the synthesis of β-chains is defective; γ-chains continue to be produced. Haemoglobin F is the principal form of haemoglobin in thalassaemia major.

The first clinical signs of β-thalassaemia major are seen at about the age of four months and consist of listlessness, fever, diarrhoea, a poor appetite and pallor due to anaemia. These signs result from the destruction of the defective Hb F-containing red cells and the great compensatory increase in the amount of haemopoietic tissue both in the bone marrow and in other tissues such as the spleen. The abdomen may be greatly enlarged (fig. 16.21) and the bones may show characteristic appearances (fig. 16.22).

The outlook is poor since the susceptibility to infection is increased. Also the repeated blood transfusions which are needed lead eventually to iron

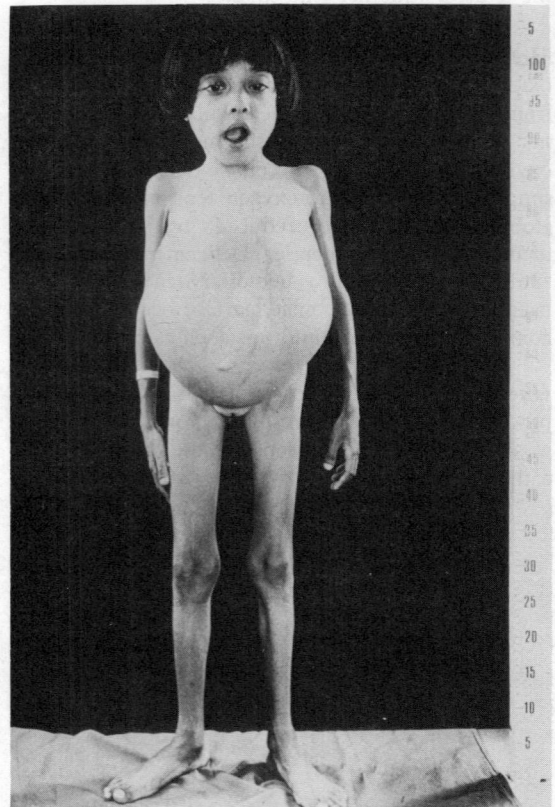

Fig. 16.21 A ten-year old girl with β-thalassaemia major. The abdominal distension was probably due partly to the enlargement of the liver and spleen and partly to the accumulation of fluid in peritoneal cavity (ascites). Her haemoglobin level at that time was 3.5 g/dl (normal 12.5 to 14 g/dl). (Courtesy of D. Higgs, D. J. Weatherall and Elizabeth A. Letsky)

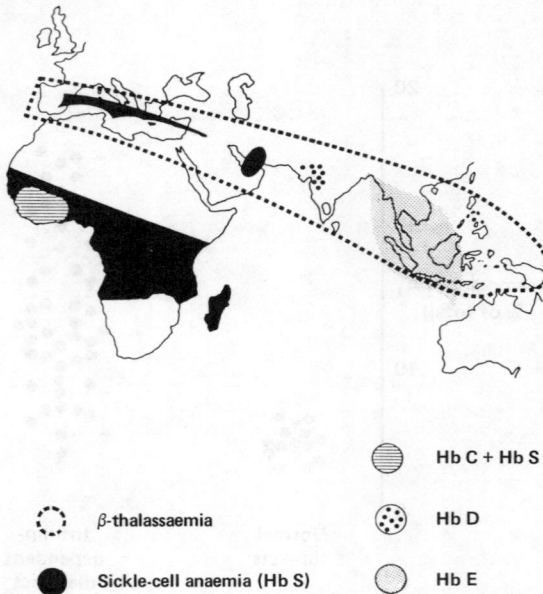

β-thalassaemia

Sickle-cell anaemia (Hb S)

Hb C + Hb S

Hb D

Hb E

Fig. 16.20 World distribution of the principal haemoglobin disorders. (From Lehmann, H. (1978) In *The Metabolic Basis of Inherited Disease*, eds Stanbury, J. C. *et al.*, 4th edn. New York: McGraw Hill)

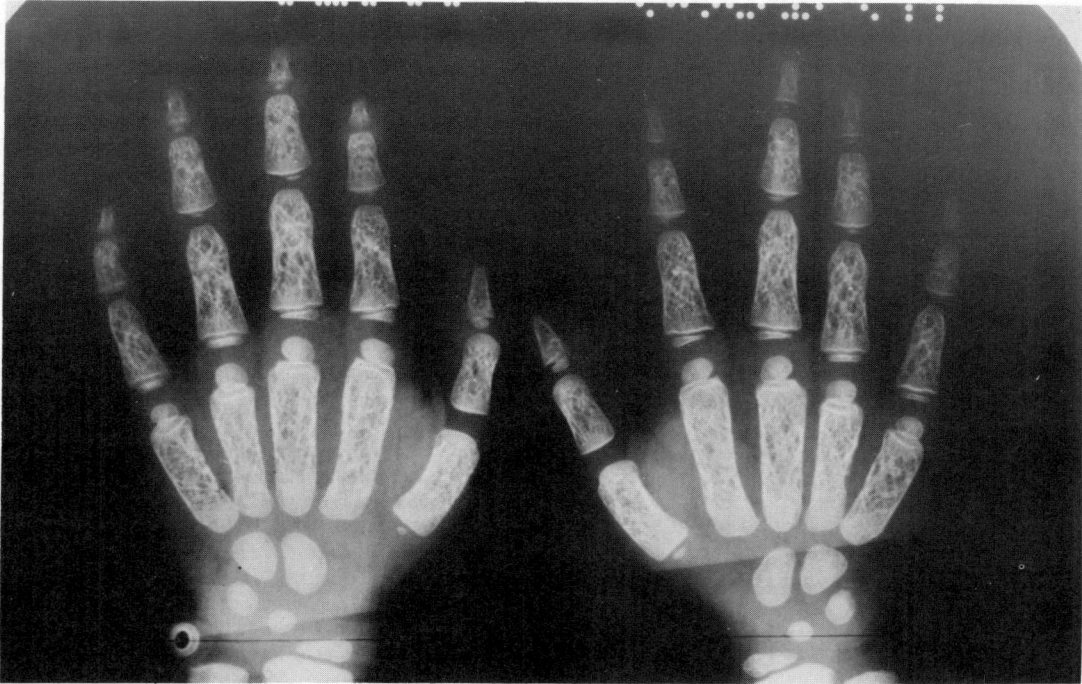

Fig. 16.22 X-rays of the hands in a child with β-thalassaemia major. The 'honeycomb' appearance of the phalanges is caused by the great increase in the amount of bone marrow. (Courtesy of D. Higgs and D. J. Weatherall)

overload (Chap. 21) and to iron deposition in tissues, particularly in cardiac muscle. Heart failure eventually follows.

The heterozygous state for β-thalassaemia is known as thalassaemia minor; it is characterised by a mild anaemia.

In α-thalassaemia α-chain synthesis is defective. Normal subjects have four genes for α-chain synthesis. Patients with only three or two functional genes are normal or have a moderate anaemia. Patients with one functional α-chain gene have severe anaemia characterised by a large excess of β-chains which combine to produce Hb H (β_4). Patients with no functional α-chain genes are stillborn with large amount of functionally useless Hb Bart's (γ_4).

Sickle-cell disease. Of all the disorders caused by structural abnormalities in haemoglobin, sickle-cell disease (Hb S) is by far the most common and the most important (fig. 16.20). In some parts of Africa it is found in as many as 40 per cent of the population; in Africa as a whole the proportion is between 10 and 20 per cent. Hb S also occurs in some Mediterranean countries, in the Middle East and in India.

In Hb A position 6 of the β-chain is occupied by glutamic acid; in Hb S by valine. The molecules are otherwise identical. The amino terminal portions of the β-chains are as follows:

Hb A ($\alpha_2\beta_2$):

NH$_2$-Val-His-Leu-Thr-Pro-Glu-Glu-Lys $\cdot\cdot\cdot$

Hb S ($\alpha_2\beta_2^s$):

NH$_2$-Val-His-Leu-Thr-Pro-Val-Glu-Lys $\cdot\cdot\cdot$

In oxyhaemoglobin this substitution has little effect; in the deoxygenated state the solubility of Hb S is reduced (to 2 per cent of normal). On deoxygenation in venous blood red cells containing Hb S take on an abnormal shape (sickle-cells, fig. 16.23) and a low pH makes this change of shape more likely. Blood containing sickle-cells has an increased viscosity and blood vessels may become blocked by a mass of distorted cells. Some of the

207

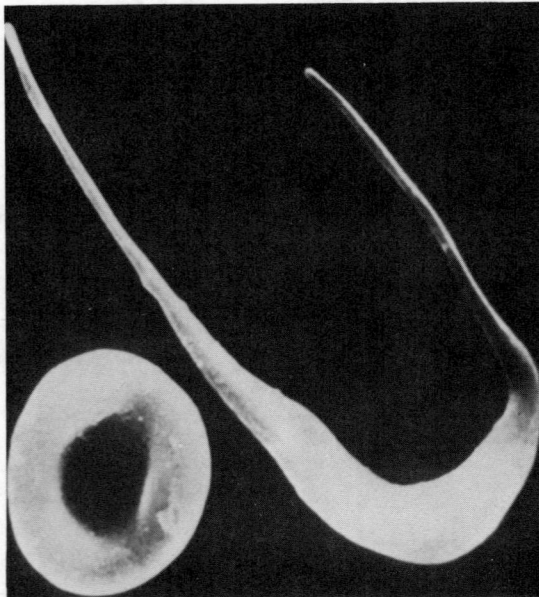

Fig. 16.23 Scanning electron micrograph to show two red cells from a patient with sickle-cell anaemia. One has 'sickled' under partial deoxygenation. (Courtesy of B. F. Cameron, D. R. Harkness and R. Zucker and the New York Academy of Sciences)

clinical features of sickle-cell anaemia are attributable to the increased blood viscosity and include abdominal pain and chronic leg ulcers. An exacerbation of the symptoms is called a 'crisis'. Death may result from the occlusion of the blood supply to the brain or the heart.

Sickle-cell trait is the name given to the heterozygous form of Hb S. Although generally symptom-free sickle-cell trait carriers may develop sickling if the P_{O_2} is particularly low, for example in unpressurised aircraft at high altitude. This heterozygous form confers resistance to falciparum malaria and is therefore an advantage in parts of the world where this form of malaria is common. There exists a *balanced polymorphism* in which the heterozygote (AS, the sickle-cell trait carrier) has a selective advantage over the two homozygotes (AA and SS). The heterozygote succumbs neither to malaria nor to sickle-cell anaemia.

Other haemoglobinopathies. Haemoglobin C also results from a substitution in position 6 of the β-chain, with lysine instead of the normal glutamic acid. The double heterozygote for Hb C and Hb S

(haemoglobin SC disease) has a milder form of sickle-cell anaemia.

Haemoglobin E is found in areas such as Malaya, Indonesia, Thailand and Burma. In the homozygote mild haemolytic anaemia results.

Unstable haemoglobins are rare but important examples of molecular pathology. In each case it is possible to relate the haemolytic anaemia directly to the particular molecular defect which causes the instability and the intracellular precipitation of globin.

Further reading

Carrell, R. W. and Lehmann, H. (1979) Abnormal haemoglobins. In *Chemical Diagnosis of Disease*, ed. Brown, S. S., Mitchell, F. L. and Young, D. S., pp. 879–926. Amsterdam: Elsevier

Elder, G. H. (1980) Haem synthesis and breakdown. In *Iron in Biochemistry and Medicine 2*, ed. Jacobs, A. and Worwood, M., pp. 245–292. London: Academic Press

Goldberg, A. and Moore, M. R. (eds) (1980) The porphyrias. *Clinics in Haematology* **9**, 225–451

Hanash, S. M. and Rucknagel, D. L. (1980) Clinical implications of recent advances in hemoglobin disorders. *Medical Clinics of North America* **64**, 775–800.

Jaffe, E. R. (1981) Methaemoglobinaemia. *Clinics in Haematology* **10**, 99–122

Kappas, A., Sassa, S., Anderson, K. E. (1983) The Porphyrias. In *Metabolic Basis of Inherited Disease*, 5th edn, eds Stanbury, J. B., Wyngaarden, J. B., Fredrickson, D. S., Goldstein, J. L., Brown, M. S. pp. 1301–1384. New York: McGraw-Hill

Lester, R. (1983) Not two but three bilirubins. *New England Journal of Medicine* **309**, 183–184

Miedema, K. and Casparie, T. (1984) Glycosylated haemoglobins: biochemical evaluation and clinical utility. *Annals of Clinical Biochemistry* **21**, 2–15

Sasaki, R., Ikura, K., Narita, H., Yanagawa, S. and Chiba, H. (1982) 2,3-Bisphosphoglycerate in erythroid cells. *Trends in Biochemical Sciences* **7**, 140–142

Weatherall, D. J. (1982) *The New Genetics and Clinical Practice*. London: Nuffield Provincial Hospitals Trust

Weatherall, D. J. (1983) The molecular genetics of the thalassaemias: practical applications and implications for other genetic diseases. In *Advanced Medicine 19*, ed. Saunders, K. B., pp. 202–215. London: Pitman

Weatherall, D. J. and Clegg, J. B. (1981) *The Thalassaemia Syndromes*, 3rd edn. Oxford: Blackwell

Wolkoff, A. W., Chowdhury, J. R., Arias, I. M., (1983) Hereditary jaundice and disorders of bilirubin metabolism. In *Metabolic Basis of Inherited Disease*, 5th edn, eds Stanbury, J. B., Wyngaarden, J. B., Fredrickson, D. S., Goldstein, J. L., Brown, M. S. pp. 1385–1420. New York: McGraw-Hill

CRP HL

17 The structural proteins

An essential role in the structure of many tissues is played by specialised extracellular proteins notably collagen and elastin. Collagen is a glycoprotein with a high tensile strength; it is found in bone, teeth, skin, tendon, cartilage and blood vessels.

Elastin is a protein with elastic properties. In most connective tissues and in bone and cartilage, collagen and elastin are associated with proteoglycans (p. 33). *Keratin* is a specialised structural protein in skin, hair and nails.

Collagen

A collagen molecule consists of three polypeptide chains, each containing about 1050 amino acid residues. The molecule is about 300 nm long and 1.5 nm in diameter; its mass is about 285 000.

The polypeptide chain of collagen is known as an α-chain and no less than seven different α-chains have now been identified. The most abundant are $\alpha 1(I)$, $\alpha 1(II)$, $\alpha 1(III)$ and $\alpha 2$. At least five distinct types of collagen are recognised; the most abundant is type I whose molecule contains two $\alpha 1(I)$ chains and one $\alpha 2$-chain. Type I collagen makes up 90 per cent of the collagen in an adult. The other collagen types are summarised in Table 17.1.

Composition

Collagen has an unusual amino acid composition. Nearly one-third of the residues are glycine and the proportion of proline is also exceptionally high. In addition collagen contains two residues seldom found in other proteins, hydroxyproline and hydroxylysine; these are formed by the hydroxylation of proline and lysine residues in the collagen precursor.

Collagen is a glycoprotein with glucose, galactose and mannose units attached to hydroxylysine residues. The proportion of carbohydrate varies be-

Table 17.1 Types of collagen

	Composition	Sites
Type I	$[\alpha 1(I)]_2\alpha 2$	Bone, skin and tendon
Type II	$[\alpha 1(II)]_3$	Hyaline cartilage, intervertebral disc, vitreous humour of eye
Type III	$[\alpha 1(III)]_3$	Blood vessels, skin in fetus and infant
Type IV	$[\alpha 1(IV)]_2\alpha_2(IV)$	Basement membranes for example in glomerulus. Lens capsule
Type V	$\alpha A(\alpha B)_2$	Placenta, blood vessels. Striated and smooth muscle

tween about six hexoses per chain for type I collagen and over 100 in type IV.

Structure. The three α-chains in collagen form left-handed helices which are wound round each other to form a right-handed triple helix. The triple helix is held together by hydrogen bonds between the α-chains. At each end there are small non-helical sections (telopeptide regions).

In mature collagen many molecules are associated to form large and extremely tough fibres. The molecules are arranged with an overlap between adjacent units to give a 'quarter-stagger' arrangement with gaps which are important in the initiation of the mineralisation of bone. This structure is reflected in the characteristic appearance of collagen on electron microscopy. The molecules are held together by intermolecular covalent linkages as described below.

Synthesis

The polypeptide precursors of the α-chains are formed in the usual way by translation (Chap. 14) in

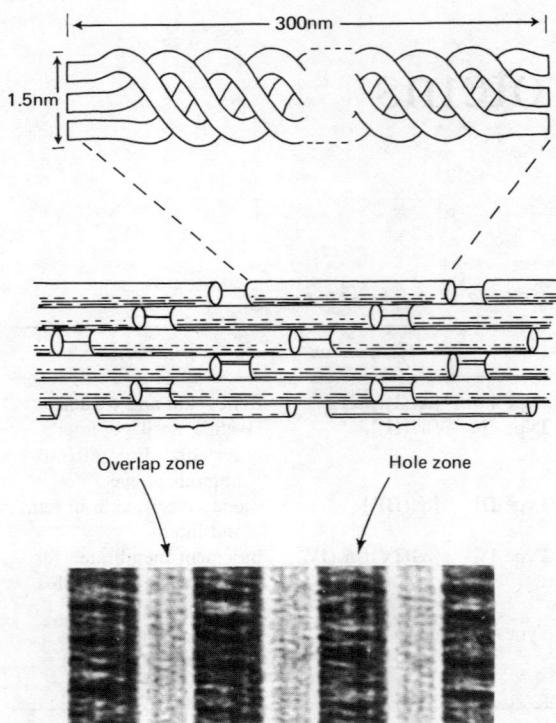

Fig. 17.1 The triple helix of a tropocollagen molecule (A), the way in which the tropocollagen molecules are associated in a fibril of mature collagen (B) and the significance of the cross-banding seen on electron microscopy (C, courtesy of J. W. Smith)

fibroblasts and osteoblasts. The synthesis of collagen is unusual in the number and variety of post-translational modifications.

Intracellular modifications. The initial polypeptide, in addition to the segment destined to become an α-chain, has globular sections at both amino and carboxyl ends. The first modification is the hydroxylation of certain proline and lysine residues to give hydroxyproline and hydroxylysine. The hydroxylation reactions require Fe^{2+}, ascorbate, molecular O_2 and α-oxoglutarate. The ascorbate acts as a reducing agent to maintain iron in the ferrous state; ascorbate deficiency causes collagen abnormalities (p. 214).

Some of the hydroxylysine residues are glycosylated in the lumen of the endoplasmic reticulum. Then, still within the cell, the modified polypeptides form triple helices and it seems likely that the

globular segments (*propeptides*) at either end are important for this process. The triple helix formed in this way is known as *procollagen*.

Extracellular modifications. Procollagen bundles are extruded from the cell and then the amino-terminal and carboxy-terminal propeptides are removed by specific peptidases to give tropocollagen (fig. 17.2).

The collagen molecules aggregate to give the fibres described in figure 17.1. The stability of these fibres increases progressively with the formation of covalent cross-links between adjacent molecules (fig. 17.3).

Breakdown

Collagen fibres can be broken down in two ways. Non-specific peptidases can attack the non-helical telopeptide regions. Since most cross-links are formed in this area the process leads to the release of molecules which can be taken up by phagocytes.

Secondly collagen can be attacked by specific collagenases which cut through collagen molecules at a specific point at about one-quarter of the distance from the carboxyl terminal. The action of collagenases, which are found in many tissues, also leads to the production of small fragments which can be taken up by phagocytes. Like many other proteolytic enzymes, collagenase is produced first as an inactive precursor which is activated by trypsin and other proteases. The activity of collagenases is regulated since their uncontrolled action could lead to inappropriate tissue destruction. Many tissues contain proteins which act as collagenase inhibitors and, in the plasma, α_2-macroglobulin (p. 20) inhibits collagenase.

Fragments produced by the action of either peptidases or collagenases can be taken up by endocytosis into phagocytes. There the vesicles fuse with lysosomes. At the very low pH ensured by the lysosomal contents the further breakdown of the fragments into small peptides is carried out by a group of enzymes known as *cathepsins*.

It will be recalled that hydroxyproline is virtually unique to collagen; it is produced by a post-translational hydroxylation of proline residues. Free hydroxyproline cannot be re-used for collagen synthesis and the hydroxyproline derived from collagen breakdown is excreted. The urinary excretion of

210

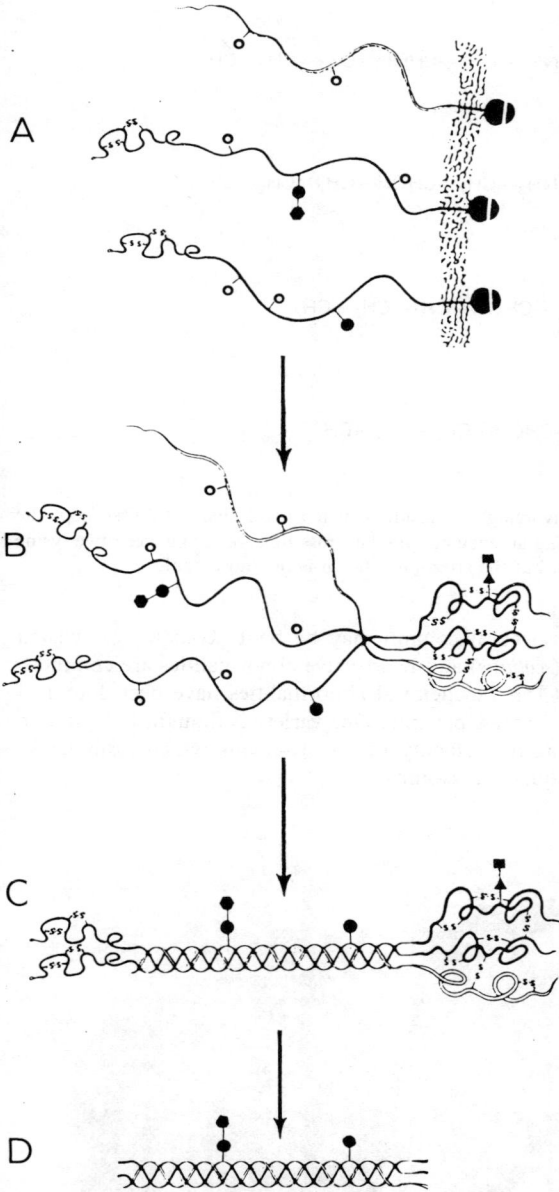

Fig. 17.2 Collagen formation. A: synthesis of the α-chain precursors on rough endoplasmic reticulum. Hydroxylation and glycosylation takes place at this stage. B and C: formation of triple helix of procollagen starting at the C-terminal ends. After extrusion of the procollagen from the cell the two propeptides are removed (D). The carbohydrate units are indicated: ● = glucose, ● = galactose, ▲ = N-acetyl glucosamine, ■ = several mannose units. (After Prockop, D. (1979) *New England Journal of Medicine* **301,** 13)

hydroxyproline provides a measure of collagen breakdown; since bone contains some 60 per cent of the body's collagen urinary hydroxyproline is sometimes used as an index of bone resorption. High values are found for example, in patients with metastatic neoplasm in bone.

Disorders of collagen

Each α-chain has some 1050 residues and to form mature collagen at least ten further polypeptides are needed as enzymes for post-translational modifications. It is not surprising therefore that a large number of disorders of collagen structure and maturation have been identified. Three main groups of diseases of collagen have been identified: *osteogenesis imperfecta*, the *Ehlers–Danlos syndrome* and *Marfan's syndrome*. In addition collagen abnormalities are recognised in vitamin C deficiency (*scurvy*) and in *homocystinuria* (p. 189).

Osteogenesis imperfecta. This is a group of at least seven inherited disorders all characterised by excessive bone fragility. In severe cases an affected child may be born with multiple fractures sustained before birth (fig. 17.4). Patients with milder varieties may have only a few fractures in chilhdood. Some patients also have excessive laxity of joints or abnormalities of the teeth, due to defective dentine collagen, or easy bruising because of defects in collagen in small blood vessels.

Collagen abnormalities found in osteogenesis imperfecta include, in some milder cases, decreased amounts of type I collagen, and in severe cases a variety of defects including lack of α_2-chains and excess mannose in the carboxyl propeptide.

Ehlers–Danlos syndrome. In this group of heritable disorders the clinical features may include hyperelastic skin with poor wound healing (figs 17.5 and 17.6), hyperextensible joints with frequent dislocations (fig. 17.7) and easy bruising.

As in osteogenesis imperfecta, clear evidence of a particular collagen defect has been obtained only in some varieties. In the most common form of the disorder the packing of collagen into fibres is probably defective. In one rare variety lysine hydroxylation is defective; in another the synthesis of

A CH—CH₂—CH₂—CHOH—CH₂—N⁺H₃   N⁺H₃—CH₂—CHOH—CH₂—CH₂—CH

B CH—CH₂—CH₂—CHOH—CH=O   N⁺H₃—CH₂—CHOH—CH₂—CH₂—CH

C CH—CH₂—CH₂—CHOH—CH=N—CH₂—CHOH—CH₂—CH₂—CH

D CH—CH₂—CH₂—C—CH₂—NH—CH₂—CHOH—CH₂—CH₂—CH
                    ‖
                    O

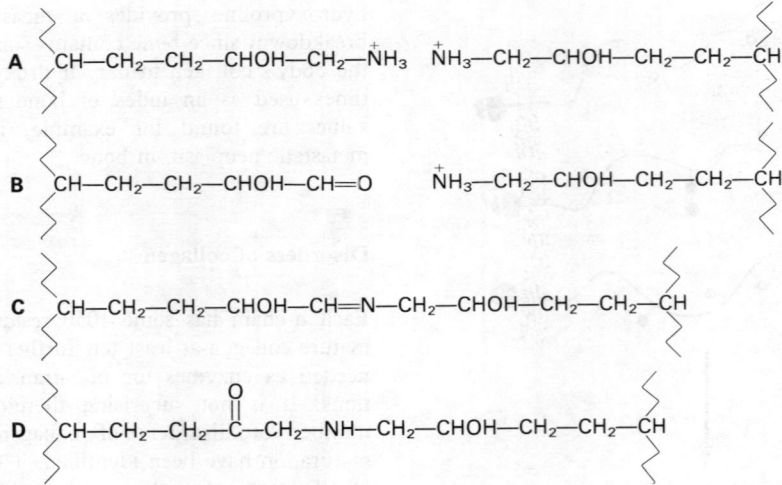

Fig. 17.3 Formation of cross-links between collagen molecules involving two residues of hydroxylysine. The first step, the oxidative deamination of one of the residues is the only one requiring an enzyme. Similar cross-links occur between two lysine residues and between a lysine residue and a hydroxylysine residue, but the cross-link shown is the most stable

type III collagen is impaired; in a third the removal of the amino-terminal part of procollagen is defective.

Marfan's syndrome. This term, too, probably includes a number of distinct disorders. Patients are often tall with unusually long fingers and toes. The vertebral column may be bent (*kyphosis*) or twisted (*scoliosis*) and heart valve abnormalities are common. Clear biochemical abnormalities have been found in very few patients. One variety is thought to be due to an abnormality of the α_2-chains which form inadequate cross-links.

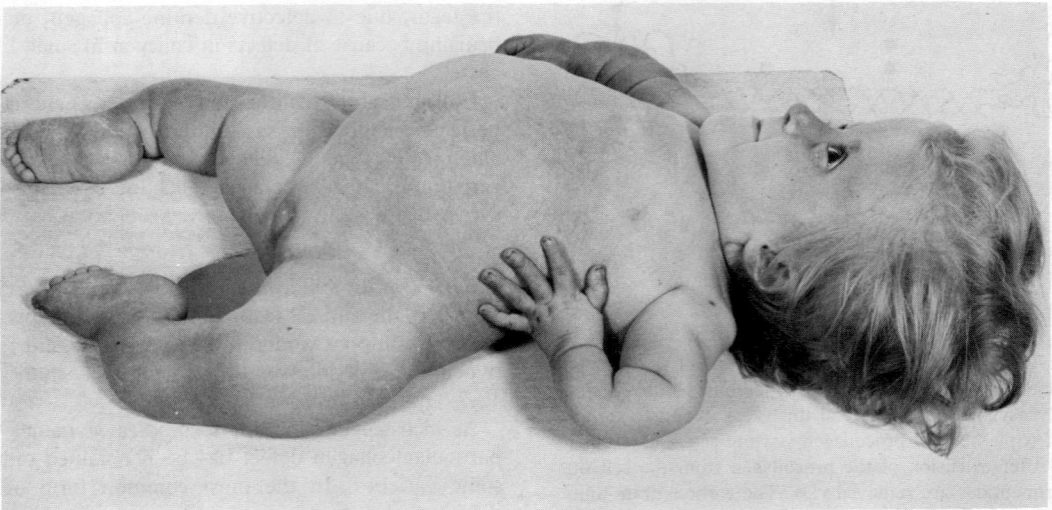

Fig. 17.4 Six-year-old child with severe osteogenesis imperfecta. The deformities of all the limb bones result from her many previous fractures. The impaired growth and the abnormally shaped head are also features of this disorder

212

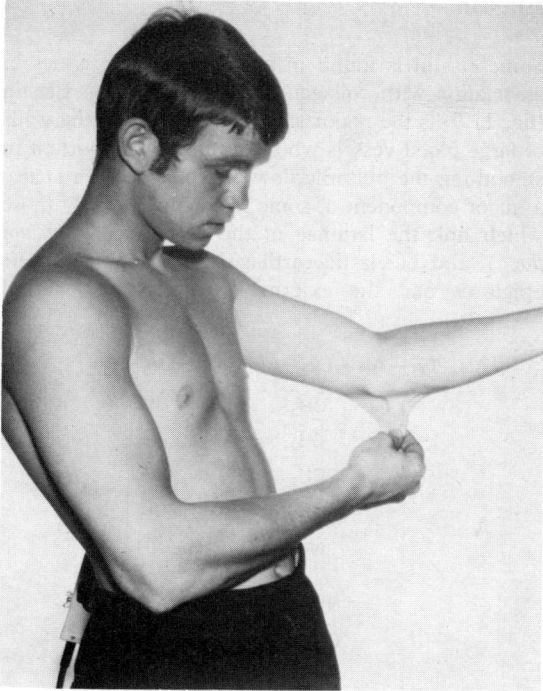

Fig. 17.5 Ehlers–Danlos syndrome to show hyperelastic skin. (Courtesy of P. Beighton)

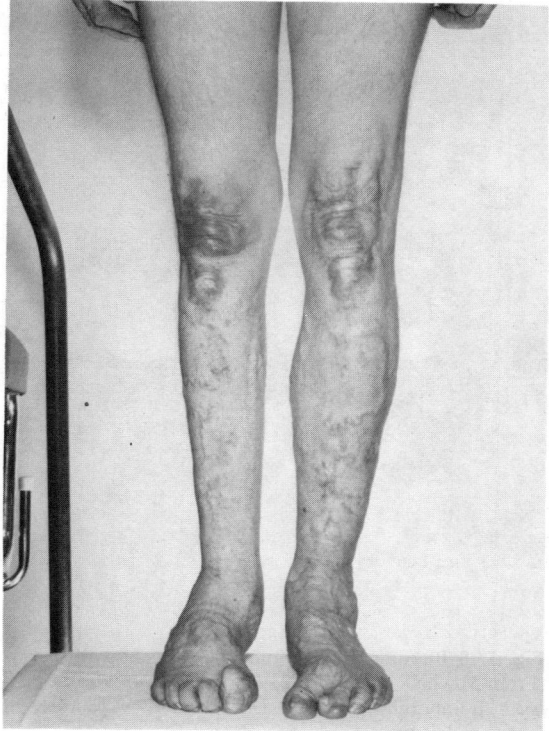

Fig. 17.6 Multiple poorly healed scars on the knees and shins in a patient with the Ehlers–Danlos syndrome. (Courtesy of P. Beighton)

Scurvy. In ascorbic acid deficiency the hydroxylation of lysine and proline is defective. The collagen formed is abnormal and patients readily develop large bruises, bleeding from the gums occurs spontaneously (fig. 17.8) and wounds in the skin fail to heal.

Gas gangrene. Infection with *Clostridium histolyticum* spreads rapidly because the organisms secrete a collagenase which breaks down the connective tissue barriers to their movement.

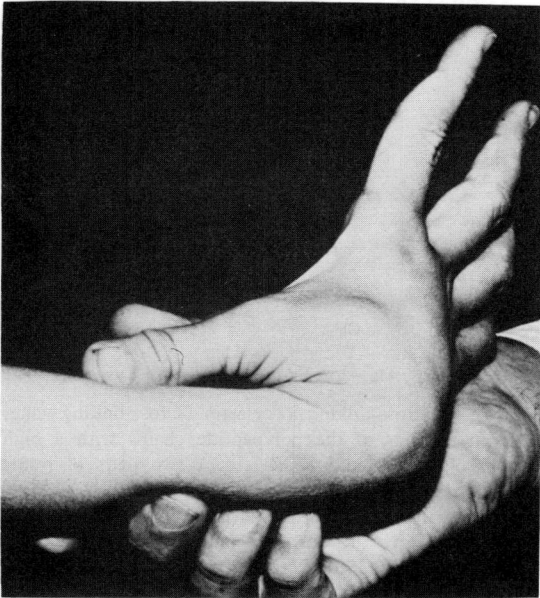

Fig. 17.7 Excessive laxity of the joints in a patient with the Ehlers–Danlos syndrome. (From Hollister, D. W. (1978) *Pediatric Clinics of North America* **25,** 575, by courtesy of author and publisher)

213

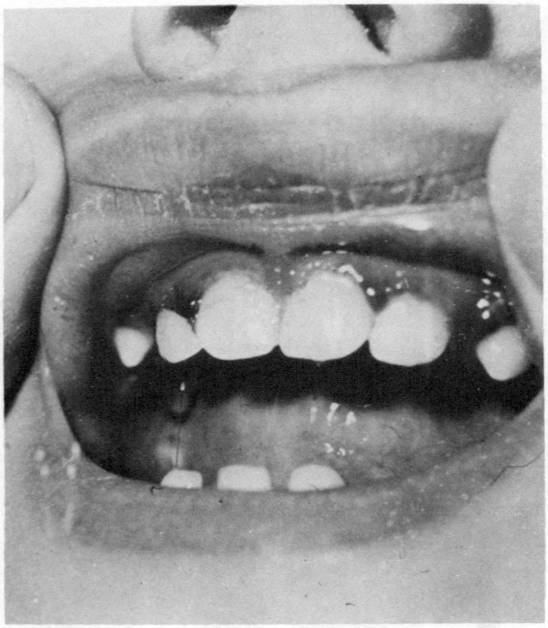

Fig. 17.8 Bleeding into the gums in a patient with scurvy. (From Mitchell, R. G. (1973) *Disease in Infancy and Childhood*. Edinburgh: Churchill Livingstone, by courtesy of author and publisher)

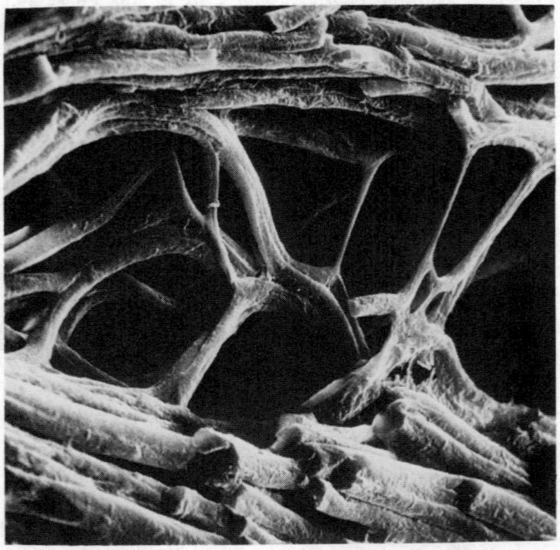

Fig. 17.9 Scanning electron micrograph of elastin to show two bundles of fibres with fibrils stretching between them (×1200). (From Gotte, L. *et al.* (1972) *Connective Tissue Research* **1**, 61, by courtesy of authors and publisher)

Elastin

Some elastin is found in most connective tissues in association with collagen and proteoglycans. Elastin (fig. 17.9) is the major structural protein in the walls of large blood vessels whose elasticity is important in smoothing the pulsatile flow of blood. Elastin is also a major component of some ligaments, notably those which link the laminae of the vertebrae (*ligamenta flava*), and of elastic cartilage, for example, in the epiglottis and the external ear. Elastin is found

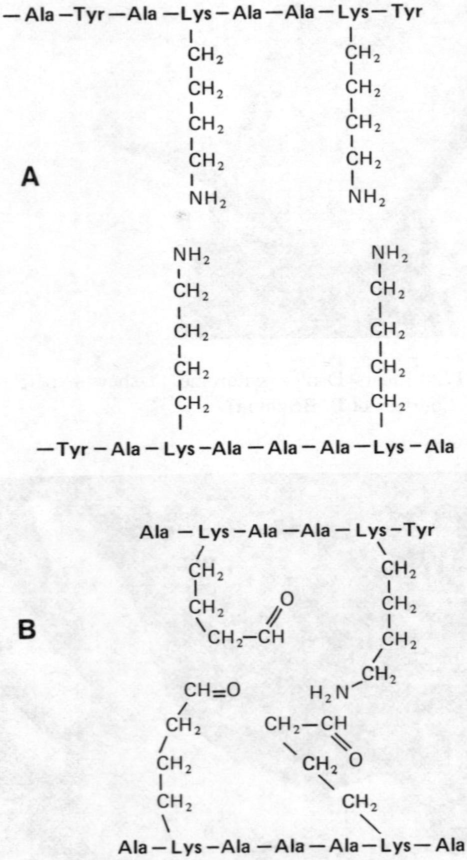

Fig. 17.10 Cross-linking of elastin. Cross-linking takes place in sections of the polypeptide chains with a high content of alanine and lysine. In A two pairs of lysine side-chains are adjacent. In B one pair has been oxidised by the enzyme lysyl oxidase. In C the four side-chains have combined to give the desmosine cross-link. (After Sandberg, L. B. *et al.* (1981) *New England Journal of Medicine* **304**, 566)

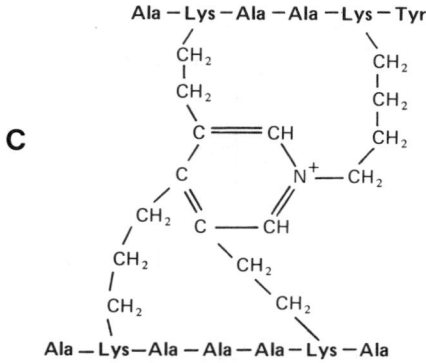

C

Fig. 17.10 (Continued)

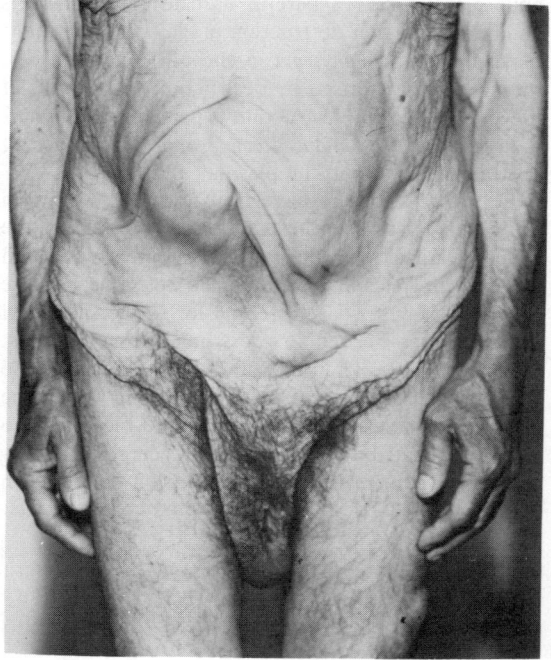

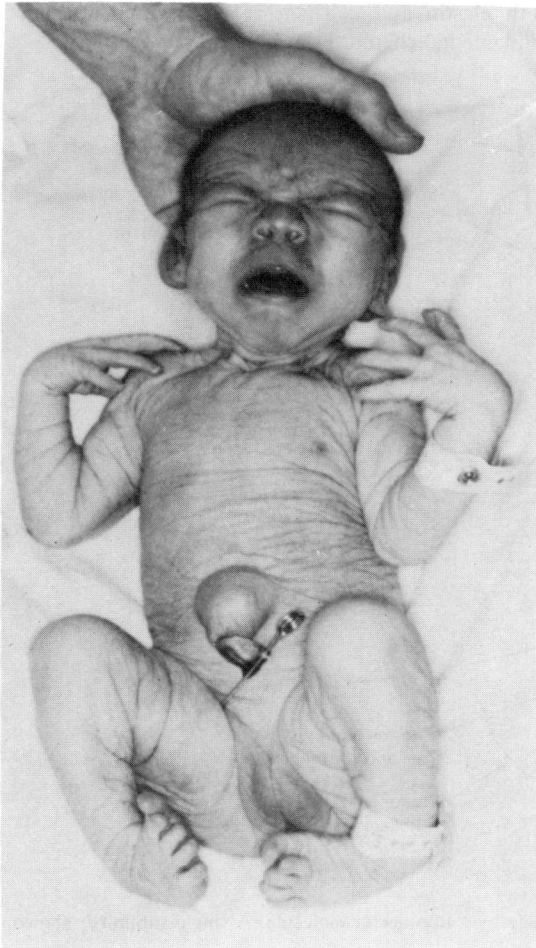

Fig. 17.12 Loose pendulous skin on the abdomen of a 55-year-old man with cutis laxa. (From Harris, R. B. *et al.* (1978) *American Journal of Medicine* **65,** 815, by courtesy of authors and publisher)

throughout the lung and it is important for its normal elasticity.

Elastin is the principal component of elastic fibres which can extend to two or three times their original length and return to their previous length when the tension is relaxed. Elastin has a high content of glycine and non-polar amino acids such as alanine, valine and proline. In addition the polypeptide chains of elastin are linked by covalent cross-links derived from pairs of lysine residues which together form the unique residue desmosine (fig. 17.10).

Disorders. Defects of elastin in major blood vessels are found in atherosclerosis ('hardening of the arteries') but it is not clear whether the elastin

Fig. 17.11 Newborn infant with cutis laxa. The thick folds of excess skin are seen over the trunk and limbs. She also had an umbilical hernia. (From Agha, A. *et al.* (1978) *Acta Paediatrica Scandinavica* **67,** 775, by courtesy of authors and editor; figure kindly provided by W. L. Nyhan)

abnormalities play a part in its causation. There is good evidence that destruction of elastin in the lungs plays a part in *emphysema*, a chronic abnormality of the lungs characterised by destruction of the walls of the alveoli. Elastin destruction is particularly obvious in cases of *elastase-induced emphysema* due to lack of protease inhibitors in the plasma (p. 20).

Very few disorders have been ascribed to inborn errors of elastin synthesis. One is type 5 of the Ehlers–Danlos syndrome (p. 211) in which both collagen and elastin maturation is defective because of lack of lysyl oxidase essential for the formation of the aldehydes needed in cross-link formation (figs. 17.3 and 17.10). The function of lysyl oxidase is also defective in *Menkes' syndrome* (p. 256) in which the elastin of arterial walls is abnormal and bleeding into the brain contributes to progressive cerebral degeneration. *Cutis laxa* is a group of inborn disorders in which the skin is loose but not hyperelastic (figs.

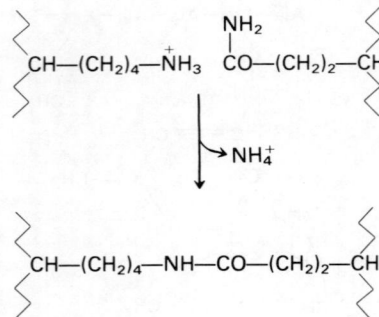

Fig. 17.13 Cross-link derived from lysine and glutamine residues linking two polypeptide chains in keratin

17.11 and 17.12). Ultrastructural abnormalities of both elastin and collagen are recognised, but the fundamental chemical defect is not yet known.

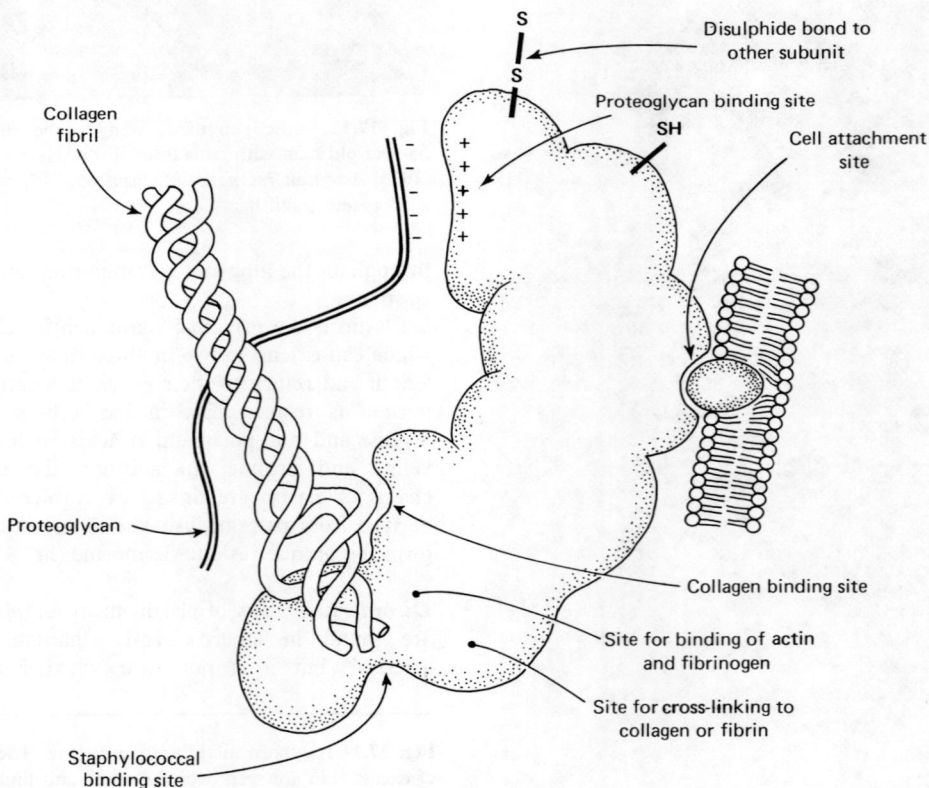

Fig. 17.14 Model to show the many attachments which may be made by a fibronectin molecule. A single subunit is shown. (After Ruoslahti, E. *et al.* (1981) *Collagen Research* **1**, 95)

Keratin

Keratins are tough fibrous proteins found in the outer layers of the skin and in nails and hair. Like collagen each keratin molecule has three polypeptide chains. Unlike collagen each chain forms a right-handed α-helix (fig. 2.8) and the chains are wound round each other to give a left-handed triple helix. Keratin can also exist in a pleated sheet structure β-keratin, p. 15). Keratin has a high content of the basic amino acids, arginine, lysine and histidine, and of cysteine. The structure of keratin is stabilised by disulphide bridges and also by covalent bonds derived from lysine and glutamine residues in adjacent chains (fig. 17.13).

Keratin is formed initially within cells which, as the surface of the skin is worn down, move outwards to be replaced by new cells. The keratin-filled cells die and form a close feltwork of keratin and other fibrous proteins. These become linked mainly by disulphide bonds.

Fibronectin

Fibronectin is a large glycoprotein found both in the plasma and in connective tissues. A fibronectin molecule has a mass of about 450 000 and consists of two polypeptide chains linked by a single disulphide bridge. It has specialised areas on its surface for binding to cell surfaces, to collagen, to proteoglycans, to actin and to fibrin (fig. 17.14). Fibronectin may serve to anchor cells in an extracellular matrix; it may also serve to bind cells in a blood clot during wound healing. The plasma fibronectin probably serves as a reserve available at any time to the tissues.

Further reading

Basu, T. K. and Schorah, C. J. (1982) *Vitamin C in Health and Disease*. London: Croom Helm

Francis, M. J. O. and Duksin, D. (1983) Heritable disorders of collagen metabolism. *Trends in Biochemical Sciences* **8**, 231–233

Mosher, D. F. (1984) Physiology of fibronectin. *Annual Review of Medicine* **35**, 561–575

Pinnell, S. R., Murad, S. (1983) Disorders of collagen. In *Metabolic Basis of Inherited Disease*, 5th edn, eds Stanbury, J. B., Wyngaarden, J. B., Fredrickson, D. S., Goldstein, J. L., Brown, M. S. pp. 1425–1449. New York: McGraw-Hill

Ryhanen, L. and Uitto, J. (1983) Elastic fibres of connective tissue. In *Biochemistry and Physiology of the Skin*, ed. Goldsmith, L. A., pp. 433–447. Oxford: University Press

Sandberg, L. B., Soskel, N. T. and Leslie, J. G. (1981) Elastin structure, biosynthesis, and relation to disease states. *New England Journal of Medicine* **304**, 566–579

Weiss, J. B. and Jayson, M. I. V. (1982) *Collagen in Health and Disease*. Edinburgh: Churchill Livingstone

Woessner, J. F. and Howel, D. S. (1980) The enzymatic degradation of connective tissue matrices. In *Scientific Foundations of Orthopaedics and Traumatology*, eds Owen, R., Goodfellow, J. and Bullough, P., pp. 232–239. London: Heinemann

Woodhead-Galloway, J. (1980) *Collagen: The Anatomy of a Protein*. London: Arnold

Yamada, K. M. (1983) Cell surface interactions with extracellular materials. *Annual Review of Biochemistry* **52**, 761–799

CRP

18 Transport across membranes

AN essential feature of all membranes is their selective permeability to solutes. Compounds such as glucose and amino acids are rapidly moved across cell membranes by specific transport mechanisms. Some substances such as urea diffuse slowly and other substances, including many harmful substances, are excluded. Membrane transport can be achieved by a number of different processes; many molecules traverse membranes by more than one of these mechanisms.

Simple diffusion

Most molecules cross membranes by simple diffusion but the rate of transfer is only significant for small uncharged molecules such as urea and certain drugs. For most solutes, the major barrier to simple diffusion is the lipid bilayer and the transport rate is governed by lipid solubility (fig. 18.1). Some small polar molecules (such as water and methanol) also cross membranes fairly easily, probably through hydrophilic channels or pores in the bilayer. Charged or highly polar molecules do not cross membranes readily by diffusion.

The driving force for simple diffusion is the *electrochemical gradient* across the membrane (fig. 18.2). It should be noted that solute molecules diffuse in both directions but that there is a net flow from the compartment with the higher to that with the lower concentration until an equilibrium is established.

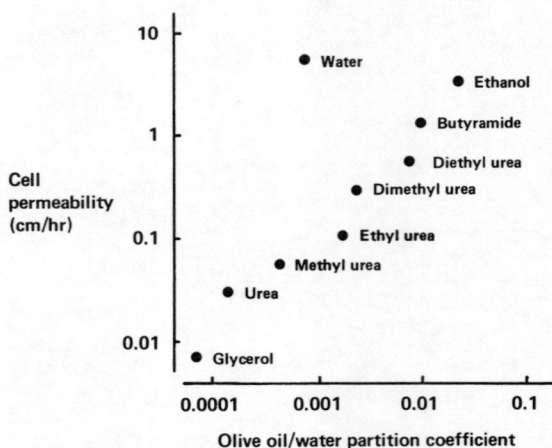

Fig. 18.1 Permeability of cell membranes to solutes in relation to the olive oil/water partition coefficient. Note that for many substances other than water the two measurements are closely related. (After Collander R. (1949) *Physiologia Plantarum* **2,** 300)

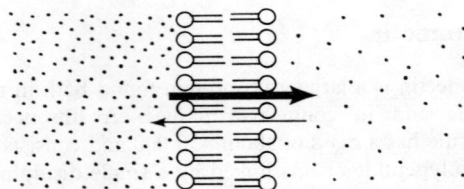

Fig. 18.2 Passive diffusion. Uncharged molecules of solute move in both directions across a membrane. The net flow is from the compartment with the higher to that with the lower concentration until the concentrations are equal

Mediated transport

Many solutes cross cell membranes much more rapidly than would be predicted from their lipid solubility and the electrochemical gradient. In these cases, *carrier mechanisms* operate to bring about specific and very rapid transport. The carriers are sometimes referred to as *permeases* because the transport processes have many of the properties of enzymic catalysis (Chap. 8).

Saturation kinetics. Whereas the net transport rate for simple diffusion increases with the concentration gradient, mediated transport has kinetics consistent

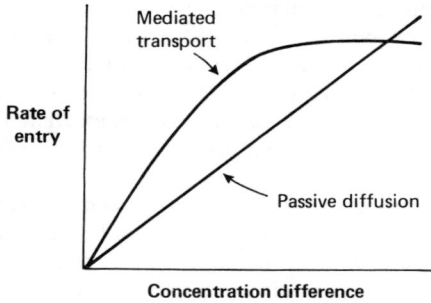

Fig. 18.3 Kinetics of membrane transport. Processes mediated by a carrier can become saturated. Transport of small uncharged molecules by diffusion is simply related to the concentration difference across the membrane

with the view that there is a fixed number of transport sites (fig. 18.3). We can define a Km and Vmax for a transport process in the same way as for an enzymic reaction.

Specificity. Carrier mechanisms are highly specific. That for D-glucose for example has only a low affinity (Km) for L-glucose (one-hundredth of that for D-glucose) or for D-galactose (one-tenth of that for D-glucose).

Temperature and pH dependence. A transport process, like an enzyme, has an optimum temperature and an optimum pH.

Inhibitor susceptibility. Like enzymes, transport processes are subject both to competitive inhibitors (substrate analogues) and to non-competitive inhibitors (such as heavy metals). As with enzymes a K_i can be determined for an inhibitor.

The nature of carriers. It seems certain that like enzymes all carriers are proteins with specific binding properties. Some of these proteins have been isolated and have even been shown to transport actively when reconstituted into artificial membranes consisting of lipids. It seems likely that most, if not all, carrier proteins are spanning proteins (p. 63).

Two distinct forms of mediated transport are recognised: *facilitated diffusion* and *active transport*.

Facilitated diffusion

This process resembles simple diffusion in that net solute transfer depends on the concentration gradient

and ceases when equilibrium is reached. However, as with enzymic reactions, equilibrium is reached rapidly because of the presence of a catalyst. Facilitated diffusion underlies the transport of many important solutes; glucose and glycerol uptake by erythrocytes are examples.

Active transport

Active transport differs from both simple and facilitated diffusion in that it can proceed against a concentration gradient. It therefore requires energy. Inhibition of cell respiration, for example by cyanide, or of oxidative phosphorylation, for example by dinitrophenol, abolishes active transport.

Examples of active transport processes include the maintenance of the Na^+ and K^+ gradients across all plasma membranes, the uptake of nutrients by the intestinal mucosa and the reabsorption of solutes by the epithelia of the renal tubules. It has been estimated that some 30 to 40 per cent of the total ATP generated in man is expended directly or indirectly for the maintenance of ion gradients. Two of the best understood active transport systems are the *sodium pump* in the plasma membrane and the *calcium pump* in the sarcoplasmic reticulum.

Sodium pump. This operates in all cells to maintain a high intracellular K^+ concentration and a low intracellular Na^+ concentration (fig. 18.4). These ion gradients are essential for the control of cell volume

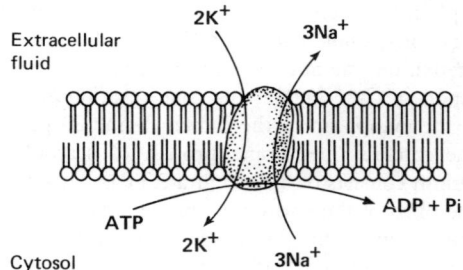

Fig. 18.4 The role of Na^+-K^+-dependent ATPase in the transport of sodium out of cells and of potassium into cells. This enzyme is specifically inhibited by the cardiac glycoside ouabain which binds to the external surface of the transport protein. Note that the exchange of $3Na^+$ ions for $2K^+$ ions leads to the generation of an electrical gradient with the cytosol negative relative to the extracellular fluid

and for other cell functions particularly in nerves and muscles.

The pump consists of a tetramer of two large and two small subunits with molecular masses of 100 000 and 50 000 respectively. The large subunits are known to span the membrane; they have sites for the binding of inhibitors such as the cardiac glycoside, ouabain, on the outside surface of the membrane; a reversible phosphorylation takes place at the cytoplasmic surface. The pump has ATPase activity and is often known as the *Na+-K+-dependent ATPase*. The ATPase activity is accompanied by the transfer of the terminal phosphate of ATP to an aspartyl residue on the large subunit. This phosphorylation and the subsequent dephosphorylation are essential features of Na+ and K+ transport. It is possible that this reversible phosphorylation of the carrier protein brings about conformational changes needed for ion translocation.

In seriously ill patients the plasma sodium level may be low. This has no obvious hormonal explanation and it now seems clear that this abnormality is caused by impaired function of the sodium pump, possibly due to hypoxia of the tissues. The intracellular sodium level rises and the disorder is sometimes known as the 'sick cell syndrome'.

Calcium pump. Ion gradients are also recognised between subcellular compartments of the ion pumps within cells; the *Ca2+-dependent ATPase* of sarcoplasmic reticulum is the best known. Muscle contraction is controlled by the intracellular levels of Ca^{2+} ions; calcium ions are moved rapidly and reversibly between the cytoplasm and the lumen of the sarcoplasmic reticulum. The Ca^{2+}-dependent ATPase or *calcium pump* embedded in the sarcoplasmic reticulum membrane is responsible for the active transport of Ca^{2+} (against its concentration gradient) from the cytoplasm to the lumen. Some 80 per cent of the integral membrane protein of sarcoplasmic reticulum consists of the calcium ATPase; it amounts to one-third of the entire membrane area.

The calcium pump is a spanning protein with a molecular mass of 100 000. In the presence of Ca^{2+} ions it is phosphorylated by ATP at the cytoplasmic face of the membrane. Calcium binding at the same surface accompanies this phosphorylation; the subsequent dephosphorylation results in the release of two calcium ions into the lumen. An equal number of phosphate ions accompany the transported calcium ions; in this way electrical neutrality is maintained.

Such a simultaneous transport of two ions is known as *symport* in contrast to the *antiport* of Na+ and K+ brought about by Na+-K+ ATPase. In both cases energy is required as ATP for the phosphorylation of the carrier protein; reversible phosphorylation/dephosphorylation is an essential part of the transport mechanism.

Highly purified isolated Ca^{2+}-dependent ATPase has been incorporated into artificial lipid bilayers and transports calcium successfully; by experiments of this sort it was shown that the lipid environment of the carrier protein is important for the pumping activity. Artificial bilayers made from phosphatidyl choline and phosphatidyl ethanolamine give maximum pump activity. In the living cell it is likely that the pump protein is surrounded by a specialised microenvironment of *boundary lipid* which does not readily exchange with the other bilayer lipids. The fatty acid composition of the boundary lipid also influences the pump's activity and plays a part in the regulation of the function of the carrier (p. 62).

Ion gradients

The two active transport processes just described generate and maintain gradients of ion concentration essential for the normal functioning of cells. Gradients across membranes may be very large; that for Na+ and K+ is usually 15:1, the Ca^{2+} gradient across the sarcoplasmic reticulum is $10^4:1$ while that for H+ across the gastric mucosa is $10^6:1$. Such gradients are essential for the function of tissues, such as nerve and muscle, but are also important in many cells for driving other transport processes (see below). A key step in the response of a cell to the binding of a hormone or neurotransmitter to its receptor is often the opening of an *ion channel* or *ionophore* in the membrane which allows the passage of an ion such as calcium down its concentration gradient.

Sodium coupled transport

The sodium and calcium pumps are unusual among active transport mechanisms in that the energy requirement is met directly by ATP. The energy needed for the active transport of several compounds is provided by the movement of sodium down its concentration gradient.

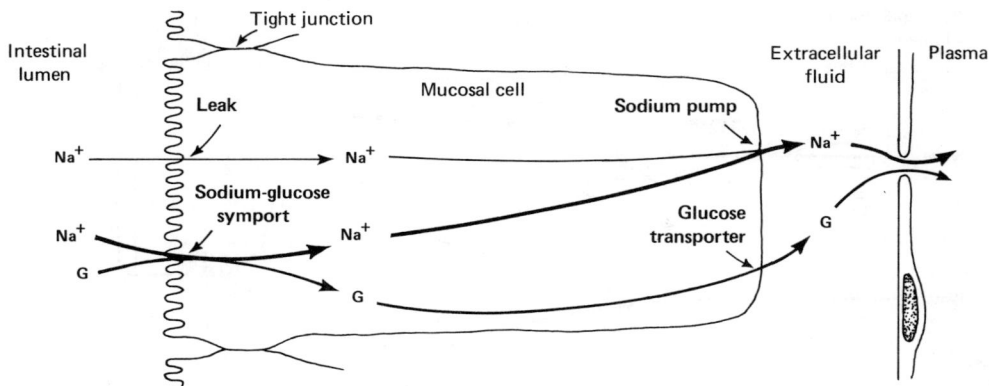

Fig. 18.5 The sodium-linked transport of glucose (G) across cells of the intestinal mucosa. The sodium gradient across the luminal surface is maintained by the action of the sodium pump at the serosal surface. Glucose passes from the cell to the extracellular fluid by facilitated diffusion

The coupled transport of glucose and Na^+ by the cells of the intestinal mucosa is an example (fig. 18.5). These cells, and those of other transporting tissues, show membrane asymmetry in that the transport functions of the mucosal surface differ from those of the serosal surface. The Na^+ gradient is maintained by the sodium pump in the serosal membrane and ensures that the concentration of Na^+ within the cell is much lower than that in the intestinal lumen. In the mucosal membrane there is a carrier protein with binding sites for both Na^+ and glucose. The binding of Na^+ increases the affinity of the carrier for glucose which is taken up from the lumen. Sodium and glucose are then transported to the cytoplasmic surface and released into the cell. Na^+ is pumped out of the cell by the sodium pump on the serosal surface and glucose crosses the serosal membrane to the extracellular fluid (ECF) by facilitated diffusion down its concentration gradient.

The same Na^+ gradient across the serosal membrane is used to drive 95 per cent of the water absorbed from the intestinal lumen by a process summarised in figure 18.6. The intercellular contacts between adjacent epithelial cells are sealed by tight junctions (Chap. 6). On the serosal side of these tight junctions are channels filled with extracellular fluid into which Na^+ ions are pumped. Cl^- ions follow passively to restore electrical neutrality. The high concentration of Na^+ and Cl^- has the effect of drawing water from the lumen both through the cells

and across the tight junctions by osmosis. The water together with Na^+ and Cl^- then pass through the capillary endothelium and enter the blood. *Cholera toxin* has the effect of making tight junctions permeable to Na^+ and Cl^- so allowing leakage of these ions back into the lumen of the gut and preventing water absorption. Massive losses of NaCl and water lead to salt and water depletion which can be fatal.

The sodium gradient is also the driving force for an antiport system which transports calcium out of cells.

Other transport mechanisms

Some important substances are transported across membranes by complex cyclical processes described elsewhere. One example is the role of carnitine in the transport of acyl coenzyme A into the matrix of mitochondria (p. 121). Similarly acetyl coenzyme A is transported out of mitochondria by a complex cycle involving citrate (p. 125).

The γ-glutamyl cycle. One of the mechanisms by which amino acids are taken up by cells involves *glutathione*. This tripeptide is found in concentrations as high as 5 to 10 mmol/l in cells. It takes part in the cyclical pathway, illustrated in figure 18.7, in which

221

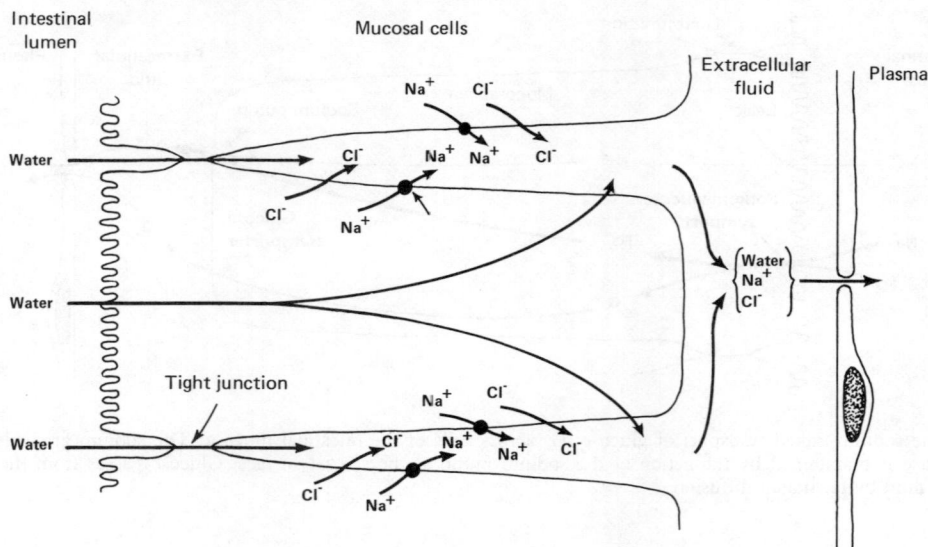

Fig. 18.6 Water absorption in the small intestine. The active transport of sodium followed by chloride leads to the generation of a high osmotic pressure in the extracellular fluid, particularly that between the cells. In turn the high osmotic pressure leads to the flow of water down the osmotic gradient through the cell and through the tight junctions

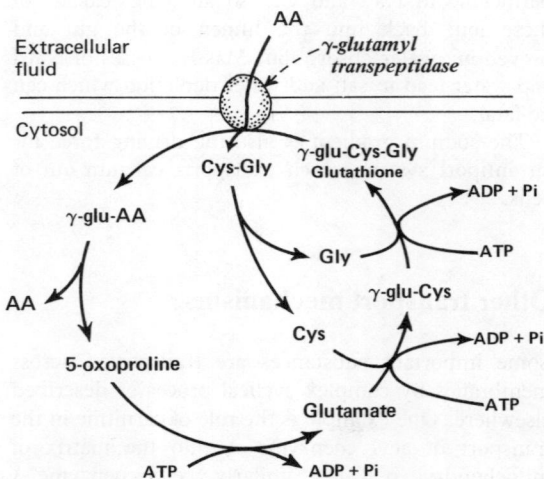

Fig. 18.7 The γ-glutamyl cycle in the active uptake of amino acids (AA) by cells. The quantitative importance of this cycle in amino acid transport is not known but it probably plays a significant role in the transport of cysteine and glutamine. The enzymes involved are particularly abundant in cells, such as those of the small intestine and renal tubules, with a special role in amino acid transport. (After Meister, A. (1974) *Annals of Internal Medicine* **81**, 247)

one amino acid is transported into the cell at the expense of the hydrolysis of three molecules of ATP.

Bulk transport

Many cells transport macromolecules, or large numbers of small molecules simultaneously, by exocytosis and endocytosis. Exocytosis occurs when an intracellular vesicle, containing for example a hormone molecule, a digestive enzyme or a collagen precursor, fuses with the plasma membrane and ruptures into the extracellular fluid (fig. 18.8).

Endocytosis (pinocytosis) is the uptake of material from the extracellular fluid by cells which engulf it to form a vesicle. One example is the uptake of bacteria by phagocytes (p. 67); another is the process whereby certain hormones are taken up by cells (fig. 18.9).

A combination of endocytosis with exocytosis leads to the movement of vesicles across cells. By this process of bulk transport, for example, cerebrospinal fluid is transferred by the arachnoid villi into the venous sinuses (fig. 18.10). A similar process is used for the reabsorption of proteins from the glomerular filtrate by cells of the renal tubules.

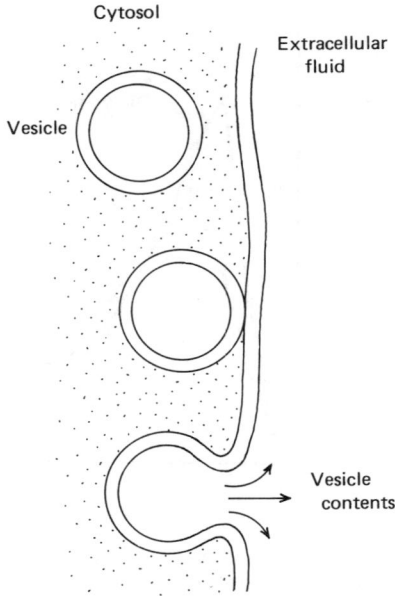

Fig. 18.8 Exocytosis of an intracellular vesicle leading to the discharge of the vesicle contents into the extracellular fluid

Disorders of transport processes

Just as inherited errors of the production or function of enzymes lead to the inherited metabolic disturbances described elsewhere, so inherited disorders of transport mechanisms also lead to disease. Like inborn errors of metabolism most of these disorders are inherited as autosomal recessive traits.

Transport of amino acids and peptides

In mammals, almost certainly including man, there are three distinct processes for the transport of amino acids.

(1) Monoamino-monocarboxylic (neutral) amino acids
(2) Dibasic amino acids and cystine
(3) Dicarboxylic (acidic) amino acids

Each of these systems operates in the intestine, in the renal tubule and in the endothelial cells of the capillaries in the brain (to transport amino acids across the 'blood brain barrier'). There is some

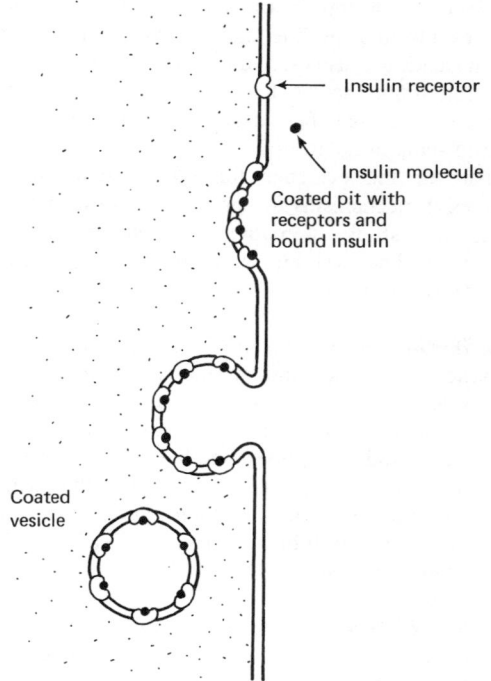

Fig. 18.9 The process whereby insulin is taken up by cells of target organs. It seems likely that receptors to which insulin has become bound cluster together and then form, in turn, a coated pit and a coated vesicle. The subsequent fate of the insulin within the cell is not yet clear

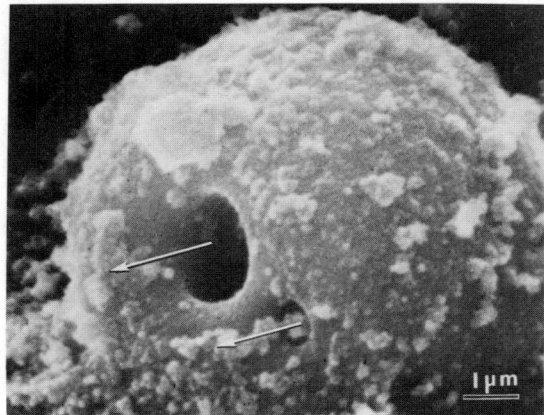

Fig. 18.10 Scanning electron micrograph of a vacuole in an arachnoid villus. Two openings are seen through which a tracer material, a colloidal suspension of thorotrast, can be seen to be passing. (From Tripathi, R. (1974) *Brain Research* **80,** 503, by courtesy of author and editor)

223

evidence of a fourth transport mechanism which carries glycine, proline and hydroxyproline; these amino acids are also carried by the mechanism for the neutral amino acids. Each process has a limited capacity and may be saturated by any one of the amino acids in its group.

Part of the evidence for the existence of the transport mechanisms for amino acids in man comes from the study of patients in whom they are defective. The best known examples are cystinuria and Hartnup disease.

Cystinuria. In this disorder the transport of the dibasic amino acids and cystine is impaired. In the intestine this defect causes no symptoms since the amino acids concerned are readily absorbed as dipeptides and tripeptides (see below). However the defect in the renal tubule means that the urinary concentration of cystine is high. Since this amino acid is very insoluble, patients develop cystine-containing renal stones (fig. 18.11).

Hartnup disease. This disorder, named after the first family in which it was described, is caused by a defect in the intestinal and renal transport of the neutral amino acids. As with cystinuria this has little

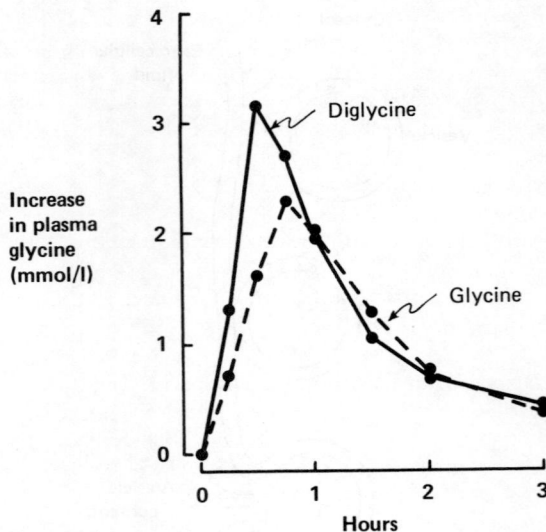

Fig. 18.12 Plasma levels of glycine after oral doses of equivalent amounts of glycine and the dipeptide diglycine. (After Croft, I. *et al.* (1968) *Gut* **9,** 425)

nutritional consequence for most of the amino acids concerned since the absorption of dipeptides is normal. However the defect in tryptophan absorption leads to low plasma tryptophan levels and skin changes similar to those of pellagra (p. 53 & p. 194). In addition patients may have intermittent neurological symptoms such as tremor and ataxia which are probably due to the absorption from the gut of toxic bacterial metabolites of the unabsorbed tryptophan.

Peptide transport. In the gut a large proportion of the amino acids are absorbed as dipeptides and tripeptides. These are taken up by the microvilli where most are broken down to the free amino acids by peptidases. Some peptides enter the portal circulation without hydrolysis. The mechanisms for the uptake of di- and tripeptides are quite different from those of the free amino acids and are highly efficient (fig. 18.12).

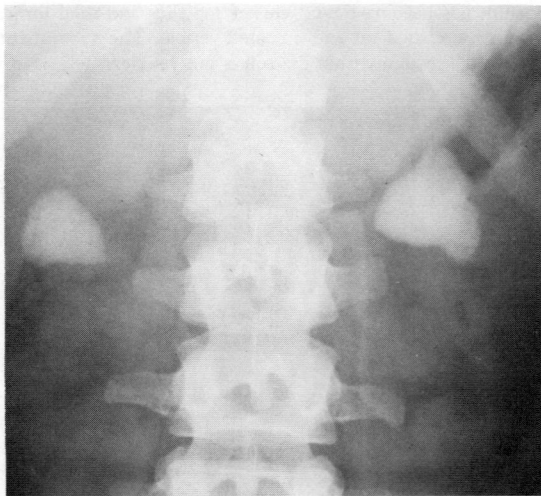

Fig. 18.11 Bilateral cystine stones in a patient with cystinuria. Since cystine contains sulphur, large cystine stones can be seen radiologically as in this example. (From Rose, G. A. (1982) *Urinary Stones: Clinical and Laboratory Aspects.* Baltimore: University Park Press, Lancaster: MTP Press, by courtesy of author and publisher)

Carbohydrate transport

The action of digestive enzymes in the gut leads to the breakdown of polysaccharides mainly to disaccharides but also, to a smaller extent, to monosaccharides.

Monosaccharide absorption. Glucose and galactose share an active transport mechanism involving simultaneous sodium movements (p. 221). It is likely that the same transport process occurs in the renal tubules and the brain. In the rare disorder, *glucose–galactose malabsorption*, patients develop abdominal pain and diarrhoea after the ingestion of any of the common carbohydrates. In addition tubular reabsorption of glucose is impaired and glucose is found in the urine.

Fructose is absorbed by facilitated diffusion, independently of sodium.

Disaccharide absorption. Disaccharides are hydrolysed by specific disaccharidases, which are membrane-bound enzymes on the microvilli. The monosaccharides produced are immediately taken up by the specific transport mechanisms which are thought to be adjacent. The disaccharidases are lactase, maltase, and sucrase–isomaltase which is a complex of two subunits, one for each enzyme activity (fig. 6.4, p. 63).

By far the most common disorder of carbohydrate absorption is lactase deficiency. Two types are recognised: congenital and acquired. Congenital lactase deficiency (*alactasia*) is characterised by profuse watery diarrhoea which begins soon after birth with the start of milk feeding (fig. 18.13). The condition is rapidly fatal unless it is recognised and lactose is withdrawn from the feeds.

Acquired lactase deficiency causes milder symptoms of abdominal distension and diarrhoea in late childhood or adult life; the patient may learn to associate his symptoms with the ingestion of milk. Acquired lactase deficiency is found in between 3 and 19 per cent of European adults and in up to 95 per cent of people of African or native American origin. In Nigeria lactase deficiency is found in most Hausa, Ibo and Yoruba people but seldom among the Fulani who habitually consume large amounts of milk. Lactase deficiency can be diagnosed by the response to an oral lactose load (fig. 18.14). A temporary lactase deficiency may follow severe gastroenteritis or other gastrointestinal diseases.

Sucrase–isomaltase deficiency is an uncommon disorder causing diarrhoea and malnutrition. It is found in some 0.2 per cent of North Americans and 10.5 per cent of Eskimos in western Greenland. The symptoms of diarrhoea diminish rapidly as soon as sucrose is removed from the diet.

Transport of inorganic substances

Nephrogenic diabetes insipidus. This is an inherited defect of water reabsorption in the distal tubules of the kidney. Unlike diabetes insipidus it does not respond to treatment with antidiuretic hormone (ADH). The urine volume is very large and severe dehydration can occur soon after birth. Many patients do well if dehydration is corrected and they are given an adequate fluid intake in feeds with a low salt and low protein content.

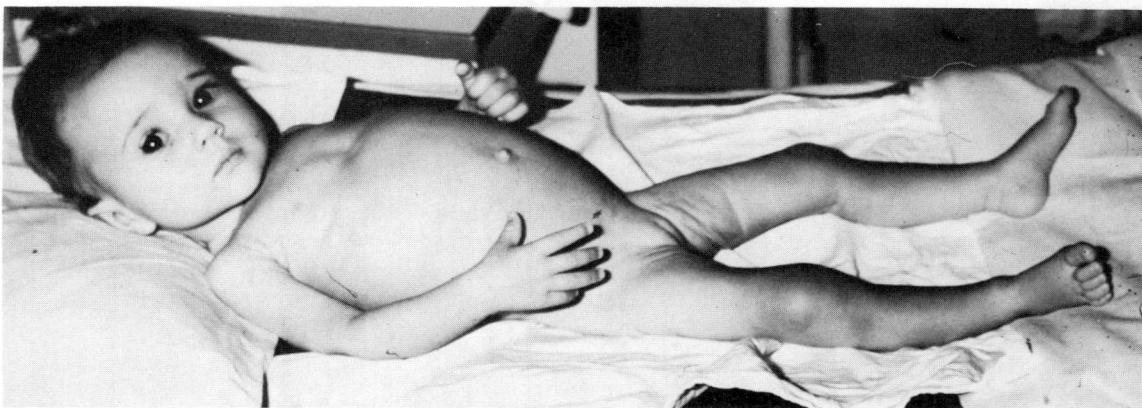

Fig. 18.13 A child with congenital lactase deficiency (alactasia). She had persistent diarrhoea, marked abdominal distension and evidence of loss of subcutaneous adipose tissue. (From Sinclair, L. (1979) *Metabolic Disease in Childhood.* Oxford: Blackwell, by courtesy of author and publisher)

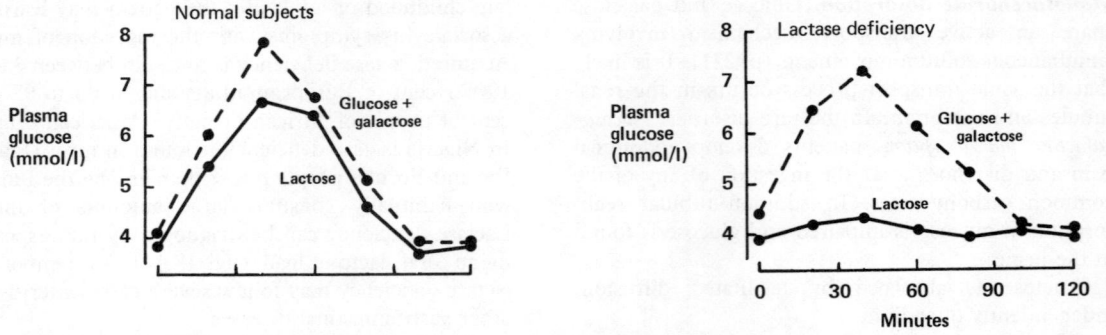

Fig. 18.14 Changes in plasma glucose after an oral dose of 50 g lactose or 25 g galactose with 25 g glucose in normal subjects and patients with lactase deficiency. (After Cuatrecasas, P. *et al.* (1965) *Lancet* **i,** 14)

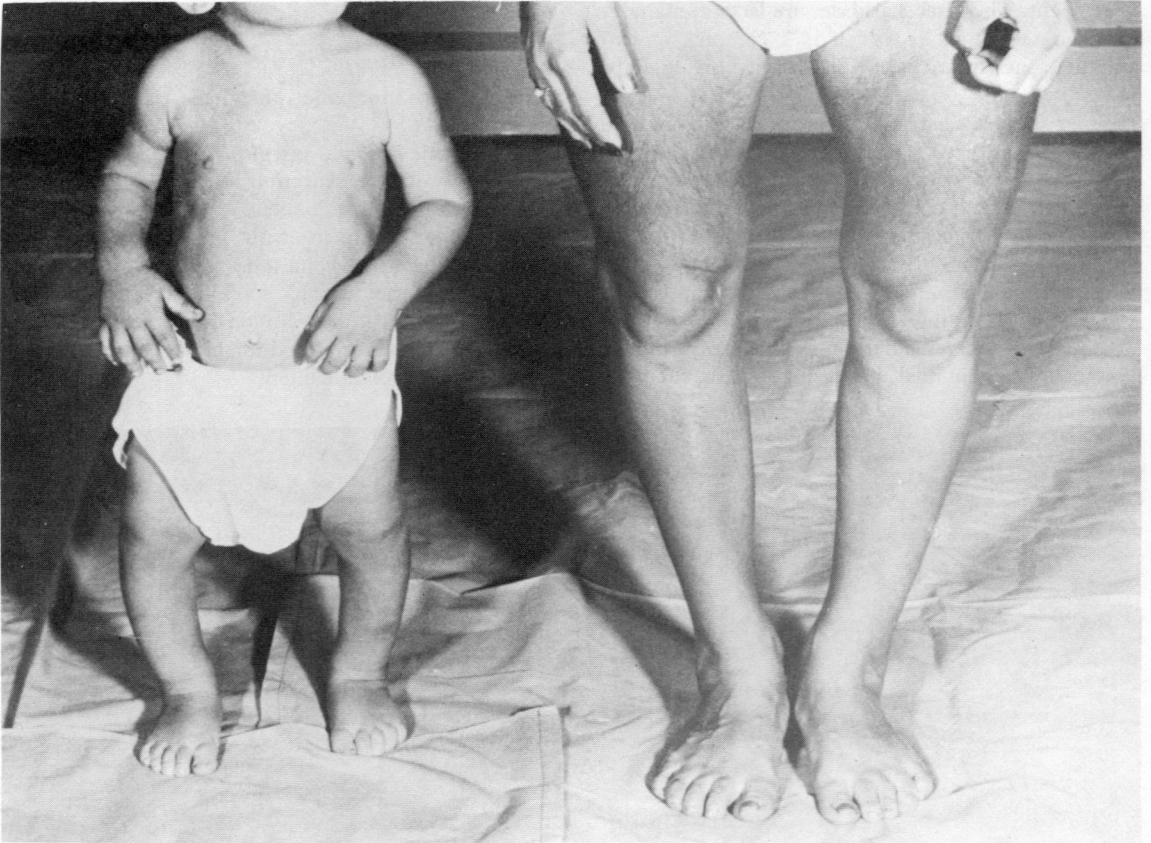

Fig. 18.15 Bow-legs in hypophosphataemic rickets in a nineteen-month-old boy and his mother. (From Evans, G. A. *et al.* (1980) *Journal of Bone and Joint Surgery* **62A,** 1130, by courtesy of authors and editor)

shipwrecked. Severe and sometimes fatal water depletion can occur in patients with deficient secretion of vasopressin (*diabetes insipidus*) or with an inborn renal insensitivity to vasopressin (*nephrogenic diabetes insipidus*, p. 225).

Intense thirst is a characteristic symptom of water depletion but can only be recognised if the patient is conscious. The other symptoms are due to reduction in the volume of cells in the central nervous system; convulsions are common. The plasma sodium is high. The disorder can be treated by the administration of water or, if the intravenous route is to be used, isotonic dextrose solutions.

Mixed disorders. A combination of saline depletion and water depletion occurs in an uncommon complication of diabetes mellitus (p. 144), *hyperosmolar non-ketotic coma.* The osmotic activity of glucose in the glomerular filtrate leads to severe losses of water and, to a smaller extent, sodium. In this condition a high plasma sodium is found despite saline depletion.

Potassium disorders

Although most of the body's potassium is within cells the plasma potassium reflects the total body potassium in many clinical disorders. There are some important exceptions. For example potassium moves into cells from the plasma when glycogen synthesis is taking place, in metabolic alkalaemia (p. 234) and in the most common variety of *familial periodic paralysis*, an inherited disorder characterised by episodes of muscle weakness. Potassium moves out of cells in metabolic acidaemia, during glycogen breakdown, during severe muscular exertion and whenever muscle breakdown is proceeding rapidly.

While methods are available for the measurement of total body potassium, the plasma potassium level is of greater clinical value. The normal function of muscles and nerves depends on a normal plasma potassium.

Hypokalaemia. While potassium levels below 3.5 mmol/l are abnormal, symptoms are not usually obvious unless the level is below 2.5 mmol/l. Patients complain of muscle weakness, tiredness, depression, thirst and an increased urine volume (*polyuria*). A complete flaccid paralysis may occur with values below 2.0 mmol/l. Longstanding hypokalaemia causes permanent damage to the renal tubules.

Hypokalaemia is most commonly caused by treatment with diuretic drugs such as frusemide (furosemide), ethacrynic acid and the thiazides. Other renal causes of hypokalaemia include aldosterone-producing tumours, uncontrolled diabetes mellitus and rare renal tubular disorders such as the *Fanconi syndrome* (p. 227). Hypokalaemia may also result from gastrointestinal losses, particularly in diarrhoea.

Hyperkalaemia. Hyperkalaemia causes few symptoms but some patients have paraesthesiae ('pins and needles'). If the K^+ level is greater than about 8.0 mmol/l, sudden cardiac arrest is likely.

The most common cause of hyperkalaemia is renal failure; it can also occur on treatment with certain diuretics such as triamterene, amiloride and the aldosterone antagonist *spironolactone*. Some patients with adrenocortical insufficiency (Addison's disease) are hyperkalaemic.

Acid–base balance

An adult on a mixed diet produces about 70 mmol H^+ daily mainly as phosphoric and sulphuric acids derived from phosphorus- and sulphur-containing compounds in food. In addition CO_2 equivalent to about 13 mol H^+ is produced daily from the metabolism of food to provide energy.

All these hydrogen ions must be excreted if the pH of the extracellular and intracellular compartments is to remain constant. The pH of the arterial blood is normally 7.36 to 7.44 (approximately 44 to 36 nmol/l). In disease pH values as low as 6.85 (140 nmol/l) or as high as 7.65 (22 nmol/l) can occur but such extreme values are associated with a high mortality.

Buffers. The nature of buffers was described in Chapter 1. In the intracellular fluid hydrogen ions are largely buffered by phosphate and by proteins particularly, in red cells, by haemoglobin. In the extracellular fluid and plasma the principal buffer is bicarbonate/carbonic acid. Bicarbonate is a particularly important buffer since the amount lost in the urine and the amount lost as CO_2 in the breath can both be regulated. The proteins also play a part in buffering in the plasma.

233

Hypophosphataemic rickets. This disorder, also known as *phosphaturic rickets* or *vitamin D-resistant rickets* is inherited as an X-linked dominant, males being more severely affected than females. The defect is in phosphate reabsorption in the proximal renal tubule; the excessive urinary loss of phosphate leads to a low plasma phosphate and bone changes (fig. 18.15) similar to those of rickets due to lack of vitamin D (p. 245).

Congenital chloride diarrhoea. This rare disorder is of interest because of the light it throws on the process of normal electrolyte transport in the intestine. Sodium and chloride absorption in the upper small intestine is normal but the exchange of chloride for bicarbonate in the ileum and colon is defective. Soon after birth patients begin to pass voluminous fluid stools with a high chloride content. Without treatment salt and water depletion progresses rapidly and death from dehydration occurs. With careful control of the fluid and electrolyte balance prolonged survival is possible although the continuing diarrhoea remains very inconvenient.

Generalised transport disorders

Defects of the intestinal absorption of a wide range of substances including fat, minerals and vitamins occur in a number of diseases. The most common cause of generalised malabsorption in western countries is *coeliac disease* resulting from damage to the intestinal mucosa caused by an inborn sensitivity to gluten, the protein of wheat.

In the *Fanconi syndrome* several functions of the proximal renal tubule are simultaneously disturbed. One well-recognised cause in childhood is cystinosis, an inherited disorder characterised by the deposition of cystine in tissues. This results in defects in the tubular reabsorption of glucose, amino acids, phosphate and potassium; phosphate depletion leads to bone changes similar to those of rickets. Renal failure causes death usually before the age of ten.

Further reading

Chapman, D. (1982) *Biomembrane Structure and Function.* London: Macmillan

Gray, G. M. (1983) Intestinal disaccharidase deficiencies and glucose-galactose malabsorption. In *Metabolic Basis of Inherited Disease*, 5th edn, eds Stanbury, J. B., Wyngaarden, J. B., Fredrickson, D. S., Goldstein, J. L., Brown, M. S. pp. 1729–1742. New York: McGraw-Hill

Harries, J. T. (1982) Disorders of carbohydrate absorption. *Clinics in Gastroenterology* **11**, 17–30

Harrison, R. and Lunt, G. G. (1981) *Biological Membranes: Their Structure and Function*, 2nd edn. Glasgow: Blackie

Meister, A and Anderson, M. E. (1983) Glutathione. *Annual Review of Biochemistry* **52**, 711–760

Racker, E. (1976) Structure and function of ATP-driven ion pumps. *Trends in Biochemical Sciences* **1**, 244–247

Rasmussen, H., Anast C. (1983) Familial hypophosphatemic rickets and vitamin D-dependent rickets. In *Metabolic Basis of Inherited Disease*, 5th edn, eds Stanbury, J. B., Wyngaarden, J. B., Fredrickson, D. S., Goldstein, J. L., Brown, M. S. pp. 1743–1773. New York: McGraw-Hill

Segal, S., Thier, S. O. (1983) Cystinuria. In *Metabolic Basis of Inherited Disease*, 5th edn, eds Stanbury, J. B., Wyngaarden, J. B., Fredrickson, D. S., Goldstein, J. L., Brown, M. S. pp. 1744–1791. New York: McGraw-Hill

Silk, D. B. A. (1982) Disorders of nitrogen absorption. *Clinics in Gastroenterology* **11**, 47–72

Tanford, C. (1983) Mechanism of free energy coupling in active transport. *Annual Review of Biochemistry* **52**, 379–409

MISH CRP

19 Metabolism of water, electrolytes & hydrogen ions

A NORMAL adult weighing 70 kg contains about 45 litres of water. This is divided into a number of compartments as shown in Table 19.1. Each compartment has a different composition and the movement of water and solutes between the compartments is not free (Chap. 18). In disease therefore the composition of the plasma, which is readily measured, does not necessarily reflect that of the other compartments.

The ionic composition of the extracellular fluid (ECF), including plasma, differs greatly from that of the intracellular fluid (ICF). Thus sodium is the principal monovalent anion in the ECF while potassium predominates in the ICF. Calcium is the most abundant divalent cation in the ECF; magnesium is the most abundant in the ICF (Table 19.2). The composition of the ICF also varies greatly between tissues. The proportion of potassium, for example, ranges between 90 per cent of the cations in erythrocytes to about 40 per cent in the liver. The intracellular pH varies between 6.90 in muscle and 7.28 in erythrocytes.

Table 19.1 Principal compartments into which the total body water is divided. Approximate figures for a 70 kg adult

		Litres
Intracellular water		30
Extracellular water		
Interstitial fluid	8.5	
Lymph	1.5	
Plasma	3.5	
* Transcellular fluid	1.5	
Total extracellular	15.0	15
Total body water		45

* The transcellular fluid comprises that, for example, within the lumen of the gut and the urinary tract.

Table 19.2 Typical figures for the composition (in mmol/l) of the interstitial fluid and an intracellular fluid

	Interstitial fluid*	Intracellular fluid
Cations		
Sodium	140	12
Potassium	4.5	165
Calcium	2.5	<0.01
Magnesium	1.0	14
Anions		
Chloride	100	<1
Bicarbonate	26	10
Phosphate (HPO_3^{2-})	1	60
Sulphate	0.5	10
Organic anions	3	5
Proteins	0.1*	8

* The plasma has a composition similar to that of the interstitial fluid except that its protein content is of the order of 2.0 mmol/l (70 g/l).

Water balance

In health the total amount of water in the body remains virtually constant from day to day since the fluid movements are balanced. The total body water at different ages is given in figure 19.1. It can be seen that the proportion of water is highest in infancy and that, in adult life, women have a smaller proportion of water than men, presumably because of their higher fat content.

The principal movements of water between the various compartments are summarised in figure 19.2. Some of these exchanges are very large. For example the glomeruli normally filter about 200 litres daily and almost all is reabsorbed. The intestinal secretions amount to about 8 litres daily and almost all is reabsorbed. Disorders which impair these reabsorptive processes rapidly lead to severe clinical disturbances.

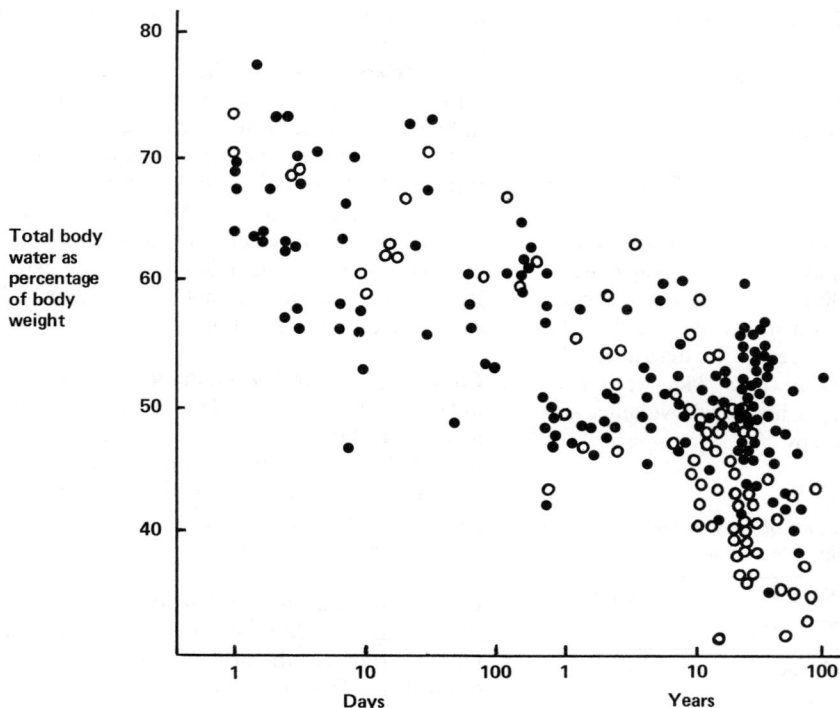

Fig. 19.1 Total body water at different ages. ● = males, ○ = females. (After Crane, C. W. (1978) In *Recent Advances in Clinical Biochemistry 1*, ed. Alberti, K. G. M. M., p. 175. Edinburgh: Churchill Livingstone)

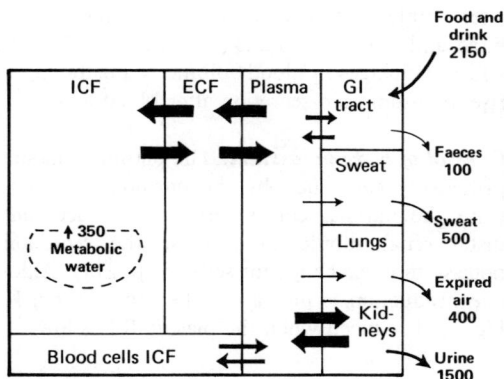

Fig. 19.2 Movements of water between different compartments and between the plasma and the exterior. The figures given are appropriate for a sedentary adult in a temperate climate. The losses of water as sweat are much higher in hot climates

Gains. The body gains water from the food and drink in the diet, and also from the metabolism of nutrients. For example the oxidation of 1 mol (284 g) of stearic acid yields 18 mol (324 g) of water. The use of proteins as a source of energy leads to the production of urea which requires water for its excretion.

Losses. A minimum volume of urine must be passed to carry the end-products of metabolism. On an average diet this is about 600 ml daily. A further 100 to 150 ml of water is lost in the faeces. Sweating amounts to about 500 ml daily in a temperate climate in a sedentary subject; in hot climates or with severe exertion the sweat volume can be as high as 10 litres per day. Water loss in urine and sweat is accompanied by loss of sodium. Water is lost by evaporation from the alveoli of the lungs. At rest the loss of water from the lungs amounts to 300 to 400 ml per day;

229

subjects who are breathless for any reason may lose much more.

Sodium and potassium balance

The sodium and potassium losses in urine, faeces and sweat for a sedentary adult in a temperate climate are given in Table 19.3. The losses in the sweat are much higher in hot climates and in subjects doing heavy manual work. After acclimatisation to a hot climate the concentration of sodium in the sweat falls. The losses in the faeces are increased in diarrhoea; the urinary losses are increased in a wide variety of disorders characterised by a high urine volume, and as a result of diuretic therapy.

Table 19.3   Intake and output of sodium and potassium by a normal adult in a temperate climate

	Sodium (mmol/day)	Potassium (mmol/day)
Intake in diet	130–300	65–140
Output		
Urine	100–250	40–120
Faeces	5	12
Sweat	10–20	5

The figures give no indication of the very large internal movements of sodium and potassium in the alimentary tract and in the kidney. For example some 600 mmol/day of sodium and 50 mmol/day of potassium are normally secreted into the gut lumen each day in the bile, the pancreatic juice and the gastric and intestinal secretions. Almost all is reabsorbed.

Control of plasma volume and composition

Several interconnected mechanisms are involved in the control of the plasma volume and of the concentration of sodium and potassium in the plasma and the ECF.

Thirst. The sensation of thirst begins soon after a person is deprived of water. At its height thirst is a far more distressing symptom than hunger; it is not caused by dryness of the mouth since diseases or drugs which impair the production of saliva do not cause thirst.

Thirst results from either an increase in plasma osmolality or from a decrease in blood volume. A fall in blood volume leads to a fall in renal perfusion and so to renin release from the juxtaglomerular cells. In turn this leads to the production of angiotensin II as shown in figure 19.3. This has a direct action on a vascular part of the brain, the *sub-fornical organ*. The mechanism of the production of thirst by hypertonicity of the plasma is not yet understood; the sub-fornical organ is not involved.

Control of aldosterone production. Figure 19.3 also indicates the role of a fall of plasma volume in increasing aldosterone production and thus sodium reabsorption in the distal tubule. The restoration of the volume of the ECF is brought about by the passive absorption of water which accompanies sodium reabsorption.

Control of plasma osmolality. The secretion of vasopressin (antidiuretic hormone) by the posterior pituitary provides the principal control of the plasma osmolality. Two types of receptor are involved: *osmoreceptors*, sensitive to plasma osmolality, in the hypothalamus, and *baroreceptors*, sensitive to blood volume, in the large blood vessels in the thorax. Vasopressin is secreted continuously. When the plasma osmolality rises, because for example of excessive sweating, vasopressin secretion is increased. When plasma osmolality falls, for example after drinking water, vasopressin secretion is inhibited and a dilute urine is produced (fig. 19.4). The effect of changes in blood volume is illustrated by the rise in vasopressin secretion after blood loss.

Control of body potassium. The plasma potassium is principally controlled by the hormone aldosterone, from the adrenal cortex. Aldosterone acts on the distal renal tubule to increase the secretion of potassium in exchange for sodium (p. 219). Aldosterone production is increased when the plasma K^+ is high and reduced when the plasma K^+ is low.

Disorders of salt and water balance

The most common disorders of salt and water balance are those in which salt and water are lost or gained in approximately equivalent amounts. These

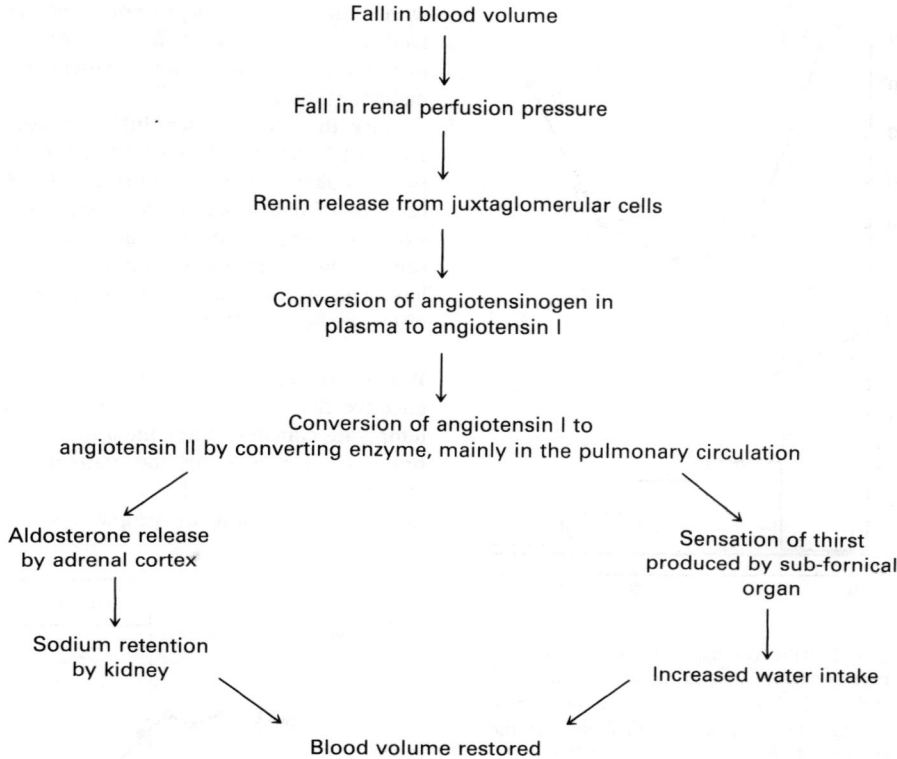

Fig. 19.3 Role of renin and angiotensin in the control of blood volume

are often called 'saline depletion' and 'saline excess'. The less common disorders, water intoxication and water depletion, are caused by changes in body water with little change in the body sodium. Disturbances of the body sodium without changes in body water are very uncommon.

Saline excess. The primary defect in saline excess is excessive renal retention of sodium. In turn the plasma osmolality rises and, because of thirst and vasopressin production, water is also retained. The plasma sodium level is frequently normal in saline excess but may be high or low. Patients with saline excess characteristically have oedema, some may have an increased blood pressure.

Saline excess occurs typically in heart failure, in the nephrotic syndrome and in some cases of renal failure. The mechanism of saline excess in heart failure is not understood but the increased venous pressure, leading to accumulation of extracellular

fluid (p. 5), is a factor. In the *nephrotic syndrome* oedema accumulates because the low plasma albumin levels lead to movement of fluid from the plasma into the interstitial space. The fall in plasma volume stimulates renin and angiotensin production and so increases renal sodium retention.

Saline depletion. The kidneys are normally able to regulate sodium reabsorption so that the amount excreted can be as high as 100 mmol/day or as low as 1 mmol/day. Loss of the ability to retain sodium is the characteristic feature of saline depletion. Saline depletion occurs therefore in a variety of disorders that affect principally the renal tubules; these include renal damage due to recurrent infections (pyelonephritis), chronic obstruction, hypercalcaemia and renal damage due to the drug phenacetin. Saline depletion occurs more commonly when loss of saline from the gut in diarrhoea, or from the skin in burns, exceeds the ability of the kidneys to conserve sodium.

231

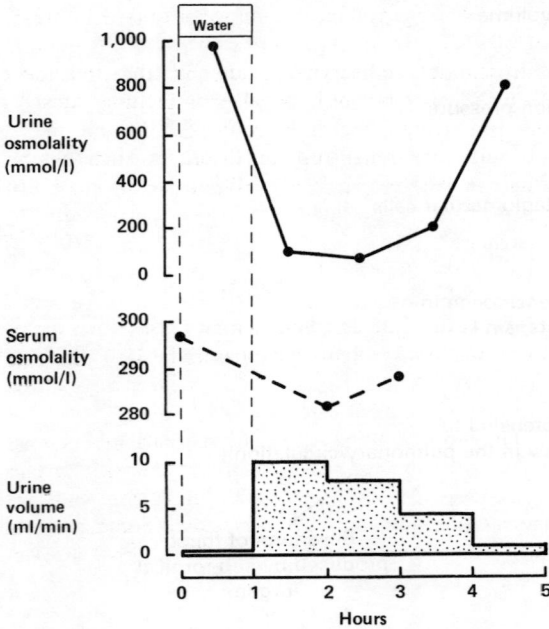

Fig. 19.4 Changes in urine volume and serum and urine osmolality in a normal subject after drinking 1375 ml water during the period marked by the vertical dotted lines. Contrast the very large change in urine osmolality with the very small change in serum osmolality. (After Bartter, F. C. (1970) *Journal of the Royal College of Physicians of London* 4, 264)

Since the cells contain little sodium the depletion is borne by the extracellular fluid. The patient shows weight loss, muscle cramps, diminished skin turgor and ocular pressure, a dry tongue and especially when standing, a low blood pressure. It is important to recognise that, since the depletion of sodium and of water is often approximately equivalent, the plasma sodium concentration may be normal.

Treatment is by giving extra salt and water by mouth, or intravenously.

Water excess. Water excess is uncommon but may occur in psychiatric patients if they drink more than 20 litres of water daily, which is more than the kidneys can secrete. More commonly water intoxication follows excessive intravenous administration of isotonic dextrose (glucose) soon after operations, when vasopressin secretion is usually increased. Water intoxication also occurs in patients with excessive secretion of vasopressin or of substances

with similar actions (*inappropriate antidiuretic hormone syndrome*). This occurs in a number of different disorders, most commonly in patients suffering from malignant disease.

Since the excess water diffuses readily into cells, the load is spread throughout the body. Each cell swells; changes in the volume of cells of the central nervous system cause nausea, vomiting, fits, confusion and ultimately death. The concentration of most solutes, but particularly sodium, in the plasma is low. The disorder can be corrected by restricting the water intake (fig. 19.5).

Water depletion. The thirst mechanism is so effective that water depletion is almost unknown in temperate climates unless the appropriate part of the brain is damaged or the patient is too ill to communicate a desire for water. Water depletion is the cause of death in people lost in deserts or

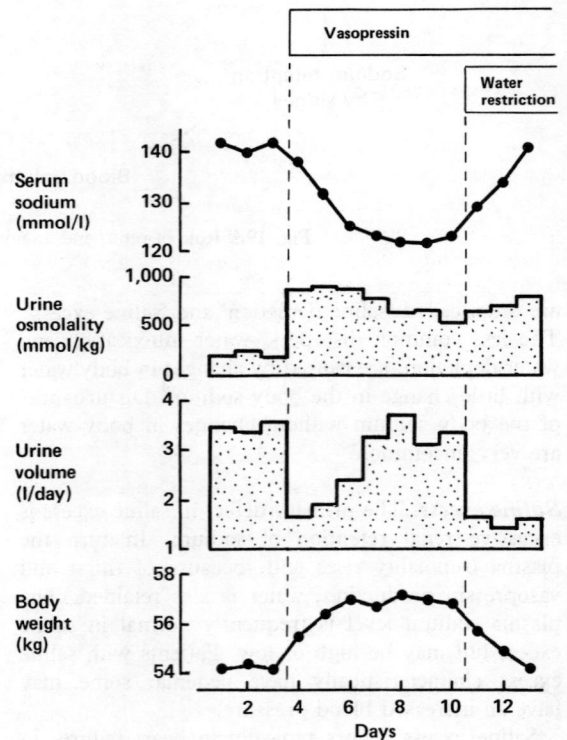

Fig. 19.5 Effect in normal subjects of (a) continuous administration of vasopressin and (b) restriction of fluid intake while vasopressin continues to be given. (After Goldberg, M. (1963) *Medical Clinics of North America* 47, 915)

Respiratory disposal of hydrogen ions

A normal adult at rest produces about 200 ml CO_2 each minute and this amount must be excreted by the lungs. In the ECF and in the plasma the dissolved CO_2 is in equilibrium with bicarbonate:

$$CO_2 + H_2O \rightleftharpoons H_2CO_3 \rightleftharpoons H^+ + HCO_3^-$$

In the tissues this reaction is driven to the right with a fall in the pH of the ECF. In the lungs the reaction is driven to the left. The rate of disposal of CO_2, and hence the loss of bicarbonate, is related to the rate and depth of respiration. Changes in pH alter these by direct effects on the respiratory centre in the medulla. The lungs thus provide a rapidly responding system which minimises fluctuations in the pH of the ECF.

Renal disposal of hydrogen ions

In the longer term control of the pH of the ECF is carried out by the cells of the renal tubules. Although this regulatory function deals with major fluctuations of pH it is slow, taking several hours to be effective.

While the glomerular filtrate has the same pH as the plasma, the urine is for most of the time more acidic with a pH as low as 5.0. The difference between glomerular filtrate and urine is accounted for by three mechanisms which, in effect, pump H^+ into the lumen of the tubules in exchange for Na^+ (fig. 19.6).

In a normal adult on an average diet 30 to 50 mmol of ammonia are produced daily. In exceptional circumstances this amount can be increased to as much as 500 mmol daily. However the stimulation of the synthesis of the enzymes involved in ammonia production (glutaminase and glutamate dehydrogenase) is a slow process so that the full renal response to an increased acid load takes several days.

This adaptive mechanism can be illustrated by the response to chronic respiratory insufficiency. If the ability of the lungs to dispose of CO_2 is impaired, the P_{CO_2} in the extracellular fluid and thus that within the tubular cell rises. In turn this leads to a rise in H^+ concentration and an increase in H^+ excretion. Most of the extra H^+ ions are excreted as NH_4^+; ammonia production increases markedly in chronic respiratory insufficiency.

Assessment of acid–base status

Current electrode methods allow the determination of pH and P_{CO_2} on a sample of blood less than 100 μl in as little as five minutes. The sample is most commonly obtained from an artery. Alternatively, worthwhile results can be obtained by skin puncture of a warm extremity so that the 'capillary sample' is largely arterial in composition. The second method is that often used for babies.

The pH indicates whether a patient has an excess or deficit of free H^+ ions—*acidaemia* or *alkalaemia*. The P_{CO_2} level in arterial blood is a measure of the respiratory component of the acid–base balance. The P_{CO_2} reflects the rate of removal of CO_2 by the lungs; high levels are found in chronic obstructive disease of the airways and low levels result from overbreathing.

There is no direct measurement to reflect the metabolic component of an acid–base disorder. In the equation

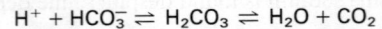

$$H^+ + HCO_3^- \rightleftharpoons H_2CO_3 \rightleftharpoons H_2O + CO_2$$

it is clear that the HCO_3^- level depends in part on the P_{CO_2}. However a measure of the metabolic component is provided by the *base excess* which is defined as the base concentration, as measured by titration to pH 7.40, at a P_{CO_2} of 40 mmHg (5.4 kPa). In practice base excess can be estimated from pH and P_{CO_2} by using standard tables or nomograms; the normal range is -3 to $+3$ mmol/l.

Disorders of acid–base balance

Metabolic acidaemia. This can be caused either by an excess of H^+ ions or by excessive losses of base. The most common causes of H^+ excess are lactic acidosis (p. 95), diabetic ketosis (p. 144) and renal failure in which acids such as H_2SO_4 and H_3PO_4 accumulate in excess. Loss of HCO_3^- occurs in diarrhoea since the fluids of the small intestine are rich in bicarbonate.

In a patient who is otherwise normal metabolic acidaemia leads to stimulation of the respiratory centre and an increase in the rate and depth of respiration, *acidotic breathing* or *Kussmaul breathing*. This leads to partial correction of the abnormality in pH at the expense of a fall in P_{CO_2}.

Metabolic alkalaemia. This may be caused by loss of H^+ ions or an increase in HCO_3^- ions. Loss of H^+

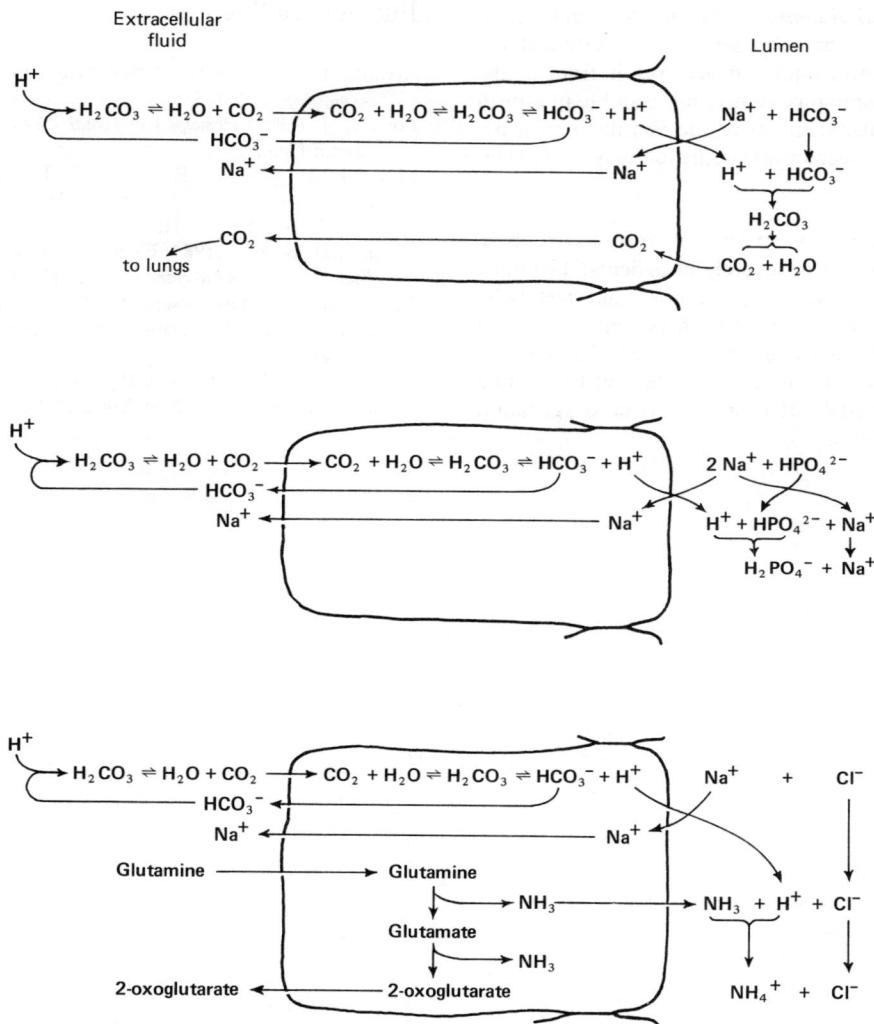

Fig. 19.6 Three processes by which the cells of the renal tubule secrete hydrogen ions into the lumen of the tubule. Carbonic anhydrase in the cell plays an important part in all three processes

ions occurs in patients who are vomiting or whose gastric secretions are being regularly aspirated. Increased HCO_3^- can result from excessive administration of alkalis. Metabolic alkalaemia is also caused by potassium depletion (p. 233); H^+ ions move into cells in replacement for the K^+ ions.

Respiratory acidaemia. A high P_{CO_2} and subsequently acidaemia result from a variety of disorders which affect the transfer of CO_2 from the blood to the expired air. The most common causes are pneumonia, chronic obstructive airways disease ('chronic bronchitis'), paralysis of the respiratory muscles, and drugs, poisons or neurological disorders which affect the respiratory centre.

In acute *asphyxia* the condition is uncompensated but on most occasions respiratory acidaemia is sufficiently chronic for renal compensation, as described above, to restore the pH towards normal at the expense of a rise in plasma HCO_3^-.

235

Respiratory alkalaemia. A fall in P_{CO_2} most commonly results from overbreathing in hysterical behaviour. Overbreathing can also result from a disorder of the respiratory centre, for example in aspirin poisoning. Maladjustment of the ventilator in a patient receiving mechanical ventilation may cause alkalaemia.

Mixed disorders. A variety of mixed acid–base disorders can occur in severely ill patients. The most common combination is respiratory and metabolic acidosis which occur together in patients who have had both respiratory and cardiac arrest. The respiratory failure leads to the accumulation of CO_2 while the cardiac arrest leads to loss of tissue oxygenation and lactic acidosis.

Further reading

Arruda, J. A. L. (ed.) (1981) Acid–base. *Seminars in Nephrology* **1,** No. 3

Lote, C. J. (1982) *Principles of Renal Physiology.* London: Croom Helm

Maxwell, M. H. and Kleeman, C. R. (1980) *Clinical Disorders of Fluid and Electrolyte Metabolism,* 3rd edn. New York: McGraw-Hill

Morgan, D. B. (ed.) (1984) Electrolyte disorders. *Clinics in Endocrinology and Metabolism* **13,** 233–434

Porter, R. and Lawrenson, G. (eds) (1982) *Metabolic Acidosis* (Ciba Foundation Symposium 87). London: Pitman

Rose, B. D. (1980) *Clinical Physiology of Acid–Base and Electrolyte Disorders.* New York: McGraw-Hill

CRP DBW

20 Calcium, phosphorus & magnesium

SOME 99 per cent of the body's calcium is in bone but calcium has many other important functions. In particular it acts as the 'second messenger' in the response by cells to certain hormones and it plays an essential part in neuromuscular transmission and muscle contraction.

About 88 per cent of the body's phosphorus is in bone and most of the remainder, in the form of organic phosphates, phospholipids or nucleic acids, is important for the structure and function of cells.

About half of the body's magnesium is in bone, almost all the rest is within cells where it is the principal divalent cation. It is an essential component of many enzymes and a co-factor in reactions involving ATP. Magnesium is also essential for the stability of intracellular organelles, such as ribosomes and mitochondria.

Bone composition

Dried bone contains about 25 per cent of organic material and 75 per cent of inorganic substances. The principal organic component is collagen (Chap. 17); others include glycosaminoglycans, glycoproteins and lipids which are together sometimes known as the 'ground substance' (fig. 20.1). The inorganic component of bone is hydroxyapatite, $Ca_{10}(PO_4)_6(OH)_2$ (fig. 20.2). Various ions, notably sodium, magnesium, potassium, fluoride, chloride, bicarbonate and citrate are substituted into the hydroxyapatite crystal. Lead and strontium, if present in the environment, may be ingested and then taken up by hydroxyapatite.

Bone owes its strength to both of its main components: the hydroxyapatite crystals giving rigidity and strength particularly in compression, while the three-dimensional lattice-work of collagen fibres gives elasticity. Compact bone is nearly as strong as cast iron but is much lighter and more flexible (Table 20.1).

A bone may be abnormally fragile either when the

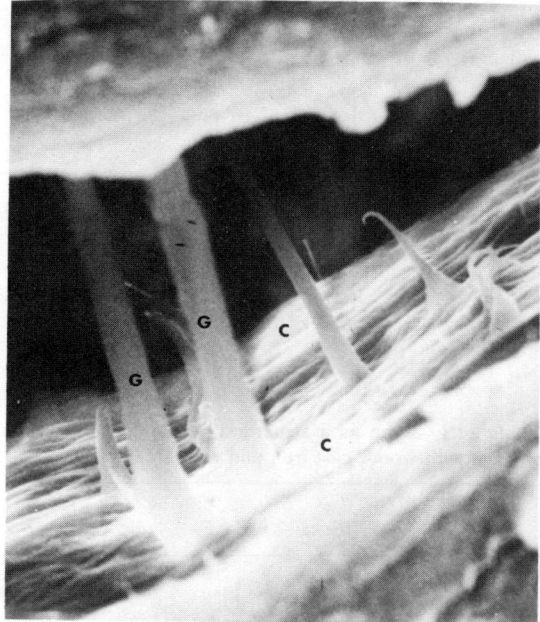

Fig. 20.1 Ground substance (G) and collagen bundles (C) seen by scanning electron microscopy in a partly disrupted fragment of human cortical bone. (From Frasca, P. (1981) *Acta Anatomica* **109**, 115, by courtesy of author and S. Karger AG, Basel)

amount of bone is abnormally low (*osteoporosis*) or when the bone material is not fully calcified (*osteomalacia*, due to vitamin D deficiency) or when the collagen is abnormal (in *osteogenesis imperfecta*, fig. 17.4 p. 212).

Calcium metabolism

The body of a young adult contains about 1.2 kg of calcium and more than 99 per cent of this is in bone.

237

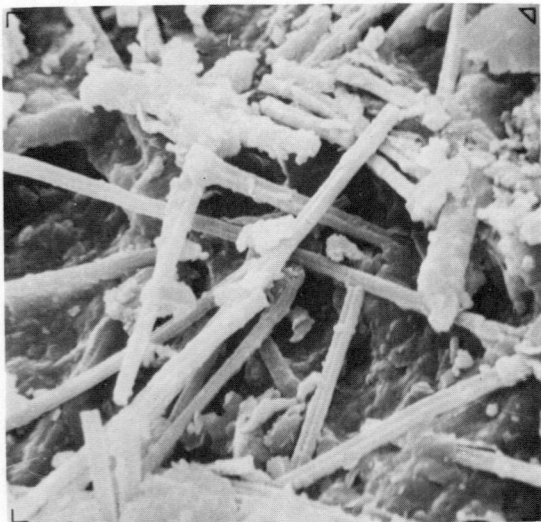

Fig. 20.2 Crystals of hydroxyapatite from human bone as seen by scanning electron microscopy. (From Mongiorgi, R. and Krajewski A. (1981) *Biomaterials* **2**, 147, by courtesy of authors and publisher)

Table 20.1 Bone as a structural material. Comparison of bone with wood, cast iron and mild steel

	Bone	Wood	Cast iron	Mild steel
Breaking stress on bending (kg/mm^2)	20	7	28	50
Breaking stress on twisting (kg/mm^2)	6	0.7	14	25
Young's modulus (kg/mm^2)	1100	1000	9000	20 000
Density (kg/m^3)	2000	600	8000	8000

Data of G. H. Bell (1970) Living bone as an engineering material. *Advancement of Science* **6**, 1–11.

In older women and, to a smaller extent, older men a progressive loss of bone and therefore of calcium takes place with increasing age.

Calcium requirements

In most western societies the daily calcium intake is around 1000 mg (25 mmol). It is derived mainly from milk and milk products, flour, which often contains added calcium, and vegetables. In Japan and some developing countries the daily calcium intake may be as low as 200 to 500 mg per day. There is little evidence that such calcium intakes do any harm in otherwise healthy people or that calcium supplements would be desirable in developing countries. In particular there is no evidence that the bones or teeth are defective in such communities. The intestinal absorption of calcium in healthy people increases in response to a fall in the dietary calcium (see below). It is debatable whether calcium lack plays a part in the bone loss in some old people, and also whether calcium deficiency can, in exceptional circumstances, contribute to a form of rickets seen in children in South Africa.

Calcium absorption

Calcium is partly absorbed by passive diffusion which takes place in all parts of the small intestine and also by an active transport mechanism whose capacity is greatest in the duodenum. This active transport requires the presence of the hormone calcitriol (p. 245) which acts by stimulating the production of a calcium-binding protein. In a normal adult in balance a large proportion of the dietary calcium is lost in the faeces but the proportion absorbed increases in response to increased needs as in growth, pregnancy and lactation. These adaptations probably depend on parathyroid hormone and calcitriol. Calcium absorption is greatly reduced in vitamin D deficiency and in renal failure in which the production of calcitriol is impaired.

Plasma calcium

Calcium is constantly being exchanged between the plasma and the bones, the gut and the kidney (fig. 20.3) but the plasma calcium remains remarkably constant in normal subjects. The physiological mechanisms underlying this homeostasis are described below.

The total plasma calcium is normally maintained within the range 2.2 to 2.6 mmol/l (8.8 to 10.4 mg/dl). Just under half of this is ionic calcium and most of the remainder is bound to plasma proteins, particularly albumin. A small amount is complexed with citrate, phosphate and other anions. It is the ionic calcium which is of physiological importance,

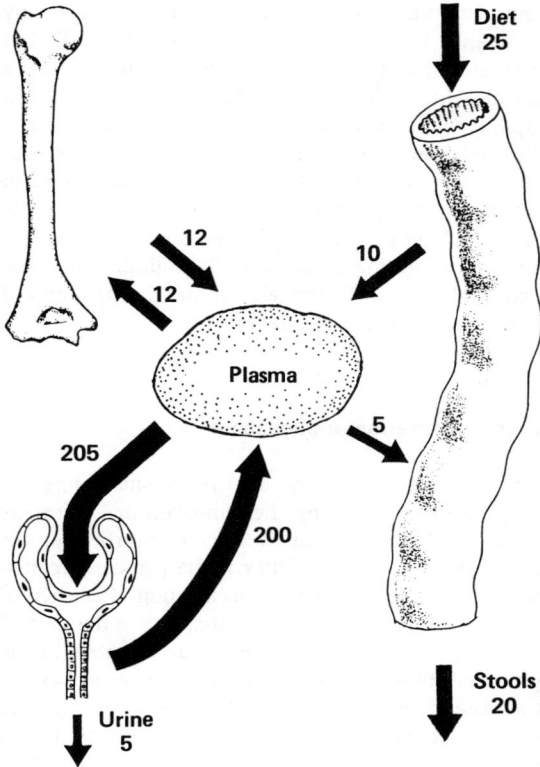

Fig. 20.3 Calcium exchanges (in mmol/day) between the plasma and the tissues in an adult

Urinary excretion of calcium

Approximately 225 mmol (9 g) of calcium, derived from the diffusible fractions in the plasma (ionic and complexed calcium), are filtered through the glomeruli each day. About 90 per cent of the calcium in the glomerular filtrate is ionised and most of this is reabsorbed in the tubules. Only about 20 per cent of the total calcium in the urine is ionised, the remainder being complexed with a variety of anions, such as sulphate, citrate and gluconate.

The renal handling of calcium can be assessed in two ways. One is by measuring the urinary calcium during a 24-hour period. Typical results for normal adults are shown in figure 20.4. It is difficult to define the upper limit of the normal range but conventionally it is taken as 10 mmol (400 mg) per day for males and 8 mmol (320 mg) per day for females. High values for calcium excretion are found in most patients with hypercalcaemia of any cause and also, in the presence of a normal plasma calcium, in patients with a poorly understood disorder, *idiopathic hypercalciuria*. This is thought to be caused by increased calcium absorption in the gut and may contribute to the formation of calcium-containing stones in the urinary tract.

A more helpful measure of calcium handling by the renal tubule is obtained by a number of 'indices' the most widely used being the *calcium excretion per*

its level being controlled very closely by the hormones described below. However it is difficult in practice to determine the ionic calcium level reliably in routine laboratories; usually it is the total calcium which is measured routinely, but in interpreting such results allowance must be made for variations in the plasma albumin level. For example the level of total calcium in the plasma is low in patients who have a low plasma albumin for any reason (p. 20). The binding of calcium to albumin is also affected by the pH of the plasma: the proportion bound is increased in alkalaemia and decreased in acidaemia.

The plasma calcium is increased in several disorders including parathyroid overactivity, vitamin D poisoning and malignant disease causing excessive bone destruction. The plasma calcium may be low in vitamin D deficiency, in parathyroid insufficiency and in chronic renal failure.

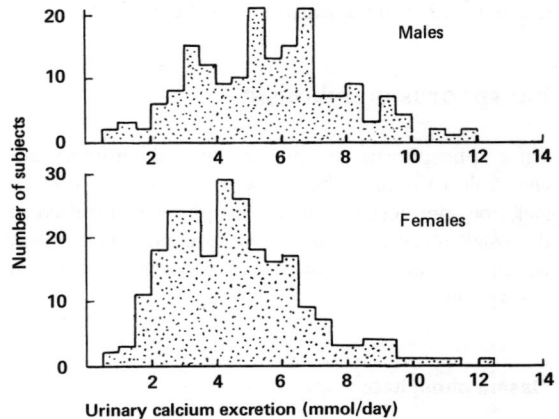

Fig. 20.4 Urinary excretion of calcium in 178 normal males and 240 normal females between the ages of 20 and 69 (After Bulusu, L. *et al.* (1970) *Clinical Science* **38**, 601, and Davis, R. H. *et al.* (1970) *Clinical Science* **39**, 1)

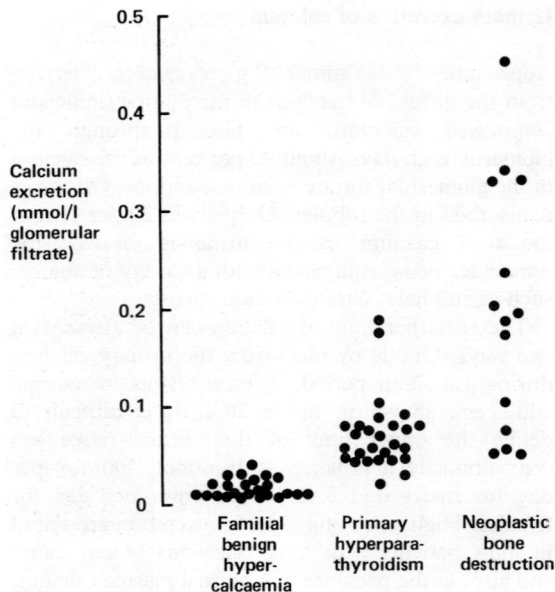

Fig. 20.5 Calcium excretion in urine, expressed as mmol per litre of glomerular filtrate, in three groups of patients with high plasma calcium levels. Familial benign hypercalcaemia is an inherited disorder in which hypercalcaemia is accompanied by an unusually low calcium excretion

ml of glomerular filtrate. This can be calculated from measurements of calcium and creatinine in simultaneous samples of urine and serum. The value of this determination is illustrated in figure 20.5.

Phosphorus metabolism

Since phosphorus is present in all animal and vegetable cells phosphorus deficiency due to dietary lack does not occur in man. Phosphate is absorbed in the small intestine by an active transport mechanism which is independent of that responsible for calcium absorption.

Plasma phosphate

Approximately three-quarters of the total plasma phosphorus is bound in organic molecules, particularly phospholipids. The inorganic phosphate concentration in the plasma is the fraction that is routinely measured. In adults this is normally between 0.8 and 1.4 mmol/l (2.5 to 4.5 mg/dl) but the value may be higher after meals. Higher values are also found in children, particularly at times of rapid growth. Almost all the inorganic phosphate is diffusible; only about 0.2 mmol/l is bound to protein.

Plasma inorganic phosphate levels are low in patients with phosphate depletion due to renal tubular abnormalities or after excessive ingestion of phosphate-binding antacids, such as aluminium hydroxide. Low values are also found in parathyroid overactivity and in renal failure.

Urinary excretion of phosphate

The diffusible fraction of the plasma inorganic phosphate is filtered by the glomeruli and normally more than 80 per cent of it is reabsorbed in the proximal tubule. The kidney is the principal regulator of plasma phosphate concentration. Parathyroid hormone (p. 242) plays an essential role in the control of phosphate reabsorption by its action both in the proximal tubule and in the terminal nephron beyond the distal convoluted tubule. There is no evidence of tubular secretion. Phosphate excretion can be varied very widely and rapidly. Within one day of dietary phosphate restriction, the urine may be free of phosphate whereas during a period of high phosphate intake almost all of the filtered phosphate load is excreted.

The tubular handling of phosphate is best expressed as the theoretical renal tubular threshhold for phosphate which is the ratio

$$\frac{\text{maximum tubular reabsorption of phosphate}}{\text{glomerular filtration rate}}$$

(TmP/GFR) (fig. 20.6).

Measurements of TmP/GFR are useful both in experimental studies and as an indication of the levels of parathyroid hormone (fig. 20.7).

Pyrophosphate

Pyrophosphate (PPi) is formed by a combination of two molecules of phosphate with a loss of one molecule of water:

$$2H_3PO_4 \rightarrow H_4P_2O_7 + H_2O$$

240

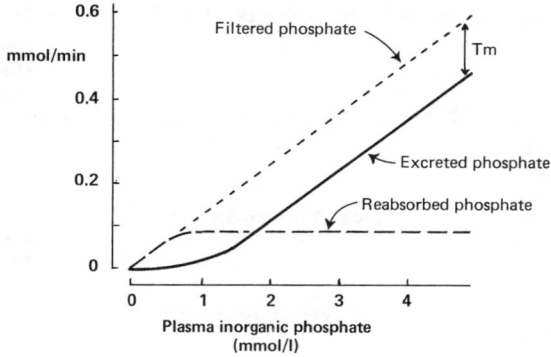

Fig. 20.6 The relationship between phosphate excretion in the urine in a normal adult male, fasting and during an intravenous infusion of phosphate. (After Bijvoet, O. L. M. and Morgan, D. B. (1971) In *Phosphate et Metabolisme Phosphocalcique*, Paris: Sandoz)

Pyrophosphate bonds are of course present in many substances of biochemical importance including ADP, ATP and uridine diphosphate glucose (UDPG). Substances such as these are probably the main biological sources of inorganic pyrophosphate.

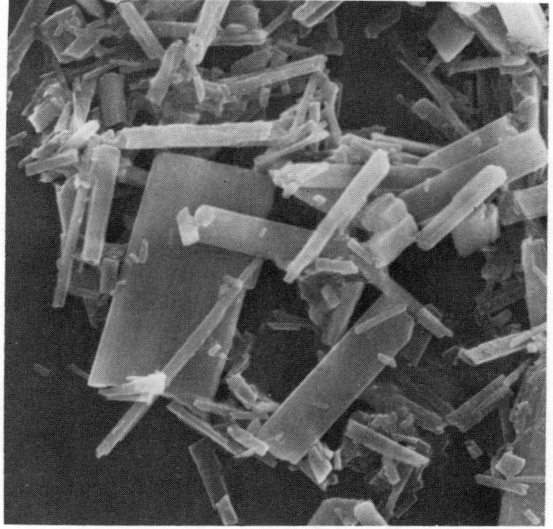

Fig. 20.8 Crystals of calcium pyrophosphate from a patient with pseudogout. (From Leisen, J. C. C. *et al.* (1980) *Journal of the American Medical Association* **244,** 1711, by courtesy of author and the American Medical Association)

Plasma levels of pyrophosphate are in the range 1 to 6 μmol/l but considerably higher values are found in serum; the extra pyrophosphate is probably derived from platelets but the function of pyrophosphate in platelets is unknown. Pyrophosphate is also found on bone surfaces where it appears to have a role in inhibiting bone deposition. While, in theory, pyrophosphate might be used to prevent inappropriate calcification it is so rapidly hydrolysed when administered that it cannot be used in this way. However a group of substances with a P–C–P core, the *bisphosphonates* (diphosphonates) which are analogues of pyrophosphate, are not broken down by phosphatases. These drugs, such as sodium etidronate (EHDP), are of value in the control of excessive bone turnover in certain diseases. A rare joint disorder, *pseudogout*, is characterised by the deposition of crystals of calcium pyrophosphate in joints (fig. 20.8).

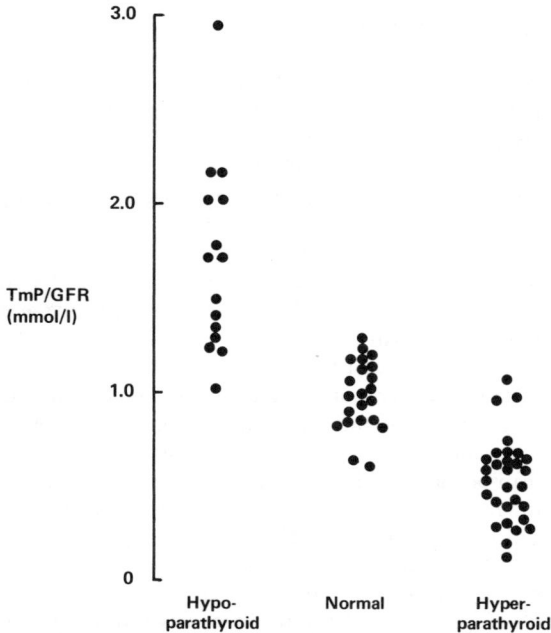

Fig. 20.7 Effect of hypoparathyroidism and hyperparathyroidism on the renal tubular handling of phosphate. (Courtesy of J. A. Gibb and J. Evans)

Control of the plasma calcium

Constant plasma calcium levels, particularly constant ionic calcium levels, are essential for normal neural

241

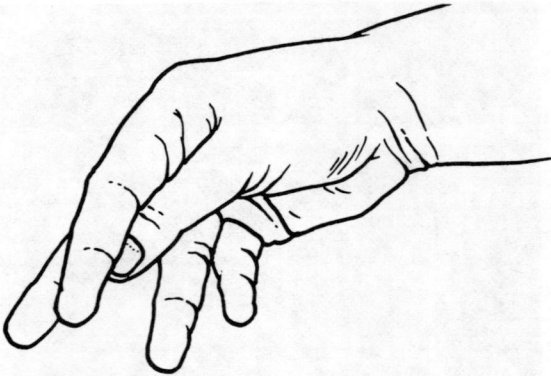

Fig. 20.9 Position of hand in tetany due to hypocalcaemia. (From Paterson, C. R. (1975) *Metabolic Disorders of Bone*. Oxford: Blackwell)

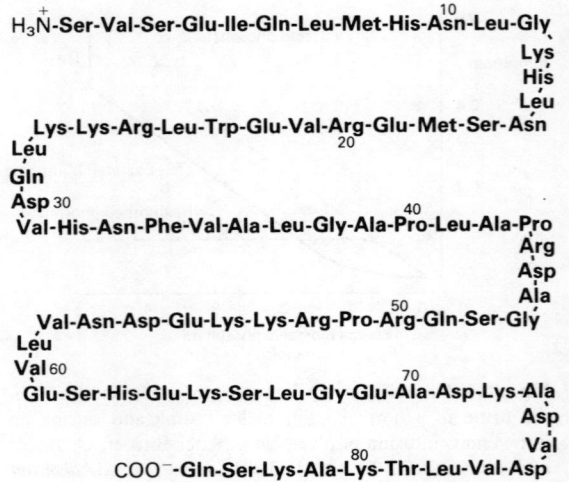

Fig. 20.10 Amino acid sequence of human parathyroid hormone (After Keutman, H. T. *et al.* (1978) *Biochemistry* 17, 5723)

and muscular function and for the activity of several enzymes. Patients with low plasma calcium values may develop spontaneous neuro-muscular activity leading to spontaneous muscle contraction (*tetany*, fig. 20.9), and abnormal sensory nerve activity leading to paraesthesiae ('pins and needles' sensations). Psychiatric disturbances, particularly irritability and anxiety, may occur. A high plasma calcium value may also lead to psychiatric disturbances and, if very high, changes in cardiac function.

The plasma calcium is controlled by the integrated action of three hormones, parathyroid hormone, calcitonin and calcitriol (1,25-dihydroxycholecalciferol).

Parathyroid hormone

Parathyroid hormone (parathyrin, PTH) is an 84-amino acid peptide produced by the parathyroid glands in the neck. The first step in its formation is the synthesis of a 115-amino acid molecule, pre-pro-PTH, by ribosomes of the rough endoplasmic reticulum. This is converted there to pro-PTH, a peptide with 90 amino acids, which is in turn split to give PTH (fig. 20.10) in the Golgi zone. PTH leaves the Golgi zone in secretory granules which are transported to the periphery of the cell. These are released into the extracellular fluid in response to a fall in the ECF calcium concentration; a fall of as little as 0.03 mmol/l is sufficient to stimulate PTH release. While the rapid response to hypocalcaemia is

due to PTH release, hypocalcaemia also stimulates PTH production probably by increasing the number of mRNA molecules produced.

Actions. PTH has direct actions on bone and on the kidney, and operates through activation of adenylate cyclase and the production of cyclic AMP. The biological activity of PTH resides in the amino-terminal portion and the minimum sequence for any biological activity is residues 2 to 27. Residues 1 to 34 are needed for full biological action and analogues of this sequence are now proving valuable in research since some have proved to be even more active and more stable than PTH itself.

PTH has at least three actions on the kidney. The first to be recognised was an inhibition of the reabsorption of phosphate in the proximal tubule. Associated with this is an inhibition of sodium reabsorption. PTH also has an effect on the distal tubule where it increases the reabsorption of calcium and magnesium. This effect of PTH is probably the most important way in which the hormone, when secreted in response to hypocalcaemia, increases the plasma calcium. The third direct action of PTH is to promote the formation of the metabolite calcitriol (see below) from 25-hydroxyvitamin D. This in turn promotes calcium absorption in the gut.

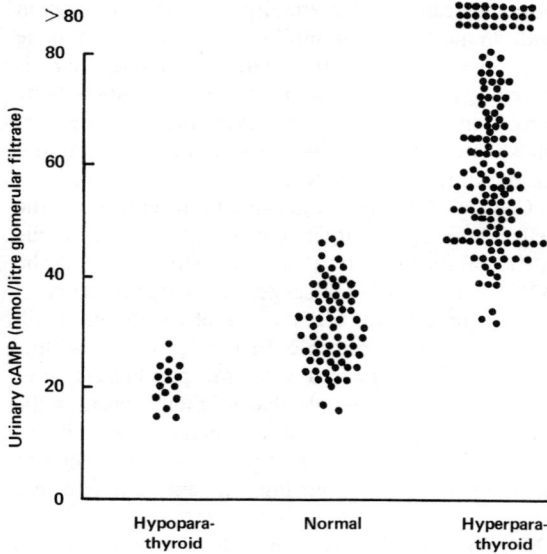

Fig. 20.11 Urinary excretion of cyclic AMP in normal subjects and in patients with hypoparathyroidism and hyperparathyroidism. (After Broadus, A. E. (1981) *Recent Progress in Hormone Research* **37,** 667)

PTH increases bone resorption both by osteoclasts at the bone surfaces and by osteocytes. In patients with parathyroid overactivity bone histology may show large numbers of active osteoclasts and also enlarged osteocyte lacunae.

In all its sites of action PTH action is mediated by adenylate cyclase and one measurable response to PTH is an increase in the urinary excretion of cyclic AMP (fig. 20.11).

Disorders. Parathyroid overactivity (*hyperparathyroidism*) is characterised by hypercalcaemia. This is due mainly to increased calcium reabsorption in the renal tubule but also to increased calcium absorption in the intestine and, particularly in the more severe cases, to increased calcium resorption from bone. A few patients have bone pain or fractures. A large number have renal calculi but nowadays most patients are identified as a result of screening procedures; some have symptoms related to the hypercalcaemia itself, particularly thirst, an increased urine volume (*polyuria*), nausea and loss of appetite.

Apart from hyperparathyroidism, hypercalcaemia can be caused by neoplastic disease, either as a result of destruction of bone by metastatic deposits or because of the production by the tumour of substances which mimic the action of PTH. Vitamin D overdosage also leads to hypercalcaemia and permanent renal damage may result. Recently a benign inherited disorder, familial benign hypercalcaemia, has been identified, the primary abnormality being an excessive calcium reabsorption in the renal tubules (fig. 20.5).

Decreased parathyroid activity (*hypoparathyroidism*) is most commonly caused by damage to the glands or their blood supply during operations on the neck. While some patients have no symptoms referable to their hypocalcaemia others have tetany or psychiatric disturbances.

Renal failure and vitamin D deficiency may also cause hypocalcaemia. In both the fundamental defect is lack of calcitriol (see below). A rare cause of hypocalcaemia is an inborn abnormality of the adenylate cyclase response to PTH, *pseudohypoparathyroidism*. In this disorder production of endogenous PTH is normal or increased but administered PTH causes smaller rises in phosphate excretion or urinary cyclic AMP than normal.

Calcitonin

Calcitonin is produced by the parafollicular cells (C cells) within the thyroid gland. Like many peptide hormones calcitonin is initially formed as a precursor molecule which is split at both carboxy- and amino-terminals to yield the main secretory product of the cells, calcitonin monomer. This has 32 amino acid residues and a molecular mass of 3500 (fig. 20.12).

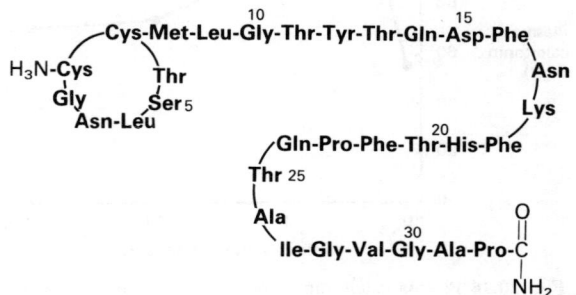

Fig. 20.12 Amino acid sequence of human calcitonin

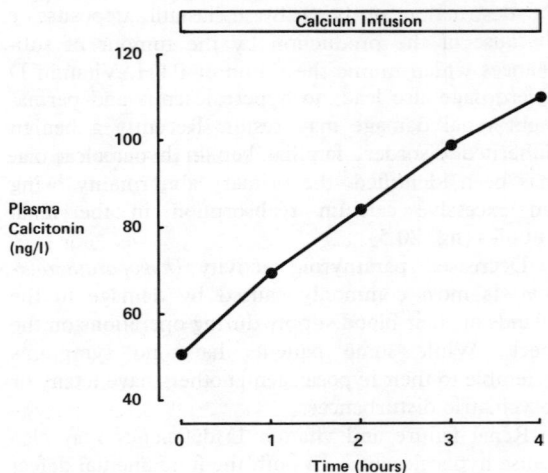

Fig. 20.13 Plasma calcitonin levels in 18 normal men who were given a constant infusion of calcium gluconate with a total dose of 15 mg Ca/kg. (Courtesy of H. Heath)

Plasma levels of calcitonin measured with the best immuno-assays currently available are about 50 pg/ml for men (15 pmol/l) and for women 30 pg/ml (10 pmol/l). A marked rise in plasma calcitonin levels follows the infusion of calcium (fig. 20.13) or the administration of certain gastrointestinal hormones notably gastrin (fig. 20.14). The response is smaller in women than in men.

Actions. While calcitonin in large doses has many actions it is not yet clear if any of these are important

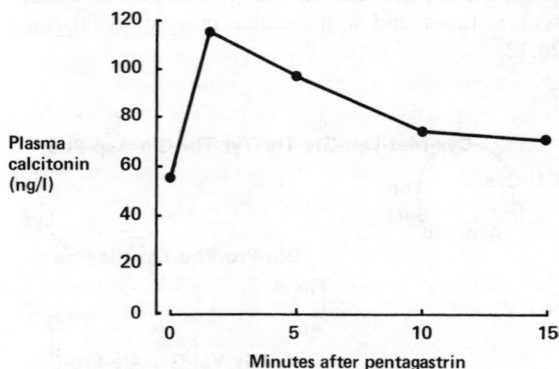

Fig. 20.14 Plasma calcitonin in normal men after an injection of pentagastrin, an analogue of gastrin (Fig. 2.5, p. 13) (Courtesy of H. Heath)

physiologically. Calcitonin inhibits bone resorption and, in particular, it inhibits the increase in bone resorption caused by parathyroid hormone; there is no evidence that calcitonin actively promotes bone formation. Calcitonin increases the excretion of phosphate as well as that of sodium, potassium and magnesium by the kidney.

Calcitonin binds specifically to receptors on the cell membrane of target tissues and increases cyclic AMP production. It is not yet certain that cyclic AMP is the second messenger of calcitonin action.

The physiological significance of calcitonin is still not clear. It is possible that it provides a 'fine adjustment' for plasma calcium, particularly after meals. It is also possible that calcitonin ensures the long-term maintenance of the bone mass and the lower calcitonin levels in females may be related to the increased rate of bone loss with ageing in females.

Disorders. No clear calcitonin deficiency disease has been described but high calcitonin levels are found in medullary carcinoma of the thyroid (fig. 20.15). In

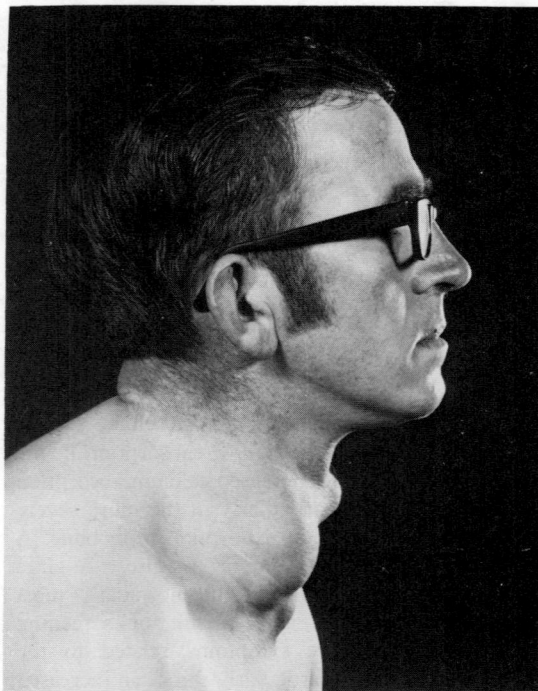

Fig. 20.15 34-year-old man with medullary carcinoma of thyroid and greatly raised plasma levels of calcitonin

addition some other tumours, notably tumours of the lung, may produce a calcitonin-like substance. Few of these patients become hypocalcaemic presumably because the parathyroid glands maintain the plasma calcium.

Calcitriol

Calcitriol (1,25-dihydroxyvitamin D_3) is a sterol produced in the kidney and derived from vitamin D.

Source. Although vitamin D can be obtained from the diet either as cholecalciferol (vitamin D_3) or as ergocalciferol (vitamin D_2) its main source in man is the action of ultraviolet radiation on 7-dehydrocholesterol in the skin. Vitamin D_3 formed in this way is partly stored in the tissues, notably, muscle and adipose tissue, and partly transformed by the smooth endoplasmic reticulum in liver to give 25-hydroxyvitamin D_3 (25-OHD$_3$). This hydroxylation involves molecular oxygen and probably cytochrome P_{450}. 25-OHD$_3$ is the principal form of vitamin D in the plasma and assays of plasma 25-OHD provide a measure of the vitamin D stores. For example, plasma levels vary with the season to reflect the increased availability of UV radiation in summer (fig. 20.16).

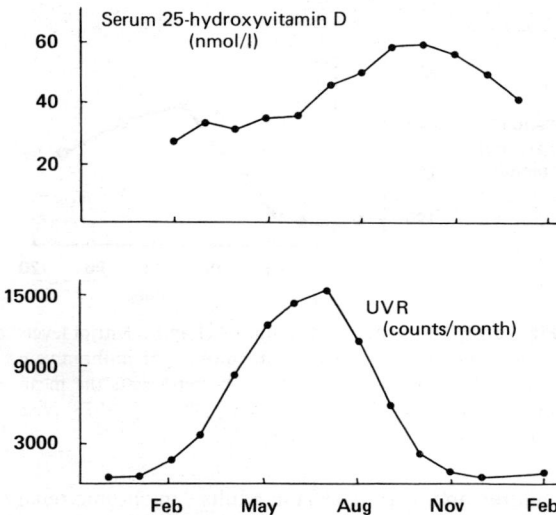

Fig. 20.16 Seasonal variation in serum levels of 25-hydroxyvitamin D in a group of laboratory staff. The environmental ultraviolet radiation in the range 290 to 330 nm for the same period is also shown. (After Devgun, M. S. *et al.* (1981) *American Journal of Clinical Nutrition* **34**, 1501)

25-OHD is transported to the kidney where a further reaction takes place in the mitochondria of the proximal convoluted tubules to give 1,25-dihydroxyvitamin D_3 (calcitriol). This hydroxylation involves cytochrome P_{450} and molecular oxygen (p. 115). This step is a key one in the regulation of the production of calcitriol. Other metabolites of 25-OHD$_3$ are known. 24,25-dihydroxyvitamin D_3 may have a physiological role but this is not yet confirmed. The pathways of vitamin D metabolism are summarised in figure 20.17.

Vitamin D and its metabolites are transported in the plasma bound to a specific α-2 globulin which is synthesised in the liver. This globulin is known as DBP or, formerly, as the group specific component (Gc).

Control of calcitriol production. Production of calcitriol is increased in hypocalcaemia and shut off by hypercalcaemia. This response depends on the function of the parathyroid glands (fig. 20.18) and is not seen after their removal. It is clear therefore that the response to hypocalcaemia is mediated by PTH. The production of calcitriol is also increased when the plasma inorganic phosphate level is low and when calcitriol levels are low. Calcitriol thus regulates its own production.

Actions. Calcitriol has a cellular mechanism of action like that of steroid hormones (fig 22.2, p. 259). On entering the cell it becomes bound to a specific protein in the cytosol. This receptor–calcitriol complex migrates into the nucleus and becomes associated with the nuclear protein. Subsequently new messenger RNA is formed leading to the production of a protein. In the intestine this is a calcium-binding protein responsible for the active transport of calcium.

The principal sites of action of calcitriol are intestine, bone and kidney. In the intestine calcitriol acts on its own but in the kidney it is likely that calcitriol and PTH act together to increase renal tubular reabsorption of calcium; in the bone PTH and calcitriol increase bone resorption. Recently some evidence has been obtained for the presence of receptors for calcitriol within the parathyroid glands; this may provide a form of feed-back control.

Disorders. The level of calcitriol in blood plasma is normally 20 to 60 pg/ml (50–150 pmol/l). Lower values are found in vitamin D deficiency (rickets in

245

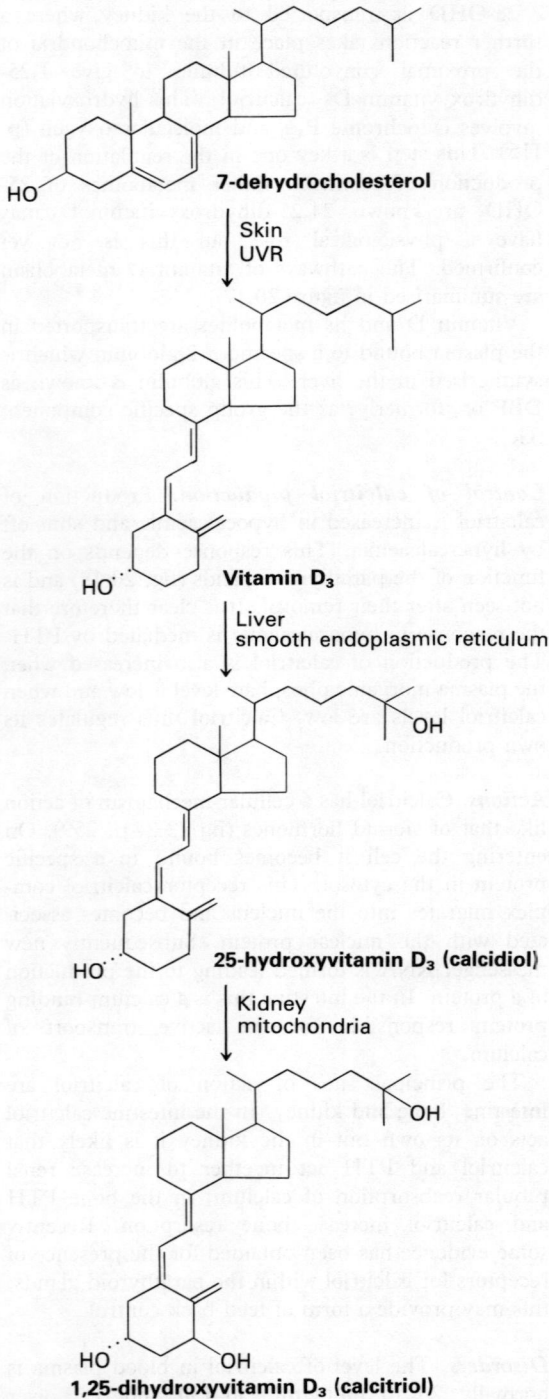

7-dehydrocholesterol

Skin
UVR

Vitamin D₃

Liver
smooth endoplasmic reticulum

25-hydroxyvitamin D₃ (calcidiol)

Kidney
mitochondria

1,25-dihydroxyvitamin D₃ (calcitriol)

Fig. 20.17 Formation and metabolism of vitamin D₃

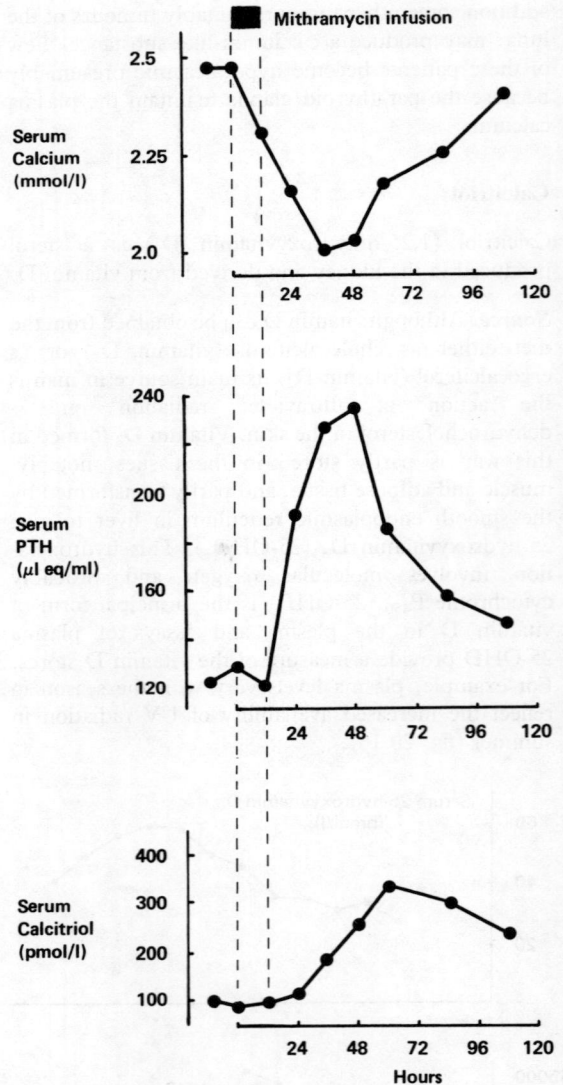

Fig. 20.18 The response of serum PTH and calcitriol levels to hypocalcaemia induced by an infusion of mithramycin which inhibits osteoclasts. Each point represents the mean for eight subjects. (After Bilezikian, J. P. *et al.* (1978) *New England Journal of Medicine* **299,** 437)

children and osteomalacia in adults), in chronic renal failure or patients without kidneys, and in a rare inherited disorder in which there is a deficiency of the 1-hydroxylase enzyme of the kidney, *vitamin D dependent rickets*. While vitamin D deficiency responds promptly to treatment with vitamin D, the other disorders do not respond to vitamin D in physiological amounts, but do respond to small doses of calcitriol given by mouth.

High values of plasma calcitriol are found in patients with parathyroid overactivity.

Magnesium

Magnesium plays a part in all enzymic reactions involving ATP, including oxidative phosphorylation, DNA transcription, RNA aggregation and protein synthesis. Many enzymes including alkaline phosphatase, hexokinase and some peptidases are dependent on magnesium for their activity. Many functions of membranes depend on an optimal concentration of magnesium. For example magnesium depletion leads to an increase in neuronal excitability.

An adult body contains approximately 20 to 30 g of magnesium. About 1 per cent is in the plasma, 45 per cent is within cells, and the remainder is bound with hydroxyapatite in bones.

Intake. Magnesium is present in foodstuffs, particularly in meat and green plants. The average daily intake of magnesium is 10 to 15 mmol (250 to 375 mg) per day. The minimum requirement for magnesium is not known but it may be as low as 4 mmol per day. Pure magnesium deficiency, because of dietary lack, does not occur.

Absorption. Magnesium is absorbed mainly in the small intestine and on an average diet some 44 per cent of the dietary magnesium is absorbed. When the magnesium content of a diet is low the proportion absorbed increases. When the dietary magnesium is high the proportion decreases. Magnesium absorption may be impaired and magnesium deficiency follow in various disorders of the small intestine, particularly chronic diarrhoea, and after extensive surgical resection of the small bowel.

Plasma magnesium. Both in infants and adults plasma magnesium levels remain remarkably constant in the range 0.7 to 1.1 mmol/l (1.5 to 2.8 mg/dl). About 55 per cent of the plasma magnesium is ionised, a further 11 per cent is diffusible, bound to phosphates and citrates and other anions, while about one-third of the plasma magnesium is bound to proteins, particularly to albumin.

Excretion. Some two-thirds of the total plasma magnesium is filtered by the glomeruli but 90 to 99 per cent of the filtered magnesium is reabsorbed, principally in the ascending limb of the loop of Henle. There appears to be an upper limit to tubular reabsorption of magnesium (TmMg) in a number of species and probably also in man. The kidney is the principal organ responsible for the control of the plasma magnesium. In magnesium depletion the kidney conserves avidly and the urinary excretion may fall as low as 0.5 mmol per day. In contrast, with a high magnesium diet, the urinary excretion of magnesium is high. A number of factors including parathyroid hormone, vitamin D, calcium and sodium, influence the renal handling of magnesium but it is not clear whether any of these have a physiological function in magnesium homeostasis.

Disorders. Magnesium deficiency is most commonly caused by prolonged loss of gastrointestinal fluids, particularly after extensive intestinal resection. Excessive renal losses of magnesium can occasionally cause magnesium deficiency. Various diuretic drugs can increase magnesium loss, as can hypercalcaemia of any cause. Magnesium deficiency is characterised clinically by muscle weakness and other neurological

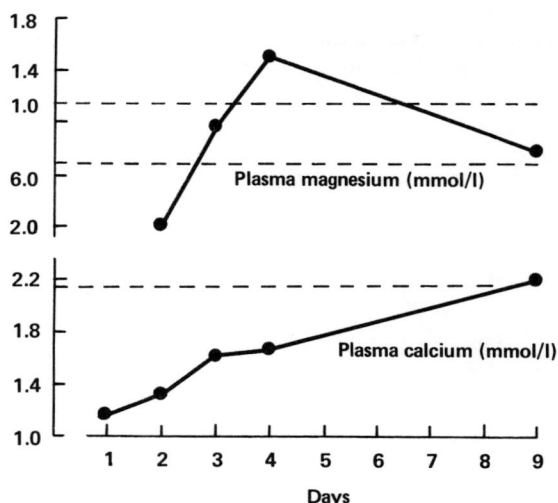

Fig. 20.19 Response to magnesium therapy in an 81-year-old woman with severe magnesium deficiency which came to light when she presented with tetany (fig. 20.9) eight months after the resection of almost all of the small intestine. The horizontal dashed lines indicate the normal range for magnesium and the lower limit of the range for calcium

247

abnormalities, tetany (p. 242) and disorders of cardiac rhythm. Most patients with hypomagnes-aemia develop hypocalcaemia which is resistant to treatment with either calcium or vitamin D. It does respond to treatment with magnesium (fig. 20.19). Magnesium deficiency leads to hypocalcaemia both because parathyroid hormone production is impaired, and because tissues are insensitive to its action.

The most frequent cause of magnesium excess is chronic renal failure. The clinical consequences of hypermagnesaemia are mainly due to the effects on nerves and muscles. Reflexes are lost and, at higher magnesium levels, patients become drowsy and the voluntary muscles progressively weaker. Eventually death may be caused by apnoea (failure to breathe) or by cardiac arrest.

Further reading

Aikawa, J. K. (1981) *Magnesium: Its Biological Significance*. Boca Raton, Florida: CRC Press

Austin, L. A. and Heath, H. (1981) Calcitonin: physiology and pathophysiology. *New England Journal of Medicine* **304**, 269–278

De Luca, H. F. and Schnoes, H. K. (1983) Vitamin D: recent advances. *Annual Review of Biochemistry* **52**, 411–439

Fraser, D. R. (1980) Regulation of the metabolism of vitamin D. *Physiological Reviews* **60**, 551–613

Habener, J. F., Rosenblatt, M., Potts, J. T. (1984) Parathyroid hormone: biochemical aspects of biosynthesis secretion action and metabolism. *Physiological Reviews* **64**, 985–1053

Kenny, A. D. (1981) *Intestinal Calcium Absorption and its Regulation*. Boca Raton, Florida: CRC Press

Norman, A. W. (1979) *Vitamin D: The Calcium Homeostatic Steroid Hormone*. New York: Academic Press

Paterson, C. R. (1975) *Metabolic Disorders of Bone*. Oxford: Blackwell

Potts, J. T., Kronenberg, H. M. and Rosenblatt, M. (1982) Parathyroid hormone: chemistry, synthesis and mode of action. *Advances in Protein Chemistry* **35**, 323–396

Raisz, L. G. and Kream, B. E. (1983) Regulation of bone formation. *New England Journal of Medicine* **309**, 29–35, 83–89

Rude, R. K. and Singer, F. R. (1981) Magnesium deficiency and excess. *Annual Review of Medicine* **32**, 245–258

Stewart, A. F. and Broadus, A. E. (1981) The regulation of calcium excretion. *Annual Review of Medicine* **32**, 457–473

Suki, W. N. and Rouse, D. (1981) Mechanisms of calcium transport. *Mineral and Electrolyte Metabolism* **5**, 175–182

Vaughan, J. (1981) *The Physiology of Bone*, 3rd edn. Oxford: University Press

Wacker, W. E. C. (1980) *Magnesium and Man*. Cambridge, Mass: Harvard University Press

Wassermann, R. H. and Fullmer, C. S. (1983) Calcium transport proteins calcium absorption and vitamin D. *Annual Review of Physiology* **45**, 375–390

CRP

21 Iron & the trace elements

OF the 25 elements now known to be essential for full health, 14 occur in the human body at concentrations of less than 100 mg per kg and are therefore described as trace elements (Table 21.1). A trace element is regarded as essential if its deficiency is associated with defects of function which can be remedied by the administration of the element. Five of the essential elements (nickel, tin, vanadium, silicon and fluorine) have been shown to be essential only within the past ten years.

Table 21.1 The essential trace elements

	Total body (mg)	Plasma concentration ($\mu mol/l$)	($\mu g/dl$)
Iron	4000–5000	9–30	50–170
Fluorine	2000–3000	5–10	10–20
Zinc	1400–2300	11–19	70–120
Copper	80–120	13–22	85–140
Selenium	20–30	1–3	10–24
Manganese	12–20	0.005–0.015	0.025–0.075
Iodine	10–20	0.25–0.55	3–7
Molybdenum	9	0.2–0.5	2–5
Chromium(trivalent)	1.7	0.06*	0.3*
Cobalt	1.5	0.005	0.03
Nickel	10	0.05	0.3
Silicon	*	15–30	40–80
Tin	<10	0.17	2
Vanadium	<10	0.2	1

*No reliable figures available.

Biological role. The function of fluorine is discussed on page 256 and that of iodine, in the thyroid hormones, in Chapter 22. The metals among the trace elements are usually complexed in the body with organic molecules, mainly proteins. In one type of metal–protein complex such as transferrin (p. 21) the element itself is biologically inactive and the complex serves for the transport or storage of the element. Such complexes shield the free metal ion from contact with the tissues.

In the second type of metal–protein complex the metal is biologically active and is essential for the function of the protein. The metal acts either in oxidation and reduction or as a means of maintaining the configuration of the active site. The best known of these complexes is haemoglobin whose function was described in Chapter 16. Most of the other trace elements are constituents of enzymes. Zinc, for example, is an essential part of alkaline phosphatase and alcohol dehydrogenase. Some enzymes require two metals, for example cytochrome c oxidase contains both copper and iron. The biochemical role of the trace elements can often be interpreted in the light of their known chemical properties. For example, transition elements such as iron and copper have more than one stable oxidation state which allows them to take part in electron-transfer (redox) reactions while zinc has only one stable oxidation state but can accept an electron pair and so participates in dehydrogenation reactions. This chapter illustrates some of the important functions of trace elements.

Iron

By far the most common cause of anaemia, both in developed countries and in the developing world, is iron deficiency. Apart from its role in haemoglobin iron has many other functions (Table 21.2).

Iron is a component of the haem molecule which forms part of both haemoglobin and myoglobin. Haemoglobin is synthesised in the bone marrow and incorporated into erythrocytes; its main function is the transport of oxygen and CO_2 (Chap. 16). In haemoglobin and in oxyhaemoglobin iron is in the ferrous form (Fe^{2+}). Within cells there is a constant tendency for the iron of haemoglobin to be oxidised to the ferric form (Fe^{3+}), yielding the inactive compound *methaemoglobin*; enzymes within the cells constantly reduce any methaemoglobin formed back

Table 21.2   Different functions of iron and the amounts in the body of a 75 kg male

Compound	Amount of iron	Function
Haemoglobin	2300 mg	Oxygen carriage in erythrocytes
Myoglobin	300 mg	Reserve supply of oxygen in muscles for use in sudden exertion
Haem-containing enzymes	80 mg	Transport of electrons in cytochromes
Other iron-containing enzymes	100 mg	For example catalase, peroxidase and succinate dehydrogenase
Ferritin and haemosiderin	1000 mg	Storage of iron within cells notably in liver, spleen and bone marrow
Transferrin	4 mg	Transport of iron in the plasma

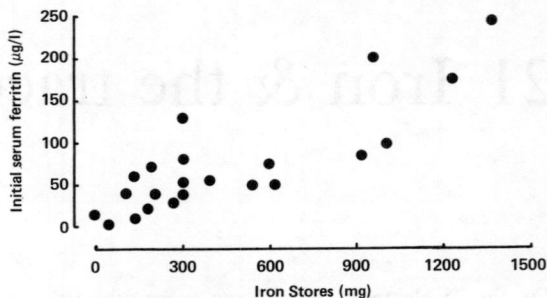

Fig. 21.1 Relationship between plasma ferritin and iron stores in 22 normal subjects. The iron stores were estimated by removing 500 ml blood weekly until the subject was unable to maintain a normal haemoglobin level and developed iron-deficiency anaemia. By calculating how much haemoglobin iron had been removed (and allowing for iron absorption) the stores of iron available for haemoglobin synthesis were determined. (After Walters, G. O. *et al.* (1973) *Journal of Clinical Pathology* **26,** 770)

to haemoglobin. Myoglobin is present in skeletal and cardiac muscle cells where it promotes the diffusion of oxygen.

Iron is stored in the liver, spleen, bone marrow and muscles bound to a protein called *apoferritin*. The complex of apoferritin and iron is known as ferritin and iron stored in ferritin is readily released. Molecules of ferritin may aggregate to form granules of *haemosiderin* visible on microscopy. The iron stored in haemosiderin can be released only slowly but it does provide a useful source of iron, for example, in patients with chronic blood loss. In iron deficiency a reduction in the storage iron can be demonstrated before anaemia develops. An approximate indication of stores can be obtained by staining with Prussian blue a bone marrow smear to show the quantity of iron present. Recently developed methods have shown that ferritin is present in very small amounts in normal human plasma (12 to 150 μg/l in females and 20 to 250 μg/l in males). Determination of plasma ferritin levels provides an indication of iron stores, low iron stores being associated with low ferritin levels (fig. 21.1). High plasma ferritin levels are found in patients with iron overload or hepatic damage.

A very small part of the body iron is present in enzyme systems within cells. Iron is an essential component of catalase, peroxidase and the cytochromes; it is essential therefore for the function of every living cell.

Only a few milligrams of iron are in the plasma at any one time. However the amounts entering and leaving the plasma are considerable (fig. 21.2). Some 20 to 25 mg daily are taken up by the bone marrow

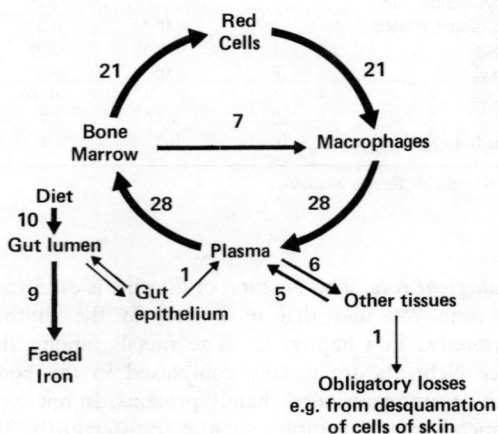

Fig. 21.2 Daily turnover of iron (in mg) between the plasma and the tissues in a male. The obligatory losses of iron are higher in menstruating women because of the additional loss of 15–20 mg monthly

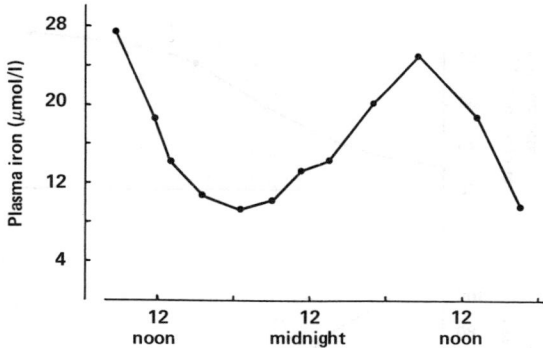

Fig. 21.3 Circadian variation in plasma iron concentration. Each point represents the average value from 19 normal subjects. (After Hamilton, L. D. *et al.* (1950) *Proceedings of the Society for Experimental Biology and Medicine* **75**, 65)

and a similar amount returns to the plasma from the macrophages. Smaller amounts of iron are exchanged between the other tissues and the plasma. While in the plasma iron is bound to the β-globulin *transferrin* (p. 21), each molecule of which has two binding sites for iron. At any one time only about one-third of the possible binding sites are occupied by iron. The plasma iron level in a normal adult is 11 to 27 μmol/l for men and 9 to 23 μmol/l for women. The amount of transferrin in plasma can be measured as the total iron-binding capacity (TIBC) of a plasma sample; the normal range is 40 to 80 μmol iron per litre. Plasma iron levels show a marked circadian variation (fig. 21.3).

Lactoferrin is an iron-binding protein which has been known for many years to exist in milk. It is also present in gastric and intestinal mucosa, in bronchial epithelium and in granulocytes. Lactoferrin in milk may be important for protecting infants against gastroenteritis due to *Escherichia coli* by depriving the organisms of iron. Similarly lactoferrin in granulocytes probably plays an important part in their function; after they disintegrate at the site of an infection they release lactoferrin into the interstitial fluid where it takes up iron.

Sources of iron. The principal foodstuffs which contain iron are green vegetables, red meat and bread. In some countries, including Britain, iron is added to white flour after milling. Milk has a very low iron content and the prolonged use of an unfortified milk diet may lead to anaemia in young children. In developed countries the typical iron intake is 10 to 20 mg daily but much less is consumed in some developing countries. The iron intake is very high in some negro males in South Africa who develop iron overload (*siderosis*) because they consume alcoholic beverages which have been stored in iron containers. Cheap wines are also rich in iron and siderosis is common in chronic alcoholics.

Iron losses. An adult male or a postmenopausal female has an iron loss of about 1 mg per day, representing iron lost in the urine, sweat, faeces and by desquamation of cells from the skin. A woman who is menstruating needs a further 0.3 to 1.0 mg iron daily to allow for the menstrual loss; women with heavy periods may need as much as 4 mg per day. A pregnant woman requires additional iron for the growing fetus, particularly during the second half of the pregnancy. Growing children need about 2 mg iron daily for the new tissues being laid down. Adolescent girls have particularly high iron requirements because they are growing rapidly and also menstruating. Lactation is accompanied by an iron loss of about 0.5 mg per day and this too needs to be allowed for in determining iron requirements.

Iron absorption. The daily iron intake in western countries is usually more than 10 mg which, were it all absorbed, would be more than enough to meet all physiological needs. However, much of this is not absorbed. A normal man on a western diet absorbs about 6 per cent of the dietary iron; the corresponding amount for a young woman is 15 per cent and higher values are found in iron deficiency. The maximum amount of iron which can be absorbed is about 3.5 mg/day (fig. 21.4).

It is important to distinguish two sorts of the dietary iron: iron bound in haem and non-haem iron. Some 25 to 50 per cent of haem iron is absorbed intact and this figure is little affected by the other constituents of the diet. On the other hand normal subjects absorb only 1 to 15 per cent of non-haem iron. The absorption of non-haem iron, largely ferric iron, is greatly reduced if the diet contains certain other components, notably phytate in cereals, tannates in tea and coffee, calcium and phosphates. Ferrous iron salts are used in the treatment of iron deficiency because ferrous iron is better absorbed than ferric iron.

Because iron deficiency and iron excess both affect health the body requires a mechanism to regulate the

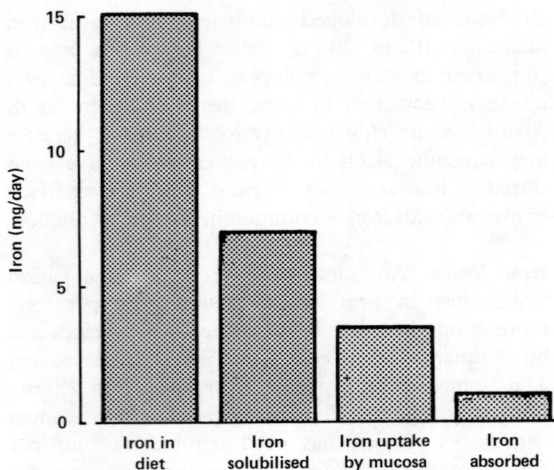

Fig. 21.4 The relative amounts of iron involved at different stages of iron absorption. Even in severe iron deficiency the amount of iron absorbed does not exceed 4 mg daily. (After Bothwell, T. H. *et al.* (1980) *Iron Metabolism in Man.* Oxford: Blackwell)

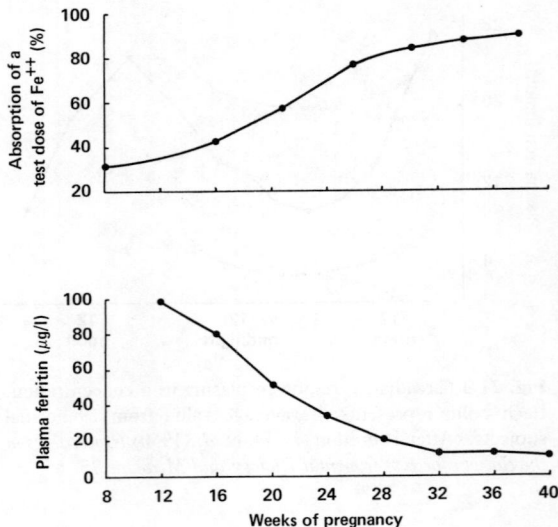

Fig. 21.5 The relationship between plasma ferritin and iron absorption illustrated by the reciprocal changes which occur during normal pregnancy. (After Fenton, V. *et al.* (1977) *British Journal of Haematology* **37,** 145 and Heinrich, H. C. *et al.* (1971) *Klinische Wochenschrift* **49,** 819)

iron stores. Since the physiological losses of iron in urine, sweat and skin cannot be adjusted, the total body iron is only controlled by regulation of the amount absorbed from food. The mechanism of this regulatory system is not yet fully understood. To be absorbed at all non-haem iron must be rendered soluble and for this the hydrochloric acid in the stomach is important. The absorption of iron takes place in two steps. First the iron has to pass from the intestinal lumen into the mucosal cells and secondly from these cells into the plasma. The first stage depends on the amount of iron in the intestine; regulation occurs at the second stage probably by varying the amount of apoferritin in the mucosal cells. In iron deficiency the apoferritin content is low and little iron is bound within the mucosal cells; most of the iron taken up passes through into the plasma. In iron overload the mucosal cells have a high apoferritin content. The iron taken up is largely bound to apoferritin and further absorption is therefore prevented. The iron remains bound as ferritin until the cells are sloughed off into the intestinal lumen. While iron absorption is clearly related in a reciprocal manner to plasma ferritin levels (fig. 21.5) it is not clear whether the apoferritin within the mucosal cell is synthesised locally or derived from the plasma.

Iron storage. The mechanism whereby iron stimulates the formation of its own storage protein apoferritin is not known. The stimulation of apoferritin synthesis by iron can be demonstrated even in isolated tissues and within 10 minutes. It is likely this mechanism provides a means for preventing the build-up of free iron in tissues, for example during the breakdown of haem.

The structure of apoferritin was described earlier (fig. 2.11, p. 17). The molecule consists of 24 subunits forming a sphere into the cavity of which iron is stored as microcrystals of ferric oxide. Before iron can be mobilised from ferritin it must be reduced to the ferrous state by flavin nucleotides. The oxidised flavin nucleotides are then reduced by NADH (fig. 21.6). Before iron can be bound to transferrin it is re-oxidised to the ferric form by the enzyme caeruloplasmin (p. 254).

Iron deficiency. It may seem surprising that, although iron is the fourth most abundant element in the earth's crust, iron-deficiency anaemia is one of the most common nutritional disorders. The explanation lies in the very limited absorption of the predominantly ferric iron of food.

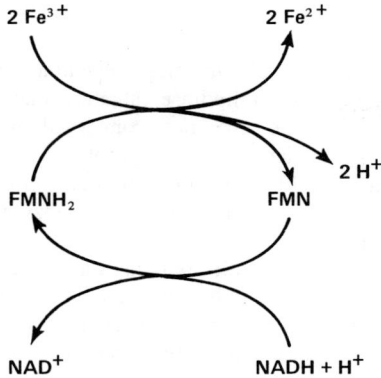

Fig. 21.6 Probable mechanism for the mobolisation of iron in ferritin. (After Osaki, S. and Sirivech, S. (1971) *Federation Proceedings* **30,** 1292)

Iron deficiency may result from excessive losses of blood, from inadequate iron intake or absorption, or from an increased need for iron, for example, in a growing child. An excessive loss of iron can be caused by excessive menstrual bleeding or by chronic blood loss from the gastrointestinal tract, for example as a result of a bleeding peptic ulcer or a hookworm infestation.

Iron deficiency is characterised by anaemia with small red cells which have a reduced haemoglobin content. Even before the anaemia develops a reduction in iron stores may be demonstrated by a fall in stainable iron in the bone marrow or by a fall in the plasma ferritin levels. Figure 21.7 shows the effects of repeated removal of blood on haemoglobin and plasma ferritin levels. The plasma iron level falls and the plasma iron-binding capacity rises as a result of increased synthesis of transferrin. Iron deficiency may also cause changes in various tissues; the mucosa of the tongue may be atrophied and the finger nails and hair may show characteristic abnormalities.

Iron-deficiency anaemia responds briskly to the administration of iron (fig. 21.8).

Iron overload. Three distinct clinical problems have to be considered: acute iron poisoning, chronic iron overload due to repeated transfusions (haemosiderosis), and haemochromatosis.

Acute iron poisoning is not uncommon in children since iron tablets are widely prescribed and since they resemble sweets. As little as 1 g (five 200 mg tablets) may cause serious symptoms in a young

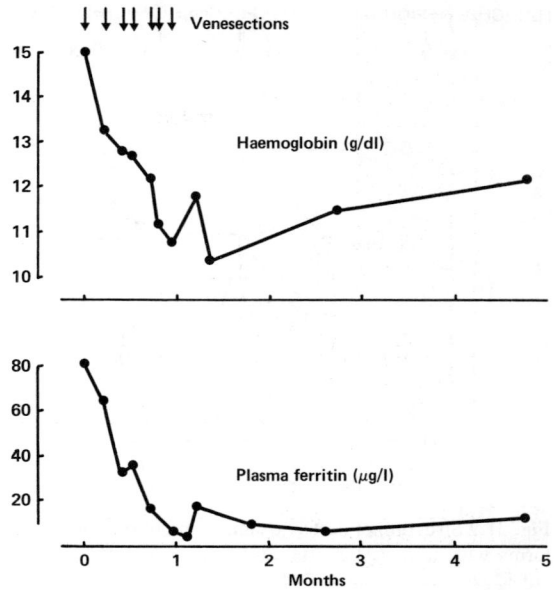

Fig. 21.7 The effect of repeated venesections on the haemoglobin concentration and the plasma ferritin concentration in a normal man. Three ferrous sulphate tablets were taken two days after the last venesection. (After Jacobs, A. *et al.* (1972) *British Medical Journal* **4,** 206)

child. Within 30 to 60 minutes of taking the tablets the child may have abdominal pain, vomiting, diarrhoea and increasing tiredness; with larger overdoses loss of consciousness and death may occur within five hours. The plasma iron level may rise to more than 500 μmol/l (3000 mg/dl) and most of the clinical consequences of iron poisoning are thought to

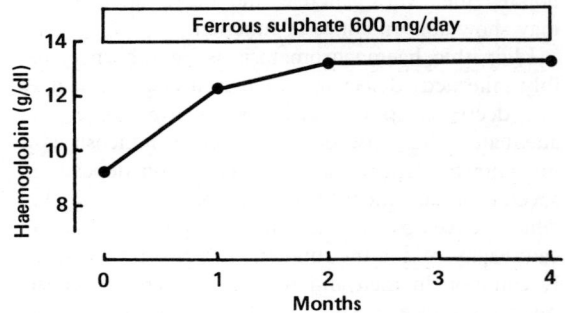

Fig. 21.8 Response of the haemoglobin level to iron therapy in a woman with iron-deficiency anaemia

$H_2N-(CH_2)_5-N-C-(CH_2)_2-C-N-(CH_2)_5-N-C-(CH_2)_2-C-N-(CH_2)_5-N-C-CH_3$

Desferrioxamine

Iron-desferrioxamine complex

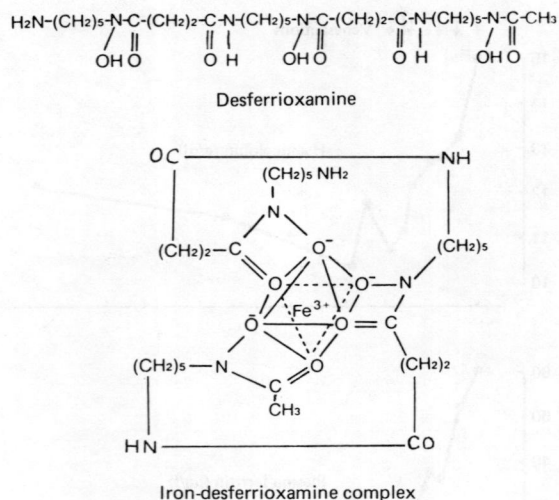

Fig. 21.9 Structure of desferrioxamine and of the complex it forms with iron. (After Meredith, T. J. and Vale, J. A. (1980) *Hospital Update* **6,** 883)

be due to the presence of free iron in the plasma, since the transferrin is fully saturated. The disorder may be treated by the administration of desferrioxamine (deferoxamine), a chelating agent (fig. 21.9).

Haemosiderosis may occur after multiple blood transfusions given in the treatment of some forms of anaemia, such as β-thalassaemia (p. 206), since the iron introduced by transfusion by-passes the normal regulatory system and is released into the body when the donor's cells are broken down. Haemosiderosis is occasionally caused, as mentioned earlier, by an excessive oral intake of iron from alcoholic drinks. In haemosiderosis iron is deposited in the tissues, notably the liver, pancreas and heart, all of which may show abnormalities of function.

Idiopathic haemochromatosis is an inborn, possibly inherited, defect in which iron absorption does not decrease appropriately when iron stores are adequate. The cause of this defect of homeostasis is not known. Patients have excessive iron deposits in several tissues including liver, pancreas and skin which develops a characteristic pigmentation in caucasians. While the fundamental disorder is equally common in men and women, the clinical consequences, with excessive iron deposition, are far more often seen in men since women lose iron in menstruation.

Zinc

Zinc has a key role in many essential enzymes including carbonic anhydrase, alkaline phosphatase and many of the enzymes concerned with the synthesis of nucleic acids and proteins. Zinc was discovered in 1940 to be an essential part of carbonic anhydrase which was the first metallo-enzyme to be recognised.

The body of an adult contains about 2 g of zinc and the plasma zinc is in the range 11 to 19 μmol/l (70 to 120 μg/dl). In normal subjects the zinc content of the body appears to be regulated by variation of zinc absorption.

A zinc-deficiency disorder was not recognised until the 1960s when it was identified in Iranian and Egyptian adolescents with impaired growth and delayed sexual maturity. These patients did not lack zinc entirely in the diet but developed the deficiency because zinc absorption was impaired by fibre and phytate which were present in unusually large amounts in village diets. Zinc deficiency can also occur in patients with impaired intestinal absorption or on total parenteral nutrition. Clinical features include depression, skin changes and loss of hair. Some of the clinical effects of zinc deficiency can be related to the impaired function of particular enzymes such as thymidine kinase involved in DNA synthesis.

The importance of zinc in human metabolism is indicated by *acrodermatitis enteropathica*, an uncommon disorder inherited as an autosomal recessive and characterised by severe skin changes, loss of hair and diarrhoea (figs. 21.10 and 21.11); untreated children die within the first four years of life. Plasma zinc levels are extremely low and the disorder can be relieved by the oral administration of zinc. This disorder is probably caused by an inherited defect in the intestinal transport of zinc.

Copper

It has been known for more than 50 years that copper is essential for the metabolism of higher animals. The body of an adult man contains about 100 mg of copper. In the plasma most of the copper is carried on the protein *caeruloplasmin* one of whose functions is the oxidation of ferrous iron to ferric iron. Iron is released from the ferritin of the tissues in the ferrous

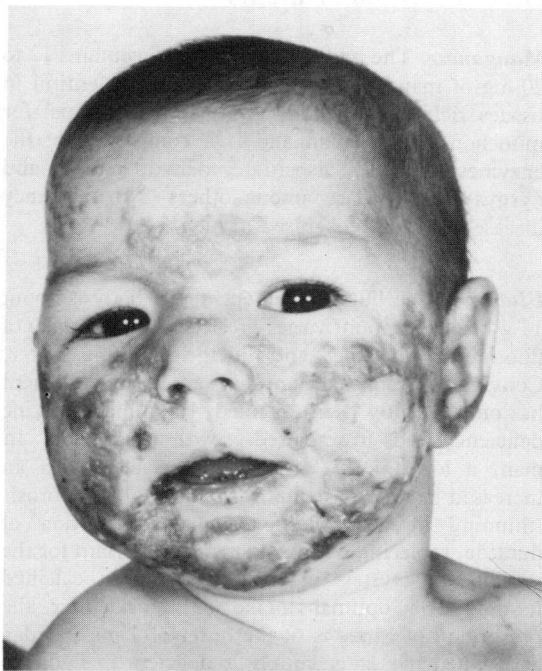

Fig. 21.10 A girl aged 4 months with acrodermatitis entero-pathica. She had an older sister with the same condition. She appeared normal at birth and was breast-fed for one month. At the age of 2 months she developed the typical skin rash and diarrhoea. (Courtesy of P. J. Aggett and E. Moynahan)

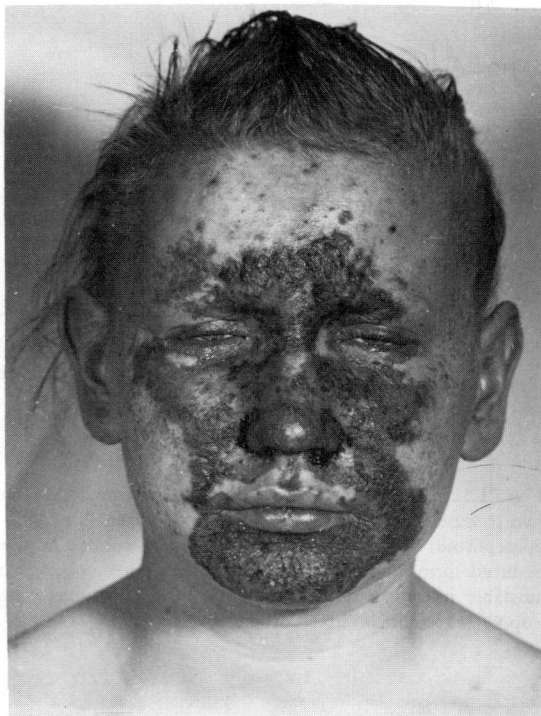

Fig. 21.11 Severe changes of acrodermatitis enteropathica in a 16-year-old girl. The photograph was taken in 1968 before the value of zinc was recognised. (Courtesy of K. Weismann)

state, but must be converted to the ferric form before stable complexes with transferrin can be formed. Because of this function caeruloplasmin is also known as ferroxidase. Copper-depleted animals may have increased iron stores in the tissues, but, at the same time, an iron-deficiency anaemia.

Lysyl oxidase is a copper-containing enzyme important for the formation of cross-links in collagen and elastin (Chap. 17). In experimental copper deficiency the cross-linking of elastin is defective and the aorta and other arteries may be weakened. Other copper-containing enzymes include cytochrome oxidase (p. 113), tyrosine hydroxylase, essential for the synthesis of melanin and the catecholamines, and dopamine hydroxylase (fig. 21.12). Erythrocytes have the copper-containing *superoxide dismutases* which dispose of the very toxic superoxide produced by the action of other enzymes:

$$2H^+ + O_2^- + O_2^- \rightarrow O_2 + H_2O_2$$

In healthy adults, the copper concentration in the plasma depends on the copper intake but is usually 13 to 22 μmol/l (85 to 140 μg/dl). Caeruloplasmin accounts for some 93 per cent of the plasma copper in man; the remainder, bound to serum albumin, is copper in transit. The daily output of copper in the urine ranges between 0.1 and 0.5 μmol (5 and 25 μg).

Disorders. Copper deficiency has not been recognised in adults but has been found in infants fed on cow's milk. Such infants have a severe anaemia with low blood levels of iron and copper.

255

HO

HO—⟨benzene ring⟩—CH_2—CH_2—NH_2

Dopamine

$2H^+ + O_2$

$2Cu^+$ → Dehydro-ascorbate + $2H^+$

$2Cu^{++}$

H_2O ← Ascorbate

HO

HO—⟨benzene ring⟩—CHOH—CH_2—NH_2

Noradrenaline

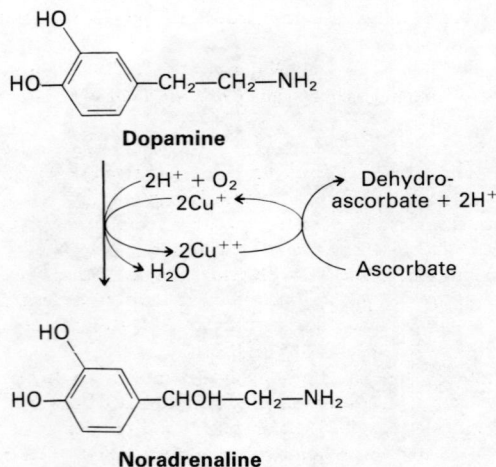

Fig. 21.12 The role of copper within the enzyme dopamine hydroxylase for the production of noradrenaline (nor-epinephrine) in the adrenal medulla. The copper in the reduced form combines with molecular oxygen to give an unstable intermediate which oxidises the dopamine. The copper is re-reduced by ascorbate

The best known disorder of copper metabolism is *Wilson's disease* (hepatolenticular degeneration) in which the concentration of caeruloplasmin in the blood is greatly reduced. The disorder is inherited as an autosomal recessive and is characterised by copper deposition in many tissues, particularly the liver, kidneys, brain and cornea. The patient may be apparently healthy until well into adult life but he eventually develops symptoms due to liver dysfunction or cerebral damage. Neurological symptoms may include, tremor, poor coordination and uncontrolled movements, and later muscular rigidity and mental deterioration. The disorder can be improved by D-penicillamine, which chelates copper and therefore promotes its excretion.

Menkes' syndrome (Menkes' kinky hair syndrome) is a rare disorder inherited as an X-linked recessive. Copper absorption is defective and the plasma levels of copper and caeruloplasmin are very low. As in experimental copper deficiency lysyl oxidase is deficient and elastin in blood vessels is abnormal. Blood vessels become excessively tortuous; bleeding and occlusion occur and a progressive cerebral degeneration takes place.

Other essential elements

Manganese. The body of an adult man contains 12 to 20 mg of manganese which is especially plentiful in tissues rich in mitochondria. It seems essential for mitochondrial function and is a component of the enzymes arginase, isocitrate dehydrogenase and pyruvate carboxylase among others. No deficiency disease has been recognised in man.

Fluorine. The fluorine content of the body is about 14 g but the amount varies greatly with the intake of fluoride in the diet and in the drinking water. Convincing proof that fluoride is an essential element has only recently been obtained. In animals fluorine deficiency leads to impaired growth and fertility. In man, a low fluoride intake is associated with an increased liability to dental caries and to osteoporosis (thinning of the bone). A small proportion of fluoride, replacing hydroxyl ions, is important for the stability of crystals of hydroxyapatite in calcified tissues. The optimal intake of fluoride for the prevention of caries is of the order of 1 mg per day and, when the water supply is deficient in fluoride, fluoridation is necessary. Fluoride excess (fluorosis) occurs in some parts of India and Africa where the soil and the water supply contain large amounts of fluoride. The teeth become discoloured, the bone becomes excessively dense, and the ligaments become calcified and ossified. Severe joint disorders and sometimes damage to the spinal cord (paraplegia) may occur eventually.

Other elements. Cobalt is essential as a component of vitamin B_{12} (p. 153). Molybdenum is a component of a number of enzymes including xanthine oxidase, which plays a part in purine catabolism (p. 158). There has so far been no convincing case report of molybdenum deficiency in man. Selenium is a component of the enzyme glutathione peroxidase in erythrocytes. Trivalent chromium is an essential element in animals and seems necessary for the action of insulin since chromium-deficient animals develop diabetes mellitus. It is however not yet clear if chromium deficiency plays any part in the causation of diabetes in man. Other elements which have been shown to be essential in animals include silicon, vanadium, nickel and tin. Silicon appears to have a role in glycosaminoglycan metabolism and is bound within the proteoglycans of bone and connective tissues. Experimental animals with silicon deficiency have unduly fragile bones.

Further reading

Bothwell, T. H., Charleton, R. W., Motulsky, A. G. (1983) Idiopathic hemochromatosis. In *Metabolic Basis of Inherited Disease*, 5th edn, eds Stanbury, J. B., Wyngaarden, J. B., Fredrickson, D. S., Goldstein, J. L., Brown, M. S. pp. 1269–1298. New York: McGraw-Hill

Bothwell, T. H., Charlton, R. W., Cook, J. D. and Finch, C. A. (1980) *Iron Metabolism in Man*. Oxford: Blackwell

Danks, D. M. (1983) Heriditary disorder of copper metabolism in Wilson's disease and Menkes' disease. In *Metabolic Basis of Inherited Disease*, 5th edn, eds Stanbury, J. B., Wyngaarden, J. B., Fredrickson, D. S., Goldstein, J. L., Brown, M. S. pp. 1251–1268. New York: McGraw-Hill

Dobbie, J. W. (1982) Silicon: its role in medicine and biology. *Scottish Medical Journal* 27, 1–2, 17–19

Fell, G. S. (1981) Essential inorganic elements in clinical nutrition. In *Recent Advances in Clinical Nutrition 1*, eds Howard, A. and Baird, I. M., pp. 75–82. London: Libbey

Hambridge, K. M. and Walravens, P. A. (1982) Disorders of mineral metabolism. *Clinics in Gastroenterology* 11, 87–117

Jacobs, A. (1982) Disorders of iron metabolism. In *Recent Advances in Haematology 3*, ed. Hoffbrand, A. V., pp. 1–24. Edinburgh: Churchill Livingstone

Jacobs, A. (ed.) (1982) Disorders of iron metabolism. *Clinics in Haematology* 11, 239–486

Jenkins, G. N. (1982) Fluoride and the fluoridation of water. In *Human Nutrition: Current Issues and Controversies*, eds Neuberger, A. and Jukes, T. H., pp. 32–72. Lancaster: MTP Press

Kay, R. G. and Knight, G. S. (1983) Trace metals. In *Surgical Nutrition*, ed. Fisher, J. E. Boston: Little Brown

Nelder, K. H. (1983) Biochemistry and physiology of zinc metabolism. In *Biochemistry and Physiology of the Skin*, ed. Goldsmith, L. A., pp. 1082–1101. Oxford: University Press

Powell, L. W. and Halliday, J. W. (1981) Iron absorption and iron overload. *Clinics in Gastroenterology* 10, 707–735

Scheinberg, I. H. and Sternlieb, I. (1984) *Wilson's Disease*. Philadelphia: Saunders

Sternlieb, I. (1983) Abnormalities of copper metabolism in disease states. In *Recent Advances in Hepatology*, eds Thomas, H. C. and MacSween, R. N. M., pp. 115–129. Edinburgh: Churchill Livingstone

CRP

22 The hormones

A WIDE variety of chemical substances act as hormones; some, such as thyroxine and the catecholamines, are small molecules derived from amino acids; others such as growth hormone, insulin, thyroid stimulating hormone and somatostatin are proteins or polypeptides; a third group, the steroid hormones, are derived from cholesterol.

Some hormones exist in a free form in plasma; growth hormone, the gonadotrophins and the catecholamines are examples. Others, including thyroxine and the steroid hormones, are carried by specific binding proteins and by albumin (p. 20).

Mode of action

The initial step in the action of any hormone consists of the interaction of the hormone with a specific recognition site, or *receptor*, in the target tissue. The thyroid hormones and the lipid-soluble steroids rapidly cross the plasma membrane; the receptors for these hormones are within the cell. The peptide hormones and the catecholamines are water-soluble and do not readily cross the cell membrane; the specific binding sites for these hormones are on the outer surface of the plasma membrane. For some peptide hormones such as ACTH, and in the action of adrenaline on β-receptors, this interaction leads to the activation of adenylate cyclase and the production of the second messenger, cyclic AMP.

Peptide and catecholamine hormones

An increase in levels of cyclic AMP within a cell activates cyclic AMP-dependent protein kinases which in turn stimulate the phosphorylation of proteins. Examples of the changes in enzyme function resulting from phosphorylation include the actions of glucagon and adrenaline in inducing increased glycogen breakdown and decreased glycogen synthesis (p. 132). The guanyl nucleotides play an important part in the activation of adenylate cyclase. When a hormone binds to a receptor protein it induces an adjacent *nucleotide binding protein* to bind GTP (fig. 22.1). In turn this activated protein stimulates adenylate cyclase, the GTP being hydrolysed to GDP at the same time, so terminating the hormone action. One effect of cholera toxin is to modify the nucleotide binding protein so that the adenylate cyclase is constantly activated. It is not yet clear how this activation leads to the profuse diarrhoea, and loss of water and electrolytes, which accompany cholera infections.

Some peptide hormones such as insulin, growth hormone, prolactin and, for some actions, adrenaline do not cause a rise in the intracellular concentration of cyclic AMP. The binding of adrenaline to α-adrenergic receptors in the liver leads to a rise in the concentration of calcium in the cytosol. The binding of calcium to a specialised protein, *calmodulin*, leads to the activation of phosphorylase kinase. It

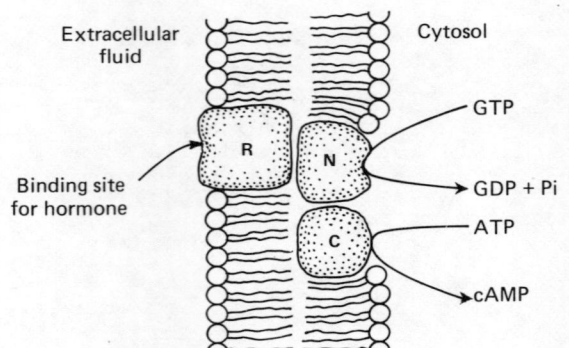

Fig. 22.1 The membrane proteins involved in the activation of adenylate cyclase by a peptide or catecholamine hormone. The binding of hormone to the receptor protein (R) leads to the binding of GTP to the nucleotide binding protein (N). In turn this activates adenylate cyclase (C)

is not yet clear whether the action of other hormones is mediated in this way.

It has recently become clear that insulin and perhaps other growth-promoting hormones act by binding to receptor proteins which are transmembrane proteins whose cytosolic component directly promotes phosphorylation of enzymes regulating metabolism (p. 138). There is no second messenger.

Intracellular receptors have been demonstrated for some peptide hormones. Such hormones enter the cell in vesicles as described earlier (fig. 18.9, p. 223). While it is possible that some polypeptide hormones act intracellularly, the uptake of the hormones could simply represent a mechanism for the degradation of both hormone and receptor to provide a means for terminating hormone action.

Steroid hormones

The steroid hormones have been widely studied. After entering a target cell a steroid hormone is bound to a receptor protein; the hormone–receptor complex then moves from the cytoplasm into the nucleus. Within the nucleus the hormone–receptor complex binds to 'acceptor sites' in the chromatin and influences the transcription of specific mRNA molecules (fig. 22.2). The usual effect of most steroid

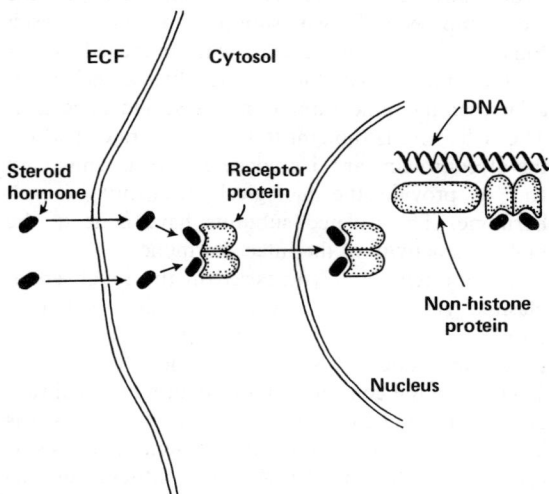

Fig. 22.2 Mechanism of action of steroid hormones. The hormone enters the cytosol, binds to a receptor protein, thought to consist of two subunits, and the hormone receptor complex migrates to the nucleus. There it associates with a non-histone protein and either stimulates or inhibits the synthesis of a particular protein

hormones is to increase mRNA synthesis but in some situations the production of mRNA is turned off. An example of this is the inhibitory effect of cortisol on the synthesis of ACTH by cells of the anterior pituitary.

Each steroid receptor is specific for a group of compounds; for example oestrogen receptors are specific for oestrogenic compounds. 'Antihormones' may act by competing for the receptor proteins or by inhibiting the transfer of the complex into the nucleus or by preventing the recycling of nuclear receptors. Anti-oestrogen drugs such as tamoxifen are important in the management of patients with oestrogen-dependent tumours.

Receptor regulation

The concentration of receptors in a cell determines its sensitivity to hormone action. Some hormones have the ability to decrease the number of their own receptors in a cell. This 'down-regulation' has been demonstrated for insulin, growth hormone, LH, TRH and TSH. Down-regulation may serve to protect target cells from over-stimulation by abnormally high hormone levels.

Some hormones such as prolactin and FSH are capable of increasing the number of receptors in their target tissue, a process known as 'up-regulation'. In addition certain hormones are able to modify the concentration of receptors for a different hormone. Insulin for example increases the number of receptors for growth hormone in the liver. LH decreases ovarian receptors for FSH and increases receptors in the breasts for prolactin.

A few disorders have been described in which tissues are insensitive to a hormone. One example is *testicular feminisation* which is caused by a severe reduction in the numbers of receptors for testosterone. Patients with this disorder are chromosomally male but have female physical features in spite of very high plasma levels of testosterone.

Synthesis and metabolism of hormones

Catecholamines

The catecholamines noradrenaline, adrenaline and dopamine are derived from the amino acid tyrosine

259

and the steps of the synthetic pathway are shown in figure 22.3. The conversion of tyrosine to dihydroxyphenylalanine is the rate limiting step. Catecholamines are degraded by two enzymatic pathways as shown in figure 22.4.

Peptide hormones

The peptide hormones vary in length between three and nearly 200 amino acid residues. Some are

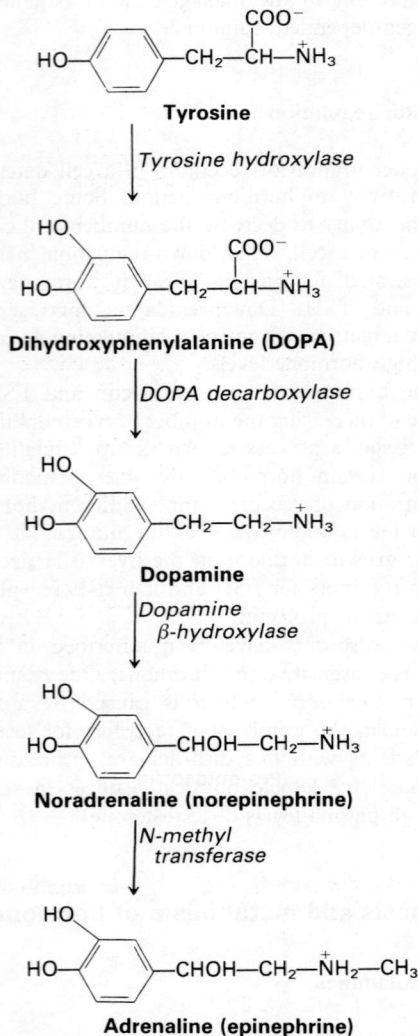

Fig. 22.3 Pathway for the synthesis of catecholamines

glycoproteins. Peptide hormones are produced by a number of tissues including the hypothalamus, pituitary, thyroid and parathyroid glands, the gastrointestinal tract, pancreas and placenta. Within the cells of these tissues the peptides to be secreted are derived from larger precursor molecules called *pro-hormones*. The example of proinsulin was described earlier (fig. 12.11, p. 139). There is evidence that some of these pro-hormones are themselves derived from even bigger molecules called pre-pro-hormones. Some peptide hormones, insulin and glucagon (Chap. 12), and parathyroid hormone and calcitonin (Chap. 20) have already been described.

Short-chain polypeptides. The hypothalamus produces a number of small peptides such as oxytocin, vasopressin, gonadotrophin releasing hormone, thyrotrophin releasing hormone and somatostatin (fig. 22.5). Somatostatin is also produced in the pancreas. These small molecules are rapidly degraded by enzymes in plasma and have a short half-life.

Glycoprotein hormones. The glycoproteins of the pituitary gland include thyroid stimulating hormone (TSH), luteinising hormone (LH) and follicle stimulating hormone (FSH). In addition chorionic gonadotrophin (HCG) derived from the placenta is structurally very similar to LH. The glycoproteins are composed of two subunits α and β each consisting of a peptide core with branched carbohydrate side-chains which include fucose, galactose, galactosamine, glucosamine, mannose and sialic acid. The sialic acid is thought to reduce the rate at which the glycoproteins are degraded. The β-subunits vary and so provide the biological specificity of each hormone. The isolated subunits have none of the biological activity of the intact hormone.

Glycoproteins are synthesised in the rough endoplasmic reticulum; the peptide chains first pass directly into the lumen where the carbohydrate groups are added by specific enzymes (fig. 14.8, p. 171). Since the concentration of α-subunits is always greater than the concentration of the β-subunits it is likely that the regulation of hormone synthesis is carried out by regulating the synthesis of the β-subunit.

ACTH and related peptides. Adrenocorticotrophin hormone (ACTH) is derived from the large molecule pro-opiocortin which splits to give ACTH and β-lipotrophin (LPH) (fig. 22.6). The function of

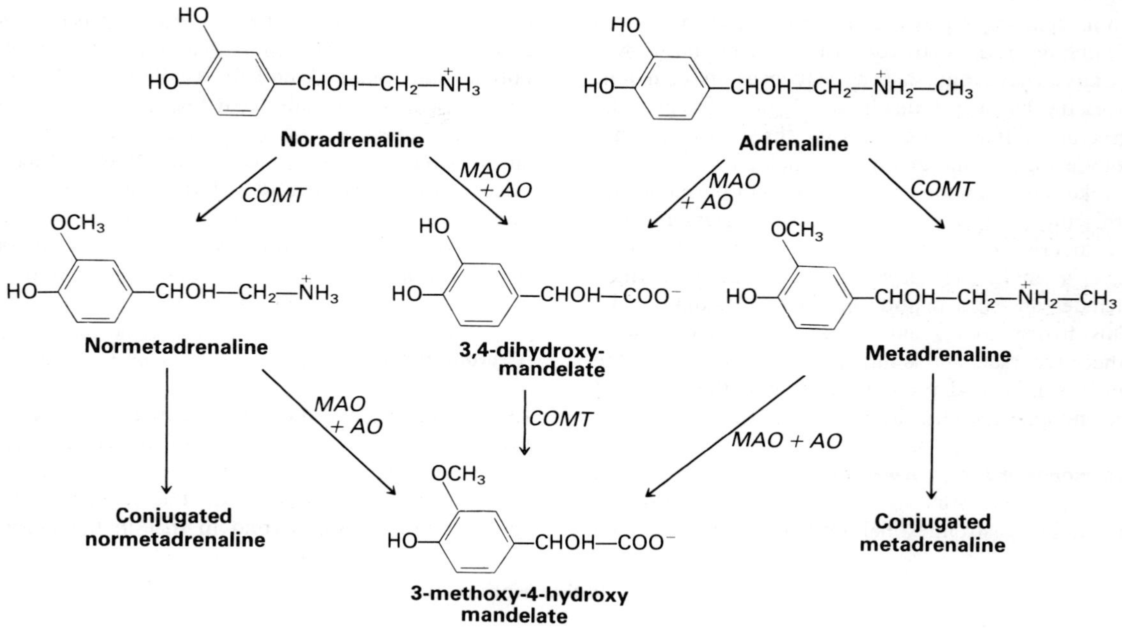

Fig. 22.4 Pathways for the metabolism of the catecholamines. All the substances in the lowest line are excreted in the urine. Measurement of the urinary excretion of 3-methoxy-4-hydroxy mandelate, often known as vanillyl mandelate (VMA = vanillyl mandelic acid), provides a screening test for the presence of catecholamine-producing tumours of the adrenal medulla, phaeochromocytomas, COMT = catechol O-methyl transferase, MAO = monoamine oxidase, AO = aldehyde oxidase

β-lipotrophin is not known; it may be a precursor of the endorphins, naturally occurring pain-relieving substances produced by the pituitary.

The enkephalins, met-enkephalin (fig. 22.5) and leu-enkephalin, are naturally occurring opiates produced mainly by the adrenal medulla. They are peptides with five residues in a sequence identical to one found in β-endorphin. They are not, however, derived from β-endorphin but from a precursor which includes at least six met-enkephalin sequences and one leu-enkephalin sequence.

Somatotrophic hormones. Growth hormone, prolactin and placental lactogen all consist of a single peptide chain with disulphide linkages between different parts of the chain. There are considerable similarities between the primary structures of growth hormone and placental lactogen; large parts of the

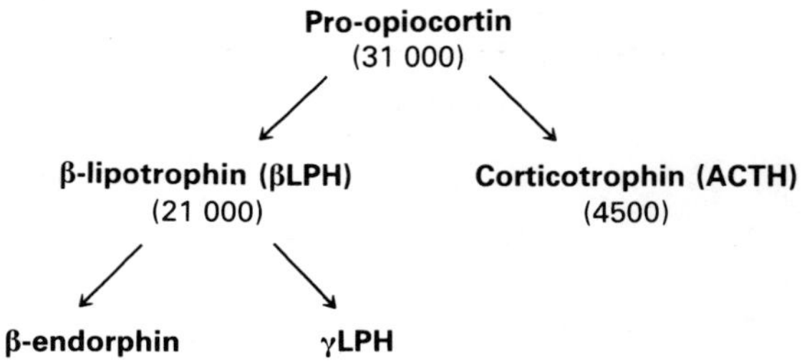

pGlu-His-Pro-NH₂
Thyrotrophin releasing hormone (TRH)

Met-Phe-Gly-Gly-Tyr-NH₂
Met-enkephalin

pGlu-His-Trp-Ser-Tyr-Gly-Leu-Arg-Pro-Gly-NH₂
Gonadotrophin releasing hormone
Luteinising hormone releasing hormone (LH-RH)

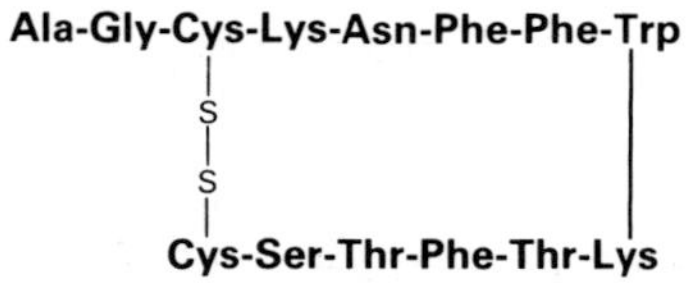

Somatostatin

Fig. 22.5 Some small peptide hormones. The structures of oxytocin and vasopressin were shown in figure 2.4 (p. 12)

Fig. 22.6 Synthesis of ACTH and related compounds. The figures in parentheses are the molecular masses

261

amino acid sequences are identical and each has two disulphide bridges. In contrast prolactin and growth hormone have only a few sequences in common; prolactin has three disulphide bonds. Despite the structural differences each of the three hormones has both lactogenic and growth-producing activity.

Like other peptide hormones growth hormone and prolactin are synthesised as larger precursors. The precursors pre-prolactin and pre-growth hormone have a molecular mass of about 28 000 which includes an extra peptide of about 30 amino acids. This hydrophobic 'signal segment' is common to other precursors and is important in ensuring that the newly synthesised protein passes into the lumen of the endoplasmic reticulum (fig. 14.8, p. 171).

Hormones of the gastrointestinal tract. The mucosa of the gastrointestinal tract is a very large endocrine organ. The gastrointestinal hormones may be divided

into two groups according to their structures: (1) gastrin and cholecystokinin which are straight-chain polypeptides (with 17 and 33 residues respectively) and (2) those structurally related to secretin. Secretin is a polypeptide containing 27 amino acid residues, all of which are required for biological activity. Gastric inhibiting peptide (GIP) and vasóactive intestinal peptide (VIP) consist of 43 and 28 residues respectively and both have sequences in common with secretin.

Thyroid hormones

In the thyroid gland inorganic iodide is converted to thyroid hormones in a series of metabolic steps shown in figure 22.7. The thyroid can concentrate inorganic iodide some 25-fold. The ratio of iodide concentration in the thyroid to that in the plasma

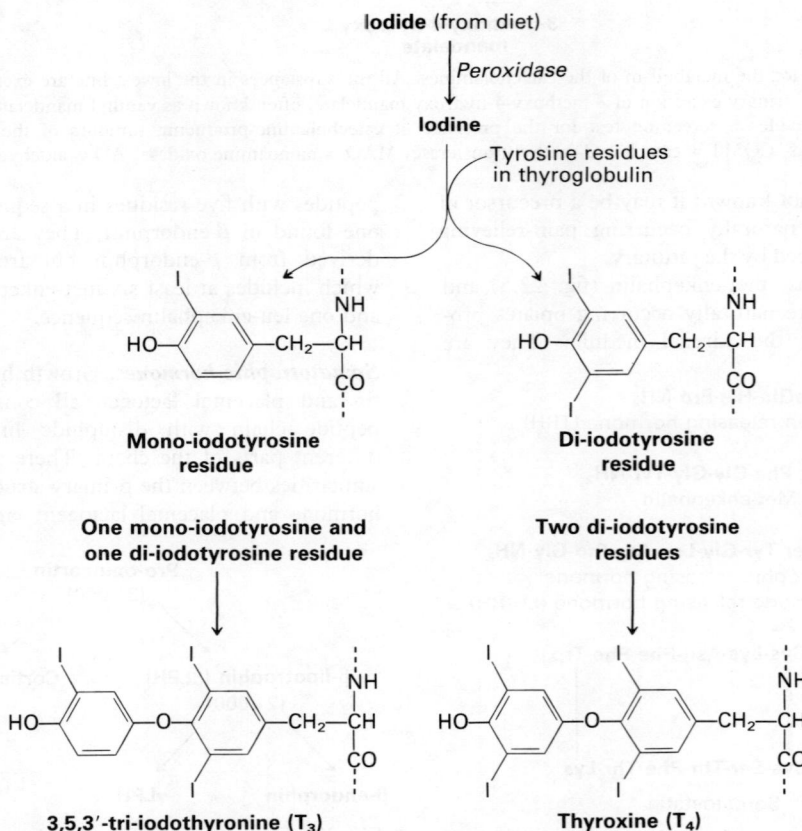

Fig. 22.7 Synthesis of the thyroid hormones

increases in subjects on an iodine-deficient diet and also after the administration of TSH. Conversely the ratio is decreased after hypophysectomy, or treatment with drugs such as thyroid hormones, thiocyanate, perchlorate or iodide. Iodide not bound to tyrosine can be discharged from the gland by thiocyanate, perchlorate or nitrate.

Within the thyroid gland iodide is oxidised and bound covalently to tyrosine residues in the protein *thyroglobulin*, giving both mono- and di-iodotyrosine residues. Still within the thyroglobulin molecule these residues in adjacent peptide chains are coupled together to give thyroxine (with two di-iodotyrosine residues) or tri-iodothyronine (with one di-iodotyrosine and one mono-iodotyrosine residue). Thyroxine and tri-iodothyronine are released from the thyroid gland by proteolysis of thyroglobulin. Some free iodotyrosines are also released but most are deiodinated within the thyroid gland and the iodine re-utilised for hormone synthesis. A number of defects in the synthesis of thyroid hormones have been described (Table 22.1).

Thyroxine and tri-iododothyronine in the blood are mainly bound to plasma proteins, principally thyroxine binding globulin (TBG), thyroxine binding pre-albumin and albumin (Table 22.2). In euthyroid adults only about one-third of the TBG binding sites contain thyroid hormones. Drugs, such as 5,5-diphenyl hydantoin and diazepam which have three-dimensional structures similar to that of thyroxine,

Table 22.1 Principal types of defect in the synthesis of the thyroid hormones

Concentration defect
  Impairment of trapping of iodide by thyroid

Organification defects
  Iodide accumulates normally within the thyroid but no iodotyrosines are formed. This is the most common inborn error of thyroid metabolism

Coupling defects
  Iodotyrosines formed but no iodothyronines

Dehalogenase defect
  Lack of the enzyme catalysing the deiodination, and therefore recovery, of iodine from iodotyrosines not used for coupling. The iodotyrosines are released and excreted in the urine

Defects in thyroglobulin synthesis
  Little thyroglobulin formed; in some cases an abnormal iodinated albumin is released

Table 22.2 Distribution of thyroid hormones in plasma

	Thyroxine	Tri-iodothyronine
Unbound	0.05%	0.3%
Bound to albumin	5–10%	25–30%
Bound to pre-albumin	15–20%	—
Bound to thyroid-binding globulin	70–75%	70–75%

can displace it from TBG. Oestrogens stimulate and androgens depress the synthesis of TBG.

Thyroid hormones may be metabolised in a number of ways: (1) by deiodination, principally in the liver, to give thyronine, (2) by deamination, mainly in the kidney to yield tetraiodothyrolactic and tetraiodothyroacetic acids which also may be deiodinated before excretion, and (3) by excretion in the faeces or urine either in the unchanged state or as glucuronide or sulphate conjugates.

Steroid hormones

The basic structure of all the steroid hormones is the ring system shown on page 41. This parent structure gives rise to three series of steroids containing 18, 19 or 21 carbon atoms (Table 4.1, p. 43) each series having very distinct biological actions.

The steroid-producing endocrine glands, the adrenals, ovaries and testes, can synthesise the steroids secreted either from two-carbon atom precursors or from cholesterol derived from the plasma. Many of the enzymes involved in steroid synthesis are common to all steroid-producing tissues. The initial conversion of acetate to cholesterol is probably identical to the corresponding process of cholesterol synthesis in the liver (p. 127).

The conversion of cholesterol to pregnenolone can occur in adrenal, ovarian and testicular tissue; it involves a complex multi-enzyme system consisting of a flavoprotein, a protein containing non-haem iron and a cytochrome P_{450} (p. 115). The conversion of cholesterol to pregnenolone seems to be the rate limiting step in steroid synthesis and is controlled by the appropriate hormones for example by ACTH in the adrenal cortex and by LH in the testis and ovary.

Adrenal steroids. The pathways involved in steroid synthesis within the adrenal cortex are shown in

263

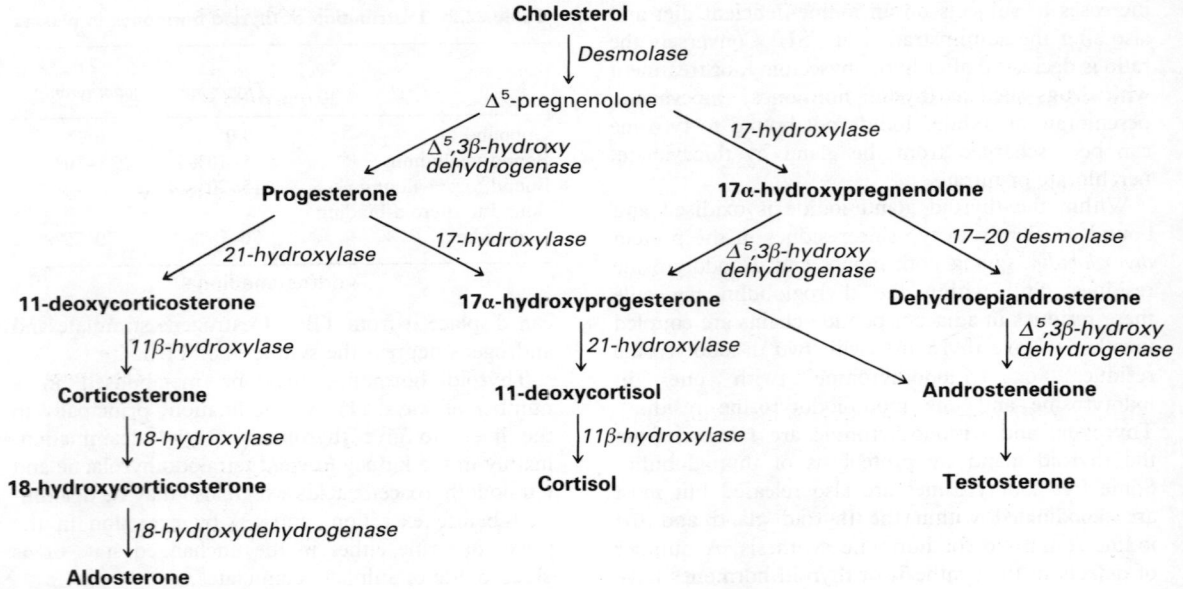

Fig. 22.8 Pathways for the synthesis of steroid hormones by the adrenal cortex

figure 22.8. The adrenal cortex has three distinct zones; the zona glomerulosa produces aldosterone while the zona fasciculata and the zona reticularis produce cortisol and the androgens. Adequate circulating levels of cortisol and aldosterone are essential for life; the androgens produced by the adrenal cortex may have a role in the initiation of puberty but their importance in adults is uncertain. Abnormalities in adrenal steroid synthesis can occur; enzyme deficiency disorders which impair the production of cortisol or aldosterone may be life threatening (fig. 22.9).

Testicular steroids. The principal androgen produced by the testis is testosterone. It is synthesised mainly within the Leydig cells (fig. 22.10). Before testosterone can exert its androgenic action

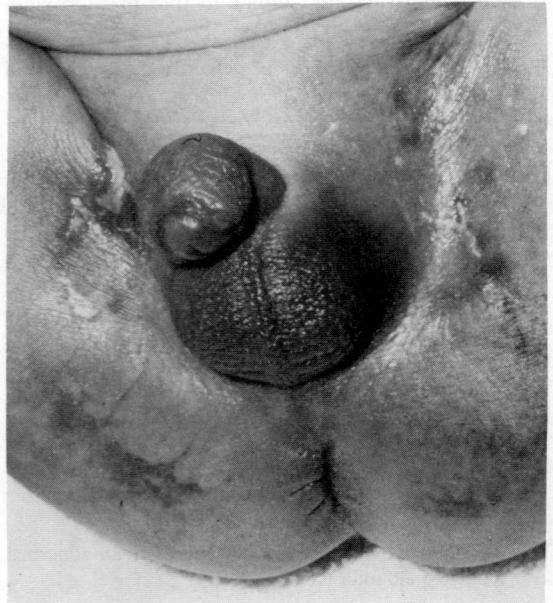

Fig. 22.9 External genitalia of a girl with 21-hydroxylase deficiency, the most common inborn error of steroid hormone synthesis. The lack of 21-hydroxylase causes impairment of the production of cortisol and aldosterone. In turn this leads to severe salt loss; if the disorder is not recognised, death may occur during an intercurrent illness. Because of the lack of cortisol, ACTH production increases and androgen production by the adrenals rises. The excess androgens leads to virilisation of the external genitalia; this child was initially thought to be a boy but her chromosome complement was 46 XX. (Courtesy of Constance C. Forsyth)

264

Pregnenolone

17α-hydroxypregnenolone 17α-hydroxyprogesterone

Dehydroepiandrosterone

Androstenediol   Androstenedione

Testosterone

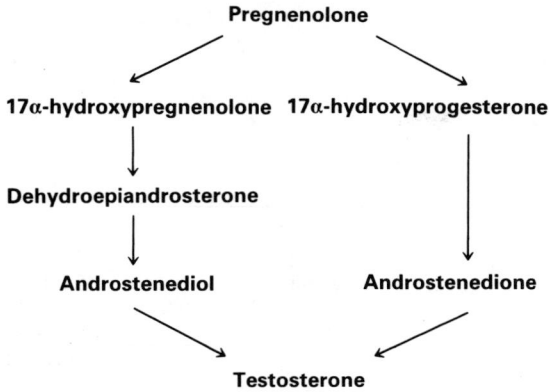

Fig. 22.10 Synthesis of testosterone by the Leydig cells of the testis

it is converted by its target tissues, such as the skin, to 5α-dihydrotestosterone.

Ovarian steroids. The enzymes needed for steroid synthesis have been found in all types of ovarian tissue: follicle, corpus luteum and stroma. Because of the cyclical changes which occur in the normal adult ovary the pathways of steroid synthesis and the amounts of each hormone produced vary throughout the menstrual cycle. During the follicular phase oestradiol-17β is the major steroid synthesised (fig. 22.11), smaller amounts of 17α-hydroxyprogesterone

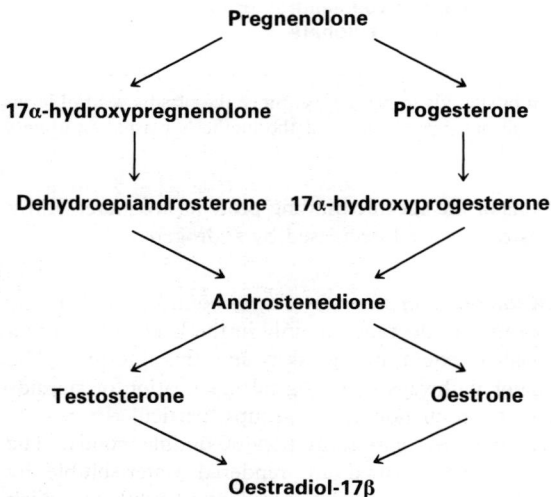

Pregnenolone

17α-hydroxypregnenolone   Progesterone

Dehydroepiandrosterone   17α-hydroxyprogesterone

Androstenedione

Testosterone   Oestrone

Oestradiol-17β

Fig. 22.11 Steroid synthesis within the ovarian follicle

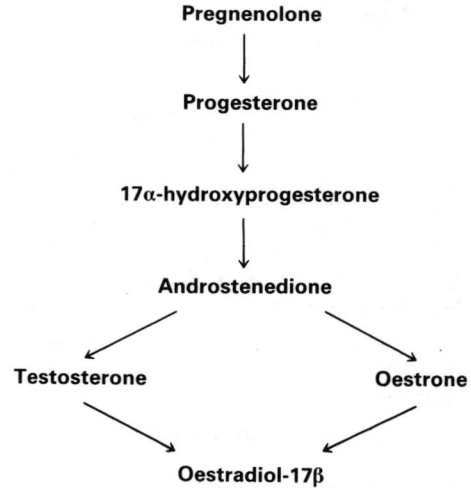

Pregnenolone

Progesterone

17α-hydroxyprogesterone

Androstenedione

Testosterone   Oestrone

Oestradiol-17β

Fig. 22.12 Steroid synthesis in the corpus luteum

and androstenedione being produced. Very little progesterone is synthesised during the follicular phase.

After ovulation the follicle differentiates to form the corpus luteum. Progesterone is the principal steroid produced at this time (fig. 22.12), smaller amounts of 17α-hydroxyprogesterone, androstenedione and oestradiol-17β being synthesised. Steroid synthesis in stromal tissue produces mainly androstenedione.

Placenta. During pregnancy the placenta synthesises large amounts of steroid hormones, mainly from precursors obtained from the maternal and fetal plasma. The placenta forms a part of the maternal–fetal–placental unit which is responsible for the increased steroid output chiefly of oestriol and progesterone during pregnancy. The role of the various components is outlined in figure 22.13.

Binding of steroids to plasma proteins. In the plasma steroid hormones are largely bound to proteins (Table 22.3). Albumin has a high capacity to bind steroids but only a low affinity so that steroids are readily displaced. Steroids bound to protein are biologically inactive and only the small unbound fraction can exert metabolic effects. The concentra-

265

Maternal liver

Acetate

↓

Cholesterol

───

Maternal plasma

↓

Cholesterol　　　　**Progesterone**　　　　**Oestriol**

───

Placenta

↓　　　　　　　　　　　　　　　　　　↑　　　　　　　↑

Cholesterol　　　　　　　　　　　　　　　　　**Oestriol**

↓　　　　　　　　　　　　　　　　　　↑

Pregnenolone ⟶ **Progesterone**　　　**16-hydroxyandrostenedione**

↑

**16α-hydroxydehydro-
epiandrosterone sulphate**

↑

───

Fetal　　　　　　　　　　　　　　　Fetal
adrenal cortex　　　　　　　　　　 liver

↓　　　　　　　　　　　　　　　　　　↑

Pregnenolone　　　　　　　　　**16α-hydroxydehydro-
epiandrosterone sulphate**

↓　　　　　　　　　　　　　　　　　　↑

**Dehydroepiandrosterone
sulphate** ─────────────⟶ **Dehydroepiandrosterone
sulphate**

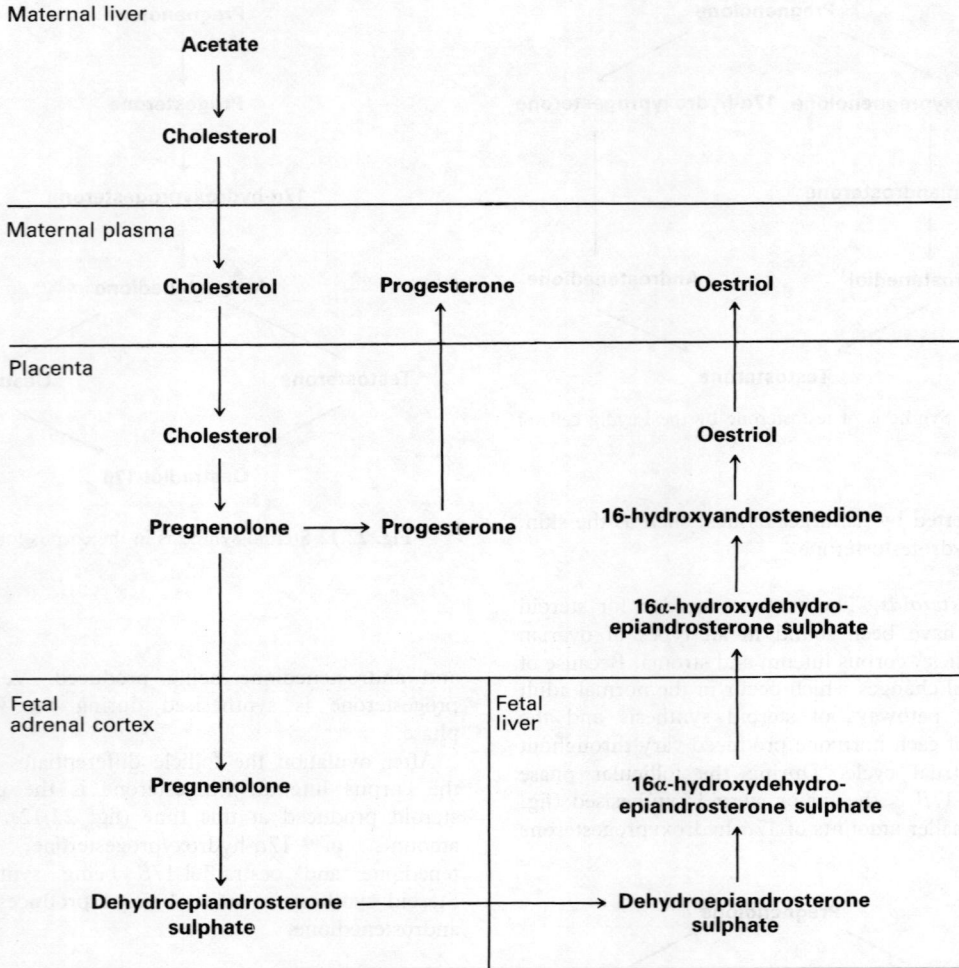

Fig. 22.13 Role of the mother, placenta and fetus in the metabolism of steroids in pregnancy. Since the plasma oestriol levels (and therefore urinary oestrogen excretion) depends on fetal metabolism, measurements of the mother's plasma or urinary oestriol provides a measure of fetal health

Table 22.3 **Binding proteins for steroids in plasma**

Plasma proteins	Steroids bound
Albumin	Most steroids
α_1-glycoprotein	Progesterone
Corticosteroid binding globulin (transcortin)	Cortisol Progesterone
Sex hormone binding globulin	Testosterone Oestradiol-17β

tions of the specific binding proteins are increased by oestrogens, and decreased by androgens.

Metabolism of steroids. The unbound steroids in the plasma are degraded mainly in the liver but also to a smaller extent in the skin and the intestine. The major pathways of biological inactivation of steroids involve reduction of oxo groups, particularly those at C_3 and C_{20}, and saturation of double bonds. The metabolites formed are rendered water-soluble by conjugation to give glucuronides and sulphates which are then excreted through the kidney.

Further reading

Brown, E. M. and Aurbach, G. D. (1982) Receptors and second messengers in cell function and clinical disorders. In *Contemporary Metabolism 2*, ed. Freinkel, N., pp. 247–299. New York: Plenum

Clayton, R. N. (1983) Hormone-receptor interactions. In *Advanced Medicine 19*, ed. Saunders, K. B., pp. 60–75. London: Pitman

Felig, P., Baxter, J. D., Broadus, A. E. and Frohman, L. A. (1981) *Endocrinology and Metabolism*. New York: McGraw-Hill

Hughes, J. (ed.) (1983) Opioid peptides. *British Medical Bulletin* **39**, 1–100

Makin, H. L. J. (ed.) (1984) *Biochemistry of Steroid Hormones 2nd edn*. Oxford: Blackwell

New, M. I. and Levine, L. S. (1984) Recent advances in 21-hydroxylase deficiency. *Annual Review of Medicine* **35**, 649–663

New, I. M., Dupont, B., Grumbach K., Levine, L. S. (1983) Congenital adrenal hyperplasia and related conditions. In *Metabolic Basis of Inherited Disease*, 5th edn, eds Stanbury, J. B., Wyngaarden, J. B., Fredrickson, D. S., Goldstein, J. L., Brown, M. S. pp. 973–1000. New York: McGraw-Hill

O'Riordan, J. L. H., Malan, P. G. and Gould, R. P. (1982) *Essentials of Endocrinology*. Oxford: Blackwell

Rees, L. H. and Smith, R. (1982) Endogenous opiates: beta endorphin and methionine enkephalin. *Recent Advances in Endocrinology and Metabolism 2*, ed. O'Riordan, J. L. H., pp. 1–15. Edinburgh: Churchill Livingstone

Tulchinsky, D. and Ryan, K. J. (ed.) (1981) *Maternal-fetal Endocrinology*. Philadelphia: Saunders

Walker, S., MacNeil, S. and Tomlinson, S. (1984) Calmodulin. *British Journal of Hospital Medicine* **32**, 198–201

MCKB DGH CRP

Index

269

Abbreviations

A	Adenine
ACTH	Adrenocorticotrophic hormone, corticotrophin
ADH	Antidiuretic hormone, vasopressin
ADP	Adenosine diphosphate
ALA	δ-amino-laevulinic acid
Ala	Alanine
AMP	Adenosine monophosphate
Asn	Asparagine
Asp	Aspartate
AST	Aspartate aminotransferase
ATP	Adenosine triphosphate
2,3-BPG	2,3-bisphosphoglycerate
C	Cytosine
C'	Complement component
cAMP	cyclic 3',5'-adenosine monophosphate
CK	Creatine kinase
CMP	Cytidine monophosphate
CoA, CoA-SH	Coenzyme A
Cys	Cysteine
DNA	Deoxyribonucleic acid
ECF	Extracellular fluid
eIF	Initiation factor (eucaryotes)
EF	Elongation factor
FAD	Flavin adenine dinucleotide (p. 54)
FSH	Follicle-stimulating hormone
G	Guanine
$\Delta G^{0\prime}$	Standard free energy change (p. 75)
GABA	γ-amino-butyric acid
GDP	Guanosine diphosphate
Gln	Glutamine
Glu	Glutamate
Gly	Glycine
GMP	Guanosine monophosphate
G6PD	Glucose 6-phosphate dehydrogenase
GFR	Glomerular filtration rate
GTP	Guanosine triphosphate
HbS	Haemoglobin S
HCG	Human chorionic gonadotrophin
HDL	High density lipoproteins

HGPRT	Hypoxanthine-guanine phosphoribosyl transferase
His	Histidine
HS-CoA	Coenzyme A
Hyl	Hydroxylysine
Hyp	Hydroxyproline
ICF	Intracellular fluid
IDL	Intermediate density lipoproteins
Ig	Immunoglobulin
Ile	Isoleucine
IMP	Inosine monophosphate
K	Equilibrium constant
Km	Michaelis constant (pp. 81, 91)
LD	Lactate dehydrogenase
LDL	Low density lipoprotein
Leu	Leucine
LHRH	Luteinising hormone releasing hormone
Lys	Lysine
Met	Methionine
mRNA	Messenger RNA
NAD	Nicotinamide adenine dinucleotide (p. 53)
NADP	Nicotinamide adenine dinucleotide phosphate (p. 53)
P_{CO_2}	Partial pressure of CO_2
PG	Prostaglandin
PGI_2	Prostacyclin
Phe	Phenylalanine
Pi	Inorganic phosphate
pKa	(see p. 4)
PPi	Inorganic pyrophosphate
Pro	Proline
PRPP	5-phosphoribosyl-1-phosphate
PTH	Parathyroid hormone
R	Side chain as in an amino acid residue
RF	Release factor
RNA	Ribonucleic acid
rRNA	Ribosomal RNA
S	Svedberg units (in ultracentrifuge) (p. 59)
Ser	Serine
T	Thymidine

THF	Tetrahydrofolate	Tyr	Tyrosine
Thr	Threonine	U	Uracil
TmP	Tubular maximum for phosphate	UDP	Uridine diphosphate
TPP	Thiamin pyrophosphate	UMP	Uridine monophosphate
TRH	Thyrotropin releasing hormone	UTP	Uridine triphosphate
tRNA	Transfer RNA	UVR	Ultraviolet radiation
Trp	Tryptophan	Val	Valine
TSH	Thyroid stimulating hormone, thyrotropin	VLDL	Very low density lipoproteins
TXA_2	Thromboxane	Vmax	Maximal velocity of an enzyme-catalysed reaction (p. 81)

Essentials of Human Biochemistry

This textbook provides a clear and concise account of biochemistry with particular emphasis on those aspects which are important for the understanding of medicine. Material irrelevant to man is omitted and the account includes several subjects important in medicine which are not found in other biochemistry texts.

The text is well illustrated with numerous new line drawings and also with clinical photographs of disorders which result from derangements of the normal pathways and control mechanisms.

While the book is designed primarily for medical and dental students in the pre-clinical part of the course, it is also likely to be useful for those working for higher qualifications in medicine, surgery and pathology, and for science graduates training in clinical biochemistry.

'This book can be strongly recommended for medical students and all biochemists with an interest in the medical aspects of their subject.' – *Nature*

'...a concise and up-to-date account of the biochemistry essential for an understanding of the molecular basis of medicine...' – *St Thomas's Hospital Gazette*

'*Essentials of Human Biochemistry* lives up to its title, and can be recommended to any preclinical student of biochemistry as an interesting and informative text' – *British Dental Journal*

'It is well illustrated with a plethora of diagrams and clinical photographs, and I should not be exaggerating to say that it is not only essential, but also compulsive, reading'. – *Nursing Mirror*

The author

Dr C R Paterson MA, DM, BSc, FRCPath, is Senior Lecturer in Biochemical Medicine at the University of Dundee, and Honorary Consultant to the Tayside Area Health Board.

ISBN 0-443-03895-3

9 780443 038952